Semiconductor Laser Photonics

This modern text provides detailed coverage of the important physical processes underpinning semiconductor devices. Advanced analysis of the optical properties of semiconductors without the requirement of complex mathematical formalism allows clear physical interpretation of all obtained results. The book describes fundamental aspects of solid-state physics and the quantum mechanics of electron–photon interactions, in addition to discussing in detail the photonic properties of bulk and quantum well semiconductors. The final six chapters focus on the physical properties of several widely used photonic devices, including distributed feedback lasers, vertical cavity surface-emitting lasers, quantum dot lasers, and quantum cascade lasers. This book is ideal for graduate students in physics and electrical engineering, and a useful reference for optical scientists.

Mauro Nisoli is Professor of Physics and Photonics at Politecnico di Milano, where he leads the Attosecond Research Center. His research is focused on attosecond science and ultrafast phenomena in matter. He has published more than 200 research papers in international journals and several physics textbooks.

Semiconductor Laser Photonics

Second Edition

MAURO NISOLI
Politecnico di Milano

CAMBRIDGE UNIVERSITY PRESS

Shaftesbury Road, Cambridge CB2 8EA, United Kingdom

One Liberty Plaza, 20th Floor, New York, NY 10006, USA

477 Williamstown Road, Port Melbourne, VIC 3207, Australia

314–321, 3rd Floor, Plot 3, Splendor Forum, Jasola District Centre, New Delhi – 110025, India

103 Penang Road, #05–06/07, Visioncrest Commercial, Singapore 238467

Cambridge University Press is part of Cambridge University Press & Assessment, a department of the University of Cambridge.

We share the University's mission to contribute to society through the pursuit of education, learning and research at the highest international levels of excellence.

www.cambridge.org
Information on this title: www.cambridge.org/9781009098748

DOI: 10.1017/9781009106153

First published 2023

A catalogue record for this publication is available from the British Library.

A Cataloging-in-Publication data record for this book is available from the Library of Congress

ISBN 978-1-009-09874-8 Hardback

To Margherita,
Matilde, and Kim

Contents

Preface

The term *photonics* was introduced in the late 1960s by Pierre Aigrain, who gave the following definition: "Photonics is the science of the harnessing of light. Photonics encompasses the generation of light, the detection of light, the management of light through guidance, manipulation and amplification and, most importantly, its utilisation for the benefit of mankind." In the last three decades impressive progress in the field of photonics has been achieved, thanks to remarkable advances in the understanding of the physical processes at the heart of light–matter interaction in photonic applications and to the introduction of crucial technological innovations. Photonics is an extremely wide field, as clearly demonstrated by the above definition, since it refers to all types of technological device and process, where photons are involved.

This book does not aim to analyze all aspects of photonics: A few excellent textbooks already exist, which present several topics relevant for photonic applications. The aim of this book is to introduce and explain important physical processes at the heart of the optical properties of semiconductor devices, such as light emitting diodes (LEDs) and semiconductor lasers. It is suitable for a half-semester (or a single-semester) course in Photonics or Optoelectronics at graduate level in engineering physics, electrical engineering, or material science. It originated from the graduate course of Photonics I have been teaching at the Politecnico di Milano since 2006. The concepts of solid-state physics and quantum mechanics, which are required to understand the subjects discussed in this book, are addressed in the introductory chapters. It is assumed that the reader has had courses on elementary quantum mechanics, solid-state physics, and electromagnetic theory at the undergraduate level.

The book presents a selection of topics, which I consider essential to understand the operation of semiconductor devices. It offers a relatively advanced analysis of the photophysics of semiconductors, trying to avoid the use of exceedingly complex formalisms. Particular attention was devoted to offer a clear physical interpretation of all the obtained results. Various worked examples are added throughout all the chapters to illustrate the application of the various formulas: The solved exercises are evidenced by the colored boxes in the text. The numerical examples are also important since they allow the reader to have a direct feeling of the order of magnitude of the parameters used in the formulas discussed in the text. The gray boxes contain concise discussions of supplementary topics or more advanced derivations of particular results reported in the main text, which may not be easily derived by the reader.

Semiconductor Laser Photonics is organized as follows. Chapter 1 focuses on the description of a few concepts of solid-state physics, which are relevant for the calculation and analysis of the band structure of semiconductors. The Bloch theorem is introduced, which describes the wavefunction of electrons in periodic structures. The tight-binding method is considered, with a few simple examples, and the $\mathbf{k} \cdot \mathbf{p}$ method, which are used to calculated the band structure of semiconductors. Chapter 2 deals with the discussion of the main properties of charged particles (electrons and holes) in intrinsic and doped semiconductors. The density of states is first calculated and the essential concepts of carrier statistics in semiconductors are discussed. Basic concepts of quantum mechanics are contained in Chapter 3. In particular, the density matrix formalism is introduced, which is used in the book for the calculation of the optical susceptibility of a semiconductor. After a very short overview of essential aspects of classical electromagnetic theory, Chapter 4 analyzes the interaction of electrons with an electromagnetic field. The expressions of the interaction Hamiltonian, which are extensively used throughout the book, are derived in this chapter. Chapters 5 and 6 build on the previous chapters. In particular, Chapter 5 deals with the optical properties of bulk semiconductors, that is, semiconductors with spatial dimensions much larger than the de Broglie wavelength of the electrons involved in the relevant physical processes. Absorption and gain coefficients are calculated and the radiative and nonradiative recombination processes in semiconductors are analyzed. Chapter 6 analyzes the principles of the photophysics in semiconductor quantum wells, that is, in semiconductor structures where the electrons are confined in one direction by a potential well, with a thickness smaller than the electron de Broglie wavelength.

In the remaining six chapters the general results obtained in the first part of the book are applied to the investigation of the main optical properties of semiconductor devices: light-emitting diodes and lasers. The general philosophy adopted in these chapters is the following: The fundamental physical processes are investigated, rather than the technological characteristics of the devices. After a short and general analysis of semiconductor lasers in Chapter 8, based on the rate equation approach, Chapter 9 is devoted to the analysis of the optical properties of quantum dots, where three-dimensional quantum confinement leads to peculiar properties, which have been used for the development of quantum dot lasers. Chapter 10 contains a detailed theoretical analysis of the distributed feedback (DFB) lasers, based on the use of the coupled-mode equations. By using a simple perturbative approach, the threshold laser conditions are obtained. Vertical cavity surface-emitting lasers (VCSELs) and Quantum cascade lasers are analyzed in the final two chapters.

1 Band Structure of Semiconductors

1.1 Crystals, Lattices, and Cells

A crystal is composed of a periodic repetition of identical groups of atoms: A group is called *basis*. The corresponding crystal lattice is obtained by replacing each group of atoms by a representative point, as shown in Fig. 1.1. A crystal can also be called *lattice with a basis*. When the basis is composed of a single atom, the corresponding lattice is called monoatomic. In a *Bravais lattice*, the position $\mathbf{R}$ of all points in the lattice can be written as :

$$\mathbf{R} = n_1\mathbf{a}_1 + n_2\mathbf{a}_2 + n_3\mathbf{a}_3, \tag{1.1}$$

where $\mathbf{a}_1$, $\mathbf{a}_2$, and $\mathbf{a}_3$ are three noncoplanar translation vectors called the *primitive vectors*, and n_1, n_2, and n_3 are arbitrary (positive or negative) integers. We note that for a given Bravais lattice, the choice of the primitive vectors is not unique. The Bravais lattice looks exactly the same when viewed from any lattice point. Not only the arrangement of points but also the orientation must be exactly the same from every point in a Bravais lattice. No rotations are needed to reach each lattice point. Therefore, two points in the lattice, whose position vectors are given by $\mathbf{r}$ and $\mathbf{r}' = \mathbf{r} + \mathbf{R}$, are completely equivalent environmentally. For example, the two-dimensional honeycomb lattice as shown in Fig. 1.2 is not a Bravais lattice. Indeed, the lattice looks the same when it is viewed from points A and C, but not when it is viewed from point B: In this case, the lattice appears rotated by 180°. We note that, for example, graphene consists of a single layer of carbon atoms arranged in a two-dimensional honeycomb structure.

A lattice can be constructed by infinite repetitions, by translations, of a single cell without any overlapping. This cell can be primitive or nonprimitive (or conventional). Primitive and conventional cells are not uniquely determined, as clearly illustrated in Fig. 1.3. The primitive cell has the minimum possible volume, given by

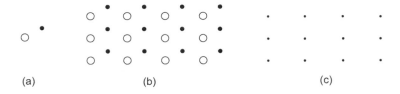

(a)　　　　　　　　(b)　　　　　　　　(c)

Figure 1.1　(a) Basis composed of two different atoms, (b) bidimensional crystal, and (c) corresponding lattice.

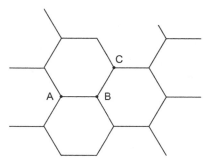

Figure 1.2   A two-dimensional honeycomb lattice is not a Bravais lattice.

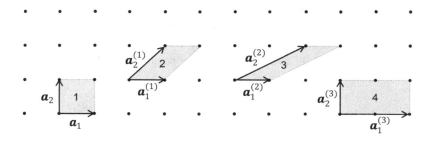

Figure 1.3   Two-dimensional lattice: cells 1, 2, and 3 are primitive cells. Cell 4 is not primitive.

$$V_u = \mathbf{a}_1 \cdot (\mathbf{a}_2 \times \mathbf{a}_3), \tag{1.2}$$

and contains exactly one lattice point. Therefore, in Fig. 1.3, which refers to a two-dimensional case, the cells 1–3 are all primitive cells: They have the same area and contain $4 \cdot \frac{1}{4} = 1$ lattice point. The primitive cell contains the minimum possible number of atoms, and there is always a lattice point per primitive cell. Cell 4 is not primitive: Its area is twice the area of the primitive cell and contains $4 \cdot \frac{1}{4} + 2 \cdot \frac{1}{2} = 2$ lattice points. Not all points in the lattice are linear combinations of $\mathbf{a}_1^{(3)}$ and $\mathbf{a}_2^{(3)}$ with integral coefficients ($\mathbf{a}_1^{(3)}$ and $\mathbf{a}_2^{(3)}$ are not the primitive vectors). In some cases, it is more convenient to consider conventional unit cells, with larger volumes (integer multiple of that of the primitive cell) but characterized by the same symmetry of the lattice.

Without entering into any detail about group theory, we can say that all the possible lattice structures are determined by the symmetry group that describes their properties. A lattice structure can be transformed into itself not only by the translations described by Eq. 1.1, which define the translational group, but also by many other symmetry operations. The symmetry operations transforming a lattice into itself keeping at least one point fixed form a group called the *point group*. In the case of three-dimensional structures, the point symmetry gives rise to 14 types of lattices, which can be classified depending on the relationships between the amplitudes of the vectors $\mathbf{a}_i$ and the angles α, β, and γ between them.

Table 1.1 The 14 lattice systems in three dimensions (the last column shows the amplitudes a_i and the angles between vectors $\mathbf{a}_i$ of the unit cell).

System	Number of lattices	Amplitudes a_i and angles
Triclinic	1	$a_1 \neq a_2 \neq a_3$ $\alpha \neq \beta \neq \gamma$
Monoclinic	2	$a_1 \neq a_2 \neq a_3$ $\alpha = \gamma = 90°,\ \beta \neq 90°$
Orthorhombic	4	$a_1 \neq a_2 \neq a_3$ $\alpha = \beta = \gamma = 90°$
Tetragonal	2	$a_1 = a_2 \neq a_3$ $\alpha = \beta = \gamma = 90°$
Cubic	3	$a_1 = a_2 = a_3$ $\alpha = \beta = \gamma = 90°$
Trigonal	1	$a_1 = a_2 = a_3$ $\alpha = \beta = \gamma < 120°, \neq 90°$
Hexagonal	1	$a_1 = a_2 \neq a_3$ $\alpha = \beta = 90°,\ \gamma = 120°$

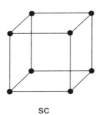
sc

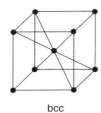
bcc

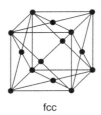
fcc

Figure 1.4 Cubic lattices: (sc) simple cubic, (bcc) body-centered cubic, and (fcc) face-centered cubic.

As reported in Table 1.1, the 14 types of lattices can be grouped in one triclinic, two monoclinic, four orthorhombic, two tetragonal, three cubic, one trigonal, and one hexagonal lattices.

Many semiconductors are characterized by a cubic lattice or by an hexagonal lattice. There are three types of cubic lattices: simple cubic (sc), body-centered cubic (bcc), and face-centered cubic (fcc), whose unit cells are shown in Fig. 1.4. Note that only the sc is a primitive cell, with volume a^3 and one lattice point per cell ($8 \times \frac{1}{8}$). The bcc lattice can be obtained from the sc by placing a lattice point at the center of the cube. The conventional cell is the cube with edge a. It has two lattice points per unit cell ($8 \times \frac{1}{8} + 1$). In terms of the cube edge a, a set of primitive vectors can be written as:

$$\mathbf{a}_1 = \frac{a}{2}(\mathbf{u}_x + \mathbf{u}_y - \mathbf{u}_z)$$
$$\mathbf{a}_2 = \frac{a}{2}(-\mathbf{u}_x + \mathbf{u}_y + \mathbf{u}_z)$$
$$\mathbf{a}_3 = \frac{a}{2}(\mathbf{u}_x - \mathbf{u}_y + \mathbf{u}_z), \tag{1.3}$$

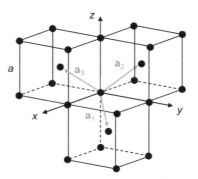

Figure 1.5 Set of primitive vectors for a body-centered cubic lattice.

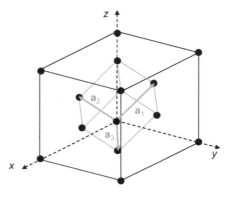

Figure 1.6 Primitive vectors and primitive cell of the face-centered cubic lattice.

where $\mathbf{u}_x$, $\mathbf{u}_y$, and $\mathbf{u}_z$ are the unit vectors of the x, y, and z axes, as shown in Fig. 1.5. The corresponding primitive cell, with volume $a^3/2$, contains by definition only one lattice point. This primitive cell does not have an obvious relation with the point symmetry (cubic) of the lattice. For this reason, it is useful to consider a unit cell larger than the primitive cell and with the same symmetry of the crystal. For a bcc lattice, the unit cell is a cube with edge a, with a volume which is twice the volume of the primitive cell. The bcc lattice can also be considered as an sc lattice with a two-point basis $0, (a/2)(\mathbf{u}_x + \mathbf{u}_y + \mathbf{u}_z)$.

The fcc Bravais lattice can be obtained from the sc lattice by adding a point in the center of each face. The fcc structure has lattice points on the faces of the cube, so that they are shared between two cells: The total number of lattice points in the cell is 4 ($8 \times \frac{1}{8} + 6 \times \frac{1}{2}$). A particular set of primitive vectors is (see Fig. 1.6):

$$\mathbf{a}_1 = \frac{a}{2}(\mathbf{u}_y + \mathbf{u}_z)$$
$$\mathbf{a}_2 = \frac{a}{2}(\mathbf{u}_z + \mathbf{u}_x)$$
$$\mathbf{a}_3 = \frac{a}{2}(\mathbf{u}_x + \mathbf{u}_y). \tag{1.4}$$

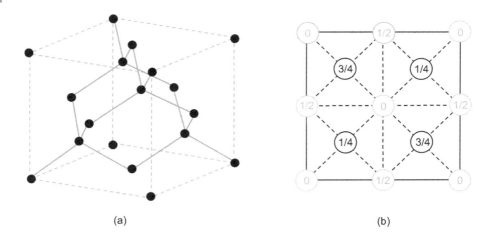

(a)   (b)

Figure 1.7   (a) Crystal structure of diamond; the solid lines show the tetrahedral bond geometry; (b) atomic position in the cubic cell projected on a cube face. The fractions correspond to the height above the cube face in the unit of the cube edge a; the black and gray circles correspond to the two interpenetrating fcc lattices, which generate the diamond structure.

The primitive cell has a volume of $a^3/4$ and contains one lattice point. Also in this case, the unit cell is generally assumed as a cube of edge a, with a volume which is four times the volume of a primitive cell. The fcc can be described as an sc lattice with a four-point basis $0, (a/2)(\mathbf{u}_x + \mathbf{u}_y), (a/2)(\mathbf{u}_y + \mathbf{u}_z), (a/2)(\mathbf{u}_z + \mathbf{u}_x)$. We recall that the numbers giving the size of the unit cell (e.g., the number a in the case of a cubic crystal) are called the lattice constants.

Many important semiconductors, for example, silicon and germanium, have a *diamond structure*, which is the lattice formed by the carbon atoms in a diamond crystal. This structure consists of identical atoms that occupy the lattice points of two interpenetrating fcc lattices, which are displaced from each other along the body diagonal of the cubic cell by one quarter the length of the diagonal, as shown in Fig. 1.7. It can be seen as an fcc lattice with a two-point basis $0, (a/4)(\mathbf{u}_x + \mathbf{u}_y + \mathbf{u}_z)$. The four nearest neighbors of each point are on the vertices of a regular tetrahedron. Note that the diamond lattice is not a Bravais lattice, since it does not look exactly the same when it is viewed from the two nearest-neighbor points. Since the unit cell of an fcc structure contains four lattice points, the unit cell of the diamond structure contains eight lattice points. In this case, it is not possible to choose a primitive cell in such a way that the basis of diamond contains only one atom. When the atoms that occupy one of the two fcc structures are different from the atoms occupying the other, the structure is called the *zinc–blende* structure. Several semiconductors are characterized by this structure, such as GaAs, AlAs, and many others.

Largely used semiconductors, such as GaN, AlN, BN, and SiC, have a hexagonal close-packet (hcp) structure (wurtzite structure) as shown in Fig. 1.8. Also, the hcp lattice is not a Bravais lattice. This lattice can be seen as two interpenetrating simple hexagonal Bravais lattices, displaced vertically by $c/2$ in the direction of the common c-axis and displaced

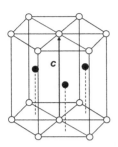

Figure 1.8   Atomic position in the hcp lattice.

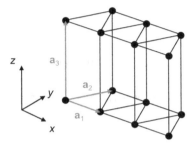

Figure 1.9   Simple hexagonal Bravais lattice.

in the horizontal plane in such a way that the points of one simple hexagonal lattice are placed above the centers of the triangles formed by the points of the other simple hexagonal lattice. Figure 1.9 shows a simple hexagonal Bravais lattice, which is obtained by stacking two-dimensional triangular Bravais lattices – one exactly above the other along a direction perpendicular to each two-dimensional lattice. This stacking direction is usually called the crystallographic c-axis. The primitive vectors can be written as:

$$\mathbf{a}_1 = a\mathbf{u}_x$$
$$\mathbf{a}_2 = \frac{a}{2}\mathbf{u}_x + \frac{\sqrt{3}}{2}a\mathbf{u}_y$$
$$\mathbf{a}_3 = c\mathbf{u}_z. \tag{1.5}$$

1.1.1 The Wigner–Seitz Cell

The primitive cell can also be chosen in such a way that it presents the full symmetry of the Bravais lattice. This can be achieved by considering the Wigner–Seitz cell. The mathematical definition is as follows: The Wigner–Seitz cell around a given lattice point is the spatial region that is closer to that particular lattice point than to any other lattice points. It can also be demonstrated that the Wigner–Seitz cell is a primitive cell. While for any given lattice there is an infinite number of possible primitive cells, there is only one Wigner–Seitz cell. The above definition does not refer to any particular choice of primitive

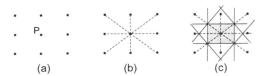

(a)   (b)   (c)

Figure 1.10   Construction of the Wigner–Seitz cell of a two-dimensional rectangular lattice.

vectors; for this reason, the Wigner–Seitz cell is as symmetrical as the Bravais lattice. The procedure for the construction of a Wigner–Seitz cell can be illustrated in the simple case of a two-dimensional rectangular lattice, as shown in Fig. 1.10(a). To determine the Wigner–Seitz cell about the lattice point P, we have first to draw the lines from P to all of its nearest neighbors (Fig. 1.10(b)) and then the bisectors to each of these lines (Fig. 1.10(c)). The Wigner–Seitz cell is the innermost region bounded by the perpendicular bisectors, as shown by the shaded region in Fig. 1.10(c). The same procedure can be applied in the case of a generic three-dimensional lattice.

1.2 The Reciprocal Lattice

In order to develop an analytic study of a crystalline solid, it is often useful to introduce the concept of a reciprocal lattice, which basically represents the Fourier transform of the Bravais lattice. The reciprocal lattice of the reciprocal lattice is the direct lattice. The reciprocal lattice provides a simple and useful basis for analyzing processes characterized by a "wave nature" in crystals, like the behavior of electrons and lattice vibrations, or the geometry of X-ray and electron diffraction patterns. Assuming a Bravais lattice defined by the primitive translation vectors $(\mathbf{a}_1, \mathbf{a}_2, \mathbf{a}_3)$, the reciprocal lattice can be defined by introducing its primitive translation vectors $(\mathbf{b}_1, \mathbf{b}_2, \mathbf{b}_3)$ in analogy with the lattice in a real space. The axis vectors of the reciprocal space can be written as:

$$\mathbf{b}_1 = \frac{2\pi}{V_u}\,\mathbf{a}_2 \times \mathbf{a}_3$$

$$\mathbf{b}_2 = \frac{2\pi}{V_u}\,\mathbf{a}_3 \times \mathbf{a}_1$$

$$\mathbf{b}_3 = \frac{2\pi}{V_u}\,\mathbf{a}_1 \times \mathbf{a}_2, \tag{1.6}$$

where V_u is the volume of the unit cell given by:

$$V_u = \mathbf{a}_1 \cdot (\mathbf{a}_2 \times \mathbf{a}_3). \tag{1.7}$$

The reciprocal lattice can be mapped by using the general translation vector $\mathbf{G}$ given by:

$$\mathbf{G} = m_1\mathbf{b}_1 + m_2\mathbf{b}_2 + m_3\mathbf{b}_3, \tag{1.8}$$

where m_1, m_2, and m_3 are integers. Vector $\mathbf{G}$ is called the reciprocal lattice vector. Note that each vector given by Eq. 1.6 is orthogonal to two-axis vectors of the crystal lattice, so that:

$$\mathbf{b}_i \cdot \mathbf{a}_j = 2\pi \delta_{ij}, \tag{1.9}$$

where δ_{ij} is the Kronecker delta symbol: $\delta_{ij} = 0$ for $i \neq j$ and $\delta_{ij} = 1$, for $i = j$. Moreover,

$$\mathbf{G} \cdot \mathbf{R} = 2\pi (n_1 m_1 + n_2 m_2 + n_3 m_3) = 2\pi \ell, \qquad \ell = 0, \pm 1, \pm 2, ..., \tag{1.10}$$

so that:

$$\exp(i\,\mathbf{G} \cdot \mathbf{R}) = 1. \tag{1.11}$$

Any function $f(\mathbf{r})$ with the periodicity of the crystal lattice, that is, $f(\mathbf{r} + \mathbf{R}) = f(\mathbf{r})$, can be expanded as:

$$f(\mathbf{r}) = \sum_{\mathbf{G}} f_{\mathbf{G}}\, e^{i\mathbf{G} \cdot \mathbf{r}}, \tag{1.12}$$

where:

$$f_{\mathbf{G}} = \frac{1}{V_u} \int_{cell} f(\mathbf{r})\, e^{-i\mathbf{G} \cdot \mathbf{r}}\, d\mathbf{r}. \tag{1.13}$$

While vectors in the direct lattice have the dimension of length, the vectors in the reciprocal lattice have the dimension of $[\text{length}]^{-1}$. The reciprocal space is therefore the most convenient space for the wave vector $\mathbf{k}$. Since each point in the reciprocal space can be reached by the translation vector $\mathbf{G}$, it is evident that we can restrict our analysis to a unit cell defined by the vector $\mathbf{b}_i$. The Wigner–Seitz cell of the reciprocal lattice is called the *first Brillouin zone*. There are also the second, third, etc. Brillouin zones, at an increasing distance from the origin, all with the same volume. These higher order Brillouin zones can be translated into the first zone by adding suitable translation vector $\mathbf{G}$. A Wigner–Seitz cell in the reciprocal space can be constructed following the same procedure used in the real space. For example, in the simple case of a linear lattice, we have a single primitive vector in the direct space: $\mathbf{a}_1 = a\mathbf{u}_x$. From Eq. 1.9, it is obvious that the primitive vector in the reciprocal space is:

$$\mathbf{b}_1 = \frac{2\pi}{a}\mathbf{u}_x. \tag{1.14}$$

The corresponding translation vector is given by:

$$\mathbf{G} = m\frac{2\pi}{a}\mathbf{u}_x, \qquad m = \pm 1, \pm 2, ... \tag{1.15}$$

Following the rules for the construction of the first Brillouin zone, we have that this region extends from $k = -\pi/a$ to $k = \pi/a$.

Following the same procedure, it is possible to construct the first Brillouin zone for three-dimensional structures. For a sc lattice, where:

$$\mathbf{a}_1 = a\mathbf{u}_x, \ \mathbf{a}_2 = a\mathbf{u}_y, \ \mathbf{a}_3 = a\mathbf{u}_z, \tag{1.16}$$

by using Eq. 1.6, where $V_u = a^3$:

$$\mathbf{b}_1 = \frac{2\pi}{a}\mathbf{u}_x, \ \mathbf{b}_2 = \frac{2\pi}{a}\mathbf{u}_y, \ \mathbf{b}_3 = \frac{2\pi}{a}\mathbf{u}_z. \tag{1.17}$$

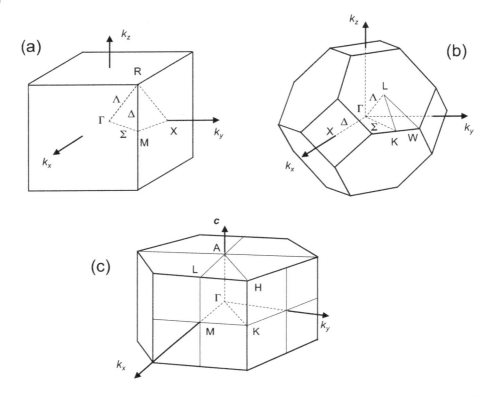

Figure 1.11 First Brillouin zone for (a) the sc lattice, (b) the fcc lattice, which forms the underlying Bravais lattice for the diamond and zinc–blende structures, and (c) the hexagonal wurtzite structure. Reported are the names of points and directions of high symmetry.

Therefore, the reciprocal lattice of an sc lattice with cubic primitive cell of edge a is an sc lattice with a cubic primitive cell of edge $2\pi/a$, and the first Brillouin zone is defined as follows:

$$-\frac{\pi}{a} \leq k_i \leq \frac{\pi}{a}, \qquad i = x, y, z. \tag{1.18}$$

This is shown in Fig. 1.11(a). The center of the first Brillouin zone is always called the Γ-point. A typical convention is to call high-symmetry points and directions inside the Brillouin zone by Greek letters and high-symmetry points on the surfaces of the Brillouin zone by roman letters. For example, in the case of the fcc structure (see Fig. 1.11(b)), the three high-symmetry directions [100], [110], and [111] are denoted by:

$$\text{[100] direction: } \dot{\Gamma} \overline{\Delta} \dot{X}$$
$$\text{[111] direction: } \dot{\Gamma} \overline{\Lambda} \dot{L}$$
$$\text{[110] direction: } \dot{\Gamma} \overline{\Sigma} \dot{K}.$$

The X-point at $(2\pi/a)(1,0,0)$ identifies the zone edge along the six equivalent [100] directions. The L-point is at $(\pi/a)(1,1,1)$, and it is at the zone edge along the eight equivalent [111] directions.

1.3 Electrons in a Periodic Crystal

After this brief overview of introductory concepts of solid-state physics, we will study the interactions of electrons with a periodic structure. We will consider bulk semiconductors, that is, semiconductors with spatial dimensions much larger than the de Broglie wavelength ($\lambda_B = h/p$, where p is the particle momentum) of the electrons involved in the interaction. In order to study the electronic and optical properties of the crystal, we have first to calculate the electronic wavefunctions and their energies inside the crystal. The system is described by the following Schrödinger equation:

$$\left[-\frac{\hbar^2}{2m_0} \nabla^2 + U(\mathbf{r}) \right] \psi(\mathbf{r}) = \mathcal{E} \psi(\mathbf{r}), \tag{1.19}$$

where $U(\mathbf{r})$ is a periodic potential due to the atoms periodically placed in the crystal lattice and to all interaction potentials between electrons. As a consequence of the crystal structure, $U(\mathbf{r})$ has the same periodicity of the lattice:

$$U(\mathbf{r} + \mathbf{R}) = U(\mathbf{r}), \tag{1.20}$$

where $\mathbf{R}$ is a translation vector of the crystal, given by Eq. 1.1. We recall that in vacuum, where $U(\mathbf{r}) = 0$, the stationary wavefunctions of the free electrons in a volume V are given by:

$$\psi_{\mathbf{k}}(\mathbf{r}) = \frac{1}{\sqrt{V}} e^{i\mathbf{k}\cdot\mathbf{r}},$$

delocalized over all the space with uniform probability density, $\mathbf{k}$ is the wavevector. The electron momentum and energy are given by the following expressions:

$$\mathbf{p} = \hbar\mathbf{k}, \qquad \mathcal{E} = \frac{\hbar^2 k^2}{2m_0}, \tag{1.21}$$

where m_0 is the free electron mass. It is possible to demonstrate that the solutions of the Schrödinger equation for a periodic potential are *Bloch–Floquet* (or simply Bloch) *functions* so that (*Bloch's theorem*):

$$\psi_{n\mathbf{k}}(\mathbf{r}) = u_{n\mathbf{k}}(\mathbf{r}) e^{i\mathbf{k}\cdot\mathbf{r}}, \tag{1.22}$$

where $u_{n\mathbf{k}}(\mathbf{r})$ has the same periodicity of the crystal:

$$u_{n\mathbf{k}}(\mathbf{r} + \mathbf{R}) = u_{n\mathbf{k}}(\mathbf{r}), \tag{1.23}$$

k is the wave vector, and n refers to the band (as will be discussed in this chapter). The Bloch functions are the product of a plane wave and a function $u_{n\mathbf{k}}(\mathbf{r})$, with the lattice periodicity. It is evident that the Bloch wavefunction is not periodic; indeed,

$$\psi_{n\mathbf{k}}(\mathbf{r} + \mathbf{R}) = \psi_{n\mathbf{k}}(\mathbf{r}) e^{i\mathbf{k}\cdot\mathbf{R}}, \tag{1.24}$$

while the electron probability density is periodic:

$$|\psi_{n\mathbf{k}}(\mathbf{r} + \mathbf{R})|^2 = |\psi_{n\mathbf{k}}(\mathbf{r})|^2. \tag{1.25}$$

1.3.1   Intuitive Proof of Bloch's Theorem in the Case of a Linear Lattice

Let us consider a linear lattice with lattice constant a. $U(x)$ is the periodic potential, such that $U(x + ma) = U(x)$, where m is an integer. We assume that the electron distribution has the same periodicity of the lattice, so that:

$$|\psi(x + ma)|^2 = |\psi(x)|^2. \tag{1.26}$$

Equation 1.26 implies that

$$\psi(x + ma) = C\psi(x), \tag{1.27}$$

where C is a quantity satisfying the condition $|C|^2 = 1$. Therefore, we may write $C = e^{ikma}$, where k is an arbitrary parameter. From Eq. 1.27, we obtain:

$$\psi(x) = e^{-ikma}\psi(x + ma). \tag{1.28}$$

Multiplying both sides of the previous equation by e^{-ikx}, we obtain

$$\psi(x)e^{-ikx} = e^{-ik(x+ma)}\psi(x + ma), \tag{1.29}$$

thus showing that the function $u(x) = e^{-ikx}\psi(x)$ is periodic with the period a. Finally, by writing $\psi(x) = e^{ikx}u(x)$, we obtain the expression of Bloch's theorem in the case of a linear lattice.   ∎

If $\mathbf{G}$ is a translation vector in the reciprocal space, the Bloch function can be written as:

$$\psi_{n\mathbf{k}}(\mathbf{r}) = u_{n\mathbf{k}}(\mathbf{r})\, e^{i\mathbf{k}\cdot\mathbf{r}} = [u_{n\mathbf{k}}(\mathbf{r})e^{-i\mathbf{G}\cdot\mathbf{r}}]e^{i(\mathbf{k}+\mathbf{G})\cdot\mathbf{r}}, \tag{1.30}$$

which is still a Bloch function since the function between the square brackets in the previous equation has the same periodicity as that of the lattice (see Eq. 1.11). Therefore, the wavefunction 1.30 is the solution of the Schrödinger equation (Eq. 1.19) for the wavevector $\mathbf{k} + \mathbf{G}$. We conclude that, for a given band n, the eigenstates, $\psi_{n\mathbf{k}}(\mathbf{r})$, and eigenvalues, $\mathcal{E}_n(\mathbf{k})$, are periodic functions of $\mathbf{k}$ in the reciprocal lattice:

$$\psi_{n\mathbf{k}}(\mathbf{r}) = \psi_{n,\mathbf{k}+\mathbf{G}}(\mathbf{r}) \tag{1.31}$$

$$\mathcal{E}_n(\mathbf{k}) = \mathcal{E}_n(\mathbf{k} + \mathbf{G}). \tag{1.32}$$

These two important expressions clearly demonstrate that we can restrict our analysis to the first Brillouin zone. In particular, it is possible to represent the dispersion relation $\mathcal{E}_n(\mathbf{k})$ only in this zone by performing a translation of the outer branches into the first Brillouin zone by the reciprocal lattice wavevector $\mathbf{G}$, so that $\mathbf{k} + \mathbf{G}$ is contained in the first zone. This procedure is called band folding.

In order to calculate the dispersion relation $\mathcal{E}_n(\mathbf{k})$, which gives the relationship among electron energy in a crystal, the wavevectors $\mathbf{k}$ in the first Brillouin zone and the band index n, one has to solve the Schrödinger equation obtained by introducing the Bloch wavefunctions 1.22 into Eq. 1.19. The set of curves $\mathcal{E}_n(\mathbf{k})$ defines the band structure of the material. The Bloch wavefunctions are labeled also with respect to the band index n since, for a given value of $\mathbf{k}$, there are many solutions of the Schrödinger equation. Indeed, if we introduce the Bloch function in the Schrödinger equation, we obtain:

$$\left[\frac{\hbar^2}{2m} (-i\nabla + \mathbf{k})^2 + U(\mathbf{r}) \right] u_{n\mathbf{k}}(\mathbf{r}) = \mathcal{E}_n(\mathbf{k}) u_{n\mathbf{k}}(\mathbf{r}). \tag{1.33}$$

Therefore, $u_{n\mathbf{k}}(\mathbf{r})$ is determined by the eigenvalue problem

$$H_B u_{n\mathbf{k}}(\mathbf{r}) = \mathcal{E}_n(\mathbf{k}) u_{n\mathbf{k}}(\mathbf{r}), \tag{1.34}$$

where:

$$H_B = \frac{\hbar^2}{2m} (-i\nabla + \mathbf{k})^2 + U(\mathbf{r}), \tag{1.35}$$

with boundary conditions:

$$u_{n\mathbf{k}}(\mathbf{r}) = u_{n\mathbf{k}}(\mathbf{r} + \mathbf{R}). \tag{1.36}$$

Due to the periodic boundary conditions, this eigenvalue problem is restricted to a single primitive cell of the crystal, and an infinite set of solutions exists with discretely spaced eigenvalues, which can be labeled by the band index n. If we fix the value of n and vary $\mathbf{k}$ in the first Brillouin zone, it is possible to obtain the dispersion curve $\mathcal{E}_n(\mathbf{k})$, which gives the nth energy band of the periodic structure.

1.3.1 Orthonormality of the Bloch Wavefunctions

The periodic functions $u_{n\mathbf{k}}(\mathbf{r})$ form a set of orthogonal wavefunctions since the operator H_B given by Eq. 1.35 is Hermitian. We can add a normalizing condition by writing:

$$\frac{1}{V_{uc}} \int_{uc} u_{m\mathbf{k}}^*(\mathbf{r}) u_{n\mathbf{k}}(\mathbf{r})\, d\tau = \delta_{mn}, \tag{1.37}$$

where V_{uc} is the volume of the unit cell and δ_{mn} is the Kronecker delta. If we now consider the Bloch function associated with the nth band corresponding to a particular $\mathbf{k}$ value, we can write the Bloch function in a slightly different way by requiring the normalization condition:

$$\int_V \psi_{n\mathbf{k}}^*(\mathbf{r}) \psi_{n\mathbf{k}}(\mathbf{r})\, d\tau = 1, \tag{1.38}$$

where $V = NV_{uc}$ is the volume of the crystal (more precisely, as will be explained in Chapter 2, V is the volume of the periodic boxes in which a crystal can be divided). If we write:

$$\psi_{n\mathbf{k}}(\mathbf{r}) = A\, e^{i\mathbf{k}\cdot\mathbf{r}} u_{n\mathbf{k}}(\mathbf{r}), \tag{1.39}$$

and use this expression in Eq. 1.38, we obtain:

$$A = \frac{1}{\sqrt{NV_{uc}}} = \frac{1}{\sqrt{V}}, \tag{1.40}$$

so that the Bloch function can be written as:

$$\psi_{n\mathbf{k}}(\mathbf{r}) = \frac{1}{\sqrt{V}} e^{i\mathbf{k}\cdot\mathbf{r}} u_{n\mathbf{k}}(\mathbf{r}). \tag{1.41}$$

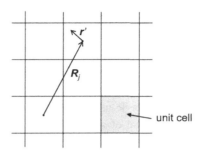

Figure 1.12 Transformation of the spatial coordinates used to calculate the integral Eq. 1.42. $\mathbf{R}_j$ is the position vector of the j-th unit cell and $\mathbf{r}'$ is confined within a single unit cell. A generic unit cell is shown as a gray square.

It is possible to demonstrate that the wavefunctions associated with different **k**-values are orthogonal. The scalar product between two Bloch states with different **k** is given by the following expression:

$$\int_V \psi^*_{m\mathbf{k}'}(\mathbf{r})\psi_{n\mathbf{k}}(\mathbf{r})\,d\tau = \frac{1}{V}\int_V e^{-i\mathbf{k}'\cdot\mathbf{r}}u^*_{m\mathbf{k}'}(\mathbf{r})\,e^{i\mathbf{k}\cdot\mathbf{r}}u_{n\mathbf{k}}(\mathbf{r})\,d\tau. \qquad (1.42)$$

We can first calculate the integral over a single unit cell and then sum the result considering all the unit cells in the crystal volume V. In order to use this approach, we first replace the spatial coordinate **r** with a new spatial coordinate:

$$\mathbf{r} \to \mathbf{r}' + \mathbf{R}_j, \qquad (1.43)$$

where $\mathbf{R}_j$ is the position vector of the j-th unit cell and $\mathbf{r}'$ is confined within a single unit cell, as schematically displayed in Fig. 1.12. Since $u_{n\mathbf{k}}(\mathbf{r})$ is periodic, we have:

$$u_{n\mathbf{k}}(\mathbf{r}' + \mathbf{R}_j) = u_{n\mathbf{k}}(\mathbf{r}'). \qquad (1.44)$$

The scalar product 1.42 can be written as:

$$\sum_{j=1}^{N}\left[\frac{1}{V}e^{i(\mathbf{k}-\mathbf{k}')\cdot\mathbf{R}_j}\int_{V_j} e^{i(\mathbf{k}-\mathbf{k}')\cdot\mathbf{r}'}u^*_{m\mathbf{k}'}(\mathbf{r}')u_{n\mathbf{k}}(\mathbf{r}')\,d\tau\right], \qquad (1.45)$$

where V_j is the volume of the j-th unit cell. Since the integral over the unit cell is the same for each unit cell, it can be extracted from the sum:

$$\left(\sum_{j=1}^{N} e^{i(\mathbf{k}-\mathbf{k}')\cdot\mathbf{R}_j}\right)\frac{1}{V}\int_{V_j} e^{i(\mathbf{k}-\mathbf{k}')\cdot\mathbf{r}'}u^*_{m\mathbf{k}'}(\mathbf{r}')u_{n\mathbf{k}}(\mathbf{r}')\,d\tau. \qquad (1.46)$$

The term within parentheses is different from zero only if $\mathbf{k} = \mathbf{k}'$ and can be written as:

$$\sum_{j=1}^{N} e^{i(\mathbf{k}-\mathbf{k}')\cdot\mathbf{R}_j} = N\delta_{\mathbf{k},\mathbf{k}'}, \qquad (1.47)$$

where $\delta_{\mathbf{k},\mathbf{k}'}$ is the Kronecker delta. Therefore, the scalar product 1.42 can be written as:

$$\int_V \psi^*_{m\mathbf{k}'}(\mathbf{r})\psi_{n\mathbf{k}}(\mathbf{r})\,d\tau = \delta_{\mathbf{k},\mathbf{k}'}\delta_{mn}. \qquad (1.48)$$

Note that the Bloch functions $\psi_{n\mathbf{k}}(\mathbf{r})$, characterized by different $\mathbf{k}$-values, are orthogonal. This property does not hold in the case of the periodic functions $u_{n\mathbf{k}}(\mathbf{r})$.

1.3.2 Momentum of an Electron in a Periodic Crystal

We note that the vector $\mathbf{k}$ used in the previous equations plays the same role of the wavevector $\mathbf{k}$ in the case of a free electron. However, there is an important difference: While for free electrons, the momentum can be written as $\mathbf{p} = \hbar\mathbf{k}$, in the case of electrons in a periodic structure $\hbar\mathbf{k}$ is not the electron momentum. Indeed, it can be easily demonstrated that the Bloch wavefunction $\psi_{n\mathbf{k}}$ is not an eigenstate of the momentum operator $\mathbf{p} = -i\hbar\nabla$. In fact, if we apply the momentum operator to a Bloch function, we obtain:

$$- i\hbar\nabla\psi_{n\mathbf{k}} = -i\hbar\nabla\left(u_{n\mathbf{k}}(\mathbf{r})e^{i\mathbf{k}\cdot\mathbf{r}}\right) = \hbar\mathbf{k}\psi_{n\mathbf{k}} - i\hbar e^{i\mathbf{k}\cdot\mathbf{r}}\nabla u_{n\mathbf{k}}(\mathbf{r}). \tag{1.49}$$

In contrast, in the case of a free electron, the application of the momentum operator leads to the following equation:

$$- i\hbar\nabla\psi_{\mathbf{k}} = \hbar\mathbf{k}\psi_{\mathbf{k}}, \tag{1.50}$$

thus showing that the wave function $\psi_{\mathbf{k}}$ is an eigenfunction of the momentum operator with the eigenvalue $\hbar\mathbf{k}$. The velocity of the electron with wavevector $\mathbf{k}$ is therefore given by $\mathbf{v} = \hbar\mathbf{k}/m$.

The physical meaning of the term $\hbar\mathbf{k}$, associated with an electron in a crystal and called *crystal momentum*, can be understood by considering the equation of motion of an electron in a periodic structure and under the action of an externally applied electromagnetic field. To study the motion of an electron in a particular band, from the quantum-mechanical point of view we have to consider the wave packet formed by wavefunctions characterized by wavenumbers $\mathbf{k}'$ distributed around a particular $\mathbf{k}$ value:

$$\begin{aligned}
\phi_{n\mathbf{k}}(\mathbf{r}, t) &= \sum_{\mathbf{k}'} g(\mathbf{k}')\, u_{n\mathbf{k}'}(\mathbf{r})e^{i(\mathbf{k}'\cdot\mathbf{r}-\omega_n t)} = \\
&= \sum_{\mathbf{k}'} g(\mathbf{k}')\, \psi_{n\mathbf{k}'}(\mathbf{r})\exp\left(-i\frac{\mathcal{E}_n(\mathbf{k}')}{\hbar}t\right),
\end{aligned} \tag{1.51}$$

where we have written $\omega_n = \mathcal{E}_n/\hbar$. It is assumed that the $\mathbf{k}'$ values are within an interval around $\mathbf{k}$ much smaller than the extension of the Brillouin zone. The group velocity of the wave packet is:

$$\mathbf{v}_{gn} = \frac{\partial\omega_n}{\partial\mathbf{k}} = \frac{1}{\hbar}\frac{\partial\mathcal{E}_n(\mathbf{k})}{\partial\mathbf{k}}, \tag{1.52}$$

which can also be written as:

$$\mathbf{v}_{gn} = \frac{1}{\hbar}\nabla_{\mathbf{k}}\mathcal{E}_n(\mathbf{k}), \tag{1.53}$$

where:

$$\nabla_{\mathbf{k}} \equiv \frac{\partial}{\partial k_x}\mathbf{u}_x + \frac{\partial}{\partial k_y}\mathbf{u}_y + \frac{\partial}{\partial k_z}\mathbf{u}_z. \tag{1.54}$$

The dispersion relation $\mathcal{E}_n(\mathbf{k})$ contains all the effects of the crystal on the motion of the electron. If $\mathbf{E}$ is the applied electric field, the work done by $\mathbf{E}$ on the electron (charge $-e$) during the time interval dt is:

$$d\mathcal{E} = -e\mathbf{E} \cdot d\mathbf{r} = -e\mathbf{E} \cdot \mathbf{v}_g dt = -e\mathbf{E} \cdot \frac{1}{\hbar}\frac{d\mathcal{E}}{d\mathbf{k}}dt. \tag{1.55}$$

Therefore, we obtain:

$$\frac{d(\hbar\mathbf{k})}{dt} = -e\mathbf{E} = \mathbf{F}_{ext}, \tag{1.56}$$

where $\mathbf{F}_{ext} = -e\mathbf{E}$ is the external force acting on the electron. Upon considering also the effects of the magnetic field, Eq. 1.56 can be written as:

$$\frac{d(\hbar\mathbf{k})}{dt} = -e[\mathbf{E}(\mathbf{r},t) + \mathbf{v}_g(\mathbf{k}) \times \mathbf{B}(\mathbf{r},t)]. \tag{1.57}$$

While in the case of free electrons $-e\mathbf{E}$ (or $-e(\mathbf{E} + \mathbf{v}_g \times \mathbf{B})$) is the total force acting on the charged particles (and therefore from Eqs. 1.56 and 1.57, we obtain that $\hbar\mathbf{k}$ is the electron momentum), for an electron in a crystal $-e\mathbf{E}$ represents only the external force, while the total force is given by $\mathbf{F}_{ext} + \mathbf{F}_{int}$, where $\mathbf{F}_{int}$ represents the resultant internal force from the crystal lattice. Therefore:

$$\frac{d\mathbf{p}}{dt} = \mathbf{F}_{ext} + \mathbf{F}_{int}. \tag{1.58}$$

1.4 The Concept of Effective Mass

In the case of free electrons, the energy is given by Eq. 1.21, so that the electron mass can be written as:

$$\frac{1}{m_0} = \frac{1}{\hbar^2}\frac{d^2\mathcal{E}}{dk^2}. \tag{1.59}$$

This expression can be generalized for an arbitrary dispersion relation $\mathcal{E}(\mathbf{k})$. If we differentiate Eq. 1.52, we obtain:

$$\frac{dv_{gn}}{dt} = \frac{1}{\hbar}\frac{\partial^2 \mathcal{E}_n}{\partial t \partial k} = \frac{1}{\hbar}\frac{\partial^2 \mathcal{E}_n}{\partial k^2}\frac{\partial k}{\partial t}. \tag{1.60}$$

By using Eq. 1.56, we have:

$$\frac{dv_{gn}}{dt} = \left(\frac{1}{\hbar^2}\frac{\partial^2 \mathcal{E}_n}{\partial k^2}\right)F_{ext}, \quad \text{or} \quad F_{ext} = \frac{\hbar^2}{\partial^2 \mathcal{E}_n/\partial k^2}\frac{dv_{gn}}{dt}. \tag{1.61}$$

Comparing Eq. 1.61 with Newton's second law, we obtain the definition of the *effective mass m^**:

$$\frac{1}{m^*} = \frac{1}{\hbar^2}\frac{\partial^2 \mathcal{E}_n}{\partial k^2}. \tag{1.62}$$

The previous equation can be generalized to take into account an anisotropic energy surface (this is the case with important semiconductors such as Si and Ge). Since the effective mass is generally dependent on direction, it is a tensor.

If we differentiate both sides of Eq. 1.53 with respect to time, we have:

$$\frac{d\mathbf{v}_{gn}}{dt} = \frac{d}{dt}\left(\frac{1}{\hbar}\nabla_{\mathbf{k}}\mathcal{E}_n\right) = \frac{1}{\hbar}\nabla_{\mathbf{k}}\left(\frac{d\mathcal{E}_n}{dt}\right) = \frac{1}{\hbar}\nabla_{\mathbf{k}}\left(\frac{d\mathbf{k}}{dt}\cdot\nabla_{\mathbf{k}}\mathcal{E}_n\right). \tag{1.63}$$

By using Eq. 1.56 in the previous expression, we obtain:

$$\frac{d\mathbf{v}_{gn}}{dt} = \frac{1}{\hbar^2}\nabla_{\mathbf{k}}\left(\mathbf{F}_{ext}\cdot\nabla_{\mathbf{k}}\mathcal{E}_n\right), \tag{1.64}$$

where:

$$\mathbf{F}_{ext}\cdot\nabla_{\mathbf{k}}\mathcal{E}_n = F_{ext,x}\frac{\partial\mathcal{E}_n}{\partial k_x} + F_{ext,y}\frac{\partial\mathcal{E}_n}{\partial k_y} + F_{ext,z}\frac{\partial\mathcal{E}_n}{\partial k_z}, \tag{1.65}$$

so that:

$$\begin{aligned}
\frac{d\mathbf{v}_{gn}}{dt} &= \frac{1}{\hbar^2}\left[F_{ext,x}\frac{\partial^2\mathcal{E}_{n\mathbf{k}}}{\partial k_x^2} + F_{ext,y}\frac{\partial^2\mathcal{E}_{n\mathbf{k}}}{\partial k_x\partial k_y} + F_{ext,z}\frac{\partial^2\mathcal{E}_{n\mathbf{k}}}{\partial k_x\partial k_z}\right]\mathbf{u}_x + \\
&= \frac{1}{\hbar^2}\left[F_{ext,x}\frac{\partial^2\mathcal{E}_{n\mathbf{k}}}{\partial k_y\partial k_x} + F_{ext,y}\frac{\partial^2\mathcal{E}_{n\mathbf{k}}}{\partial k_y^2} + F_{ext,z}\frac{\partial^2\mathcal{E}_{n\mathbf{k}}}{\partial k_y\partial k_z}\right]\mathbf{u}_y + \\
&= \frac{1}{\hbar^2}\left[F_{ext,x}\frac{\partial^2\mathcal{E}_{n\mathbf{k}}}{\partial k_z\partial k_x} + F_{ext,y}\frac{\partial^2\mathcal{E}_{n\mathbf{k}}}{\partial k_z\partial k_y} + F_{ext,z}\frac{\partial^2\mathcal{E}_{n\mathbf{k}}}{\partial k_z^2}\right]\mathbf{u}_z.
\end{aligned} \tag{1.66}$$

From the previous equation, we obtain a simple expression for a given component of the temporal derivative of the group velocity:

$$\left(\frac{d\mathbf{v}_{gn}}{dt}\right)_i = \frac{1}{\hbar^2}\frac{\partial^2\mathcal{E}_{n\mathbf{k}}}{\partial k_i\partial k_j}F_{ext,j}, \tag{1.67}$$

where i and j refer to the x, y, and z coordinates and the right-hand side of the previous equation is summed over the double subscript. Equation 1.66 can be written in a very simple form by using the concept of effective mass, m^*:

$$\frac{d\mathbf{v}_{gn}}{dt} = \frac{1}{m^*}\mathbf{F}_{ext}, \tag{1.68}$$

where $1/m^*$ is a second-rank symmetric tensor, whose elements are:

$$\frac{1}{m_{ij}^*} = \frac{1}{\hbar^2}\frac{\partial^2\mathcal{E}_{n\mathbf{k}}}{\partial k_i\partial k_j}. \tag{1.69}$$

A symmetric second-rank tensor can always be diagonalized by a proper axis rotation: The directions for which the matrix is diagonal are called the principal axis. In general, the values of the effective mass along each of the three principal axes are different.

1.5 Energy Bands

It is well known that, in atoms, bound electrons occupy discrete energy levels; in solids, such discrete levels broaden to form allowed bands, which can be separated by band gaps, as shown in Fig. 1.13. Indeed, when the distance between two initially isolated atoms

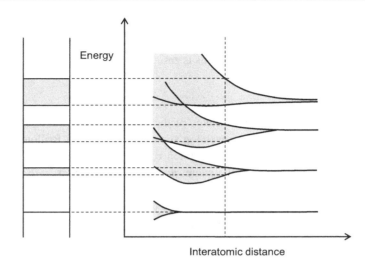

Interatomic distance

Figure 1.13   Formation of energy bands.

is decreased, their electronic orbitals overlap and the final wavefunction is described by two combinations, $\psi_1 \pm \psi_2$, of the two atomic wavefunctions, which correspond to the bonding and antibonding states of the final diatomic molecule. In the case of N atoms, N different orbitals are generated from each orbital of the isolated atom. In a given crystal lattice, there are several energy bands, which correspond to the atomic energy levels of the atoms composing the crystal, as schematically shown in Fig. 1.13, and the width of the band is proportional to the strength of the overlap interaction between neighboring atoms. The energy bands that derive from inner atomic shells are completely filled by electrons, which are more or less localized. In general, many important physical properties of the crystal are determined by electrons contained in bands, corresponding to the uppermost atomic shells, occupied by the valence electrons. If the uppermost band is not completely filled by electrons at 0 K, it is called the conduction band. If it is completely filled at 0 K, it is called the valence band and the empty band just above it is called the conduction band. The material in which the uppermost band is not completely filled is a metal, while in the second case, we have an insulator or a semiconductor, depending on the width of the energy gap, $\mathcal{E}_g$, between the valence and the conduction bands. As a rule of thumb, we have a semiconductor when $0 < \mathcal{E}_g \leq 4$ eV and an insulator when $\mathcal{E}_g > 4$ eV. This discrimination is not so sharp: Indeed, diamond, with an energy gap $\mathcal{E}_g = 5.5$ eV, is still considered a semiconductor. The materials with $\mathcal{E}_g = 0$ are called semimetals. When a band is completely filled by electrons, such electrons cannot conduct any current.

1.5.1 Electrons and Holes in a Semiconductor

In the case of a semiconductor, upon increasing the temperature, some electrons can be excited from the valence to the conduction band. In this way, the valence band is left with a few empty states. Empty states in a band are generally called holes. When an external electromagnetic field is applied to a semiconductor, a hole behaves as a particle with a positive charge e. Let us consider a completely filled band. In this case, the sum of all the wave vectors of the electrons is zero:

$$\sum \mathbf{k} = 0, \tag{1.70}$$

the sum being over all states of the considered band in the first Brillouin zone. This is a consequence of the inversion symmetry of the first Brillouin zone, which follows from the symmetry under the inversion $\mathbf{r} \rightarrow -\mathbf{r}$ about any lattice point of the crystal lattice. As a result, there is the same number of filled states with positive wavenumber $\mathbf{k}$ and with negative wavenumber $-\mathbf{k}$. If an electron with wavenumber $\mathbf{k}_e$ is promoted from the valence to the conduction band, the total wavevector is now $\sum \mathbf{k} = -\mathbf{k}_e$. This wavevector is assigned to the hole, so that the wavevector of the hole, $\mathbf{k}_h$, is given by:

$$\mathbf{k}_h = -\mathbf{k}_e. \tag{1.71}$$

The equation of motion of the missing electron in an electromagnetic field is given by Eq. 1.57:

$$\frac{d(\hbar \mathbf{k}_e)}{dt} = -e(\mathbf{E} + \mathbf{v}_g \times \mathbf{B}). \tag{1.72}$$

Therefore, the corresponding equation of motion of the hole (with $\mathbf{k}_h = -\mathbf{k}_e$) is:

$$\frac{d(\hbar \mathbf{k}_h)}{dt} = e(\mathbf{E} + \mathbf{v}_g \times \mathbf{B}), \tag{1.73}$$

which is the equation of motion of a particle with positive charge e.

1.6 Calculation of the Band Structure

Various techniques can be employed to calculate the dependence of energy versus wavevector, that is the dispersion relation $\mathcal{E}(\mathbf{k})$, which gives the band structure of the material. It is not the aim of this section to describe in detail these band structure calculations, which are extensively treated in many textbooks. Each method must solve two main problems: (i) the determination of the periodic one-electron potential, $U(\mathbf{r})$, of the crystal, which is contained in the Schrödinger equation; (ii) find the eigenvalues and eigenfunctions of this equation. The various methods differ in the procedures used to solve these two problems. In the following, the tight-binding method (TBM) will be briefly discussed, mainly for didactic reasons, since, in particular situations, it allows one to derive analytic expressions for the dispersion relation $\mathcal{E}(\mathbf{k})$, which are very useful to get a simple physical understanding of the band structure. We will also mention the so-called $\mathbf{k} \cdot \mathbf{p}$ method. Many other methods can be employed: for example, the pseudopotential method, the cellular (Wigner–Seitz) method, the so-called muffin-tin methods, which include the augmented plane wave (APW) method, and others.

1.6.1 Tight-Binding Method

TBM is an empirical approach since experimental inputs have to be used to fit the actual band structure. In the previous section, we saw how energy bands are created upon progressively reducing the distances between the atoms composing the crystal. In the final configuration, the atomic wavefunctions of neighboring atoms can be more or less overlapped: The TBM can be properly applied when the overlap of the atomic wavefunctions cannot be neglected, but it is not too large, so that the atomic description can still be used. In this case, the atomic wavefunctions can be used as a basis set for the Bloch functions. Strictly speaking, in order to obtain the eigenstates of the valence electrons of a crystal, all atomic wavefunctions of the atoms composing the crystal are required, since only all atomic orbitals compose a complete basis set. However, the largest contribution is given by the atomic orbitals of the valence shells of the free atoms: Within the TBM, only these orbitals are considered. For example, for the elemental semiconductors of group IV of the periodic table, the valence shell orbitals are formed by the $2s$ and $2p$ states in the case of C, by the $3s$ and $3p$ states for Si, and by the $4s$ and $4p$ states for Ge. In the case of semiconductors formed by different elements, the valence shell orbitals of the various atoms have to be considered. For example, in the case of GaAs, the $4s$ state and the $4p$ states of Ga and the $4s$ and $4p$ states of As must be considered. In the above examples, the valence shells of the atoms are formed by s-type or p-type orbitals.

We will assume a lattice with a single atom basis. We will further assume that the bound states of the atomic Hamiltonian, H_{at}, are well localized so that the wavefunction $\psi_n(\mathbf{r})$ for an atom placed at the origin ($\mathbf{r} = 0$) becomes very small at a distance from the origin of the order of the lattice constant. $\psi_n(\mathbf{r})$ is given by:

$$H_{at}\psi_n = \mathcal{E}_n\psi_n. \tag{1.74}$$

We assume that close to each lattice point the crystal Hamiltonian, H, can be approximated by H_{at}. The crystal Hamiltonian can be written as:

$$H = H_{at} + \Delta U(\mathbf{r}), \tag{1.75}$$

where $\Delta U(\mathbf{r})$ is the correction to the atomic potential, which has to be added to obtain the periodic potential of the crystal, as shown schematically in Fig. 1.14. If $\Delta U(\mathbf{r})$ vanishes in the region where the atomic orbital $\psi_n(\mathbf{r})$ is localized ($\mathbf{r} = 0$ in Fig. 1.14), $\psi_n(\mathbf{r})$ will also be a solution of the Schrödinger equation with the Hamiltonian given by Eq. 1.75. In this case, the crystal wavefunction $\psi_{\mathbf{k}}(\mathbf{r})$ can be written as a linear combination of atomic wavefunctions. Since $\psi(\mathbf{r})$ must satisfy the Bloch condition:

$$\psi_{\mathbf{k}}(\mathbf{r} + \mathbf{R}_0) = e^{i\mathbf{k}\cdot\mathbf{R}_0}\psi_{\mathbf{k}}(\mathbf{r}), \tag{1.76}$$

the linear combination can be written as follows:

$$\psi_{\mathbf{k}}(\mathbf{r}) = \sum_{\mathbf{R}} e^{i\mathbf{k}\cdot\mathbf{R}}\psi_n(\mathbf{r} - \mathbf{R}), \tag{1.77}$$

where the wavevector $\mathbf{k}$ varies within the first Brillouin zone and the sum over $\mathbf{R}$ spans all the crystal lattice. This wavefunction satisfies the requirement of a Bloch wavefunction, indeed, considering a generic lattice vector $\mathbf{R}_0$:

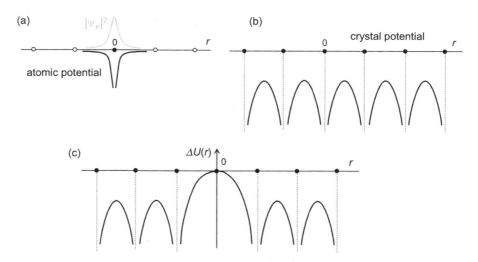

Figure 1.14 (a) Atomic potential of a single atom (black curve), the gray curve schematically represents the squared modulus of an atomic wavefunction, $|\psi_n|^2$, localized at the origin ($r = 0$). (b) Periodic crystal potential. (c) $\Delta U(r)$: correction to the atomic potential to obtain the periodic crystal potential.

$$
\begin{aligned}
\psi_{\mathbf{k}}(\mathbf{r} + \mathbf{R}_0) &= \sum_{\mathbf{R}} e^{i\mathbf{k}\cdot\mathbf{R}} \psi_n(\mathbf{r} - \mathbf{R} + \mathbf{R}_0) = \\
&= e^{i\mathbf{k}\cdot\mathbf{R}_0} \sum_{\mathbf{R}} e^{i\mathbf{k}\cdot(\mathbf{R}-\mathbf{R}_0)} \psi_n[\mathbf{r} - (\mathbf{R} - \mathbf{R}_0)].
\end{aligned}
\tag{1.78}
$$

$\mathbf{R} - \mathbf{R}_0$ is simply another crystal translation vector, and since the sum over $\mathbf{R}$ spans all the crystal, we can replace it by another translation vector, $\mathbf{R}'$, so that:

$$
\psi_{\mathbf{k}}(\mathbf{r} + \mathbf{R}_0) = e^{i\mathbf{k}\cdot\mathbf{R}_0} \sum_{\mathbf{R}'} e^{i\mathbf{k}\cdot\mathbf{R}'} \psi_n(\mathbf{r} - \mathbf{R}') = e^{i\mathbf{k}\cdot\mathbf{R}_0} \psi_{\mathbf{k}}(\mathbf{r}).
\tag{1.79}
$$

It is evident that the energy $\mathcal{E}(\mathbf{k})$ associated with the wavefunction Eq. 1.77 is simply the energy of the atomic level, $\mathcal{E}_n$. This is related to our initial assumption of a $\Delta U(\mathbf{r})$ term which vanishes where $\psi_n(\mathbf{r})$ is localized: This assumption is not realistic. Instead, we have to assume that $\Delta U(\mathbf{r})$ is small but not precisely zero where $\psi_n(\mathbf{r})$ becomes small, so that the product $\Delta U(\mathbf{r})\psi_n(\mathbf{r})$ is very small but not zero. Therefore, a more realistic solution for the crystal Schrödinger equation, still satisfying the Bloch conditions, can be written as:

$$
\psi(\mathbf{r}) = \sum_{\mathbf{R}} e^{i\mathbf{k}\cdot\mathbf{R}} \phi(\mathbf{r} - \mathbf{R}),
\tag{1.80}
$$

where $\phi(\mathbf{r})$ is very close, but not exactly equal, to the atomic wavefunction $\psi_n(\mathbf{r})$. We can assume that $\phi(\mathbf{r})$ can be written as a linear combination of a small number of localized atomic wavefunctions:

$$
\phi(\mathbf{r}) = \sum_n b_n \psi_n(\mathbf{r}).
\tag{1.81}
$$

For this reason, the TBM is also known as the *linear combination of atomic orbitals (LCAO) method*.

The Schrödinger equation is now:

$$H\psi(\mathbf{r}) = [H_{at} + \Delta U(\mathbf{r})]\psi(\mathbf{r}) = \mathcal{E}(\mathbf{k})\psi(\mathbf{r}).\tag{1.82}$$

We then multiply this equation by the atomic wave function $\psi_m^*(\mathbf{r})$, integrate over all $\mathbf{r}$, and use the fact that:

$$\int \psi_m^*(\mathbf{r})H_{at}\psi(\mathbf{r})d\mathbf{r} = \int (H_{at}\psi_m(\mathbf{r}))^*\psi(\mathbf{r})d\mathbf{r} = \mathcal{E}_m \int \psi_m^*(\mathbf{r})\psi(\mathbf{r})d\mathbf{r}.\tag{1.83}$$

We obtain:

$$\mathcal{E}_m \int \psi_m^*(\mathbf{r})\psi(\mathbf{r})d\mathbf{r} + \int \psi_m^*(\mathbf{r})\Delta U(\mathbf{r})\psi(\mathbf{r})d\mathbf{r} = \mathcal{E}(\mathbf{k}) \int \psi_m^*(\mathbf{r})\psi(\mathbf{r})d\mathbf{r},\tag{1.84}$$

which can be rewritten as:

$$[\mathcal{E}(\mathbf{k}) - \mathcal{E}_m] \int \psi_m^*(\mathbf{r})\psi(\mathbf{r})d\mathbf{r} = \int \psi_m^*(\mathbf{r})\Delta U(\mathbf{r})\psi(\mathbf{r})d\mathbf{r}.\tag{1.85}$$

Placing Eqs. 1.80–1.81 into Eq. 1.85 and using the orthonormality of the atomic wave functions:

$$\int \psi_m^*(\mathbf{r})\psi_n(\mathbf{r})d\mathbf{r} = \delta_{mn}.\tag{1.86}$$

We can write an eigenvalue equation from which it is possible to obtain the coefficients b_n and the Bloch energies $\mathcal{E}(\mathbf{k})$. In the sum over $\mathbf{R}$ we separate the terms with $\mathbf{R} = 0$ and $\mathbf{R} \neq 0$:

$$[\mathcal{E}(\mathbf{k}) - \mathcal{E}_m]b_m = -[\mathcal{E}(\mathbf{k}) - \mathcal{E}_m]\sum_n \left(\sum_{\mathbf{R}\neq 0}\int \psi_m^*(\mathbf{r})\psi_n(\mathbf{r}-\mathbf{R})e^{i\mathbf{k}\cdot\mathbf{R}}d\mathbf{r}\right)b_n +$$

$$+\sum_n \left(\int \psi_m^*(\mathbf{r})\Delta U(\mathbf{r})\psi_n(\mathbf{r})d\mathbf{r}\right)b_n +$$

$$+\sum_n \left(\sum_{\mathbf{R}\neq 0}\int \psi_m^*(\mathbf{r})\Delta U(\mathbf{r})\psi_n(\mathbf{r}-\mathbf{R})e^{i\mathbf{k}\cdot\mathbf{R}}d\mathbf{r}\right)b_n.\tag{1.87}$$

In a semiconductor, we have to consider the valence shell of the atoms from which the crystal is formed; therefore, we need to consider an s-level and triply degenerate p-levels. For this reason, Eq. 1.87 gives rise to a set of four homogeneous equations, whose eigenvalues give the dispersion relations $\mathcal{E}(\mathbf{k})$ for the s-band and the three p-bands and whose solutions $b(\mathbf{k})$ give the amplitudes of the LCAO (see Eq. 1.81) forming the wavefunction $\phi(\mathbf{r})$ in the first Brillouin zone. In this case, we have to solve a 4×4 secular problem. The calculations are rather complex. To get some physical insight into the problem, we will solve a simple problem of one atom basis with only an s-function.

1.6.2 Crystal with One-Atom Basis and Single Atomic Orbital

Since we have to consider only one atomic level, the coefficients $\{b_m\}$ are zero except for the s-level, where $b_s = 1$. A single equation results in this case. If $\mathcal{E}_s$ is the energy of the atomic s-level and we write $\psi_s(\mathbf{r}) = \phi(\mathbf{r})$, the following functions can be introduced:

$$\alpha(\mathbf{R}) = \int \phi^*(\mathbf{r})\phi(\mathbf{r} - \mathbf{R})d\mathbf{r} \tag{1.88}$$

$$\beta = -\int \phi^*(\mathbf{r})\Delta U(\mathbf{r})\phi(\mathbf{r})d\mathbf{r} = -\int \Delta U(\mathbf{r})|\phi(\mathbf{r})|^2 d\mathbf{r} \tag{1.89}$$

$$\gamma(\mathbf{R}) = -\int \phi^*(\mathbf{r})\Delta U(\mathbf{r})\phi(\mathbf{r} - \mathbf{R})d\mathbf{r}. \tag{1.90}$$

Equation 1.87 can be rewritten in terms of the previous quantities:

$$\mathcal{E}(\mathbf{k}) = \mathcal{E}_s - \frac{\beta + \sum \gamma(\mathbf{R})e^{i\mathbf{k}\cdot\mathbf{R}}}{1 + \sum \alpha(\mathbf{R})e^{i\mathbf{k}\cdot\mathbf{R}}}. \tag{1.91}$$

Since we are considering an s-level, $\alpha(-\mathbf{R}) = \alpha(\mathbf{R})$, therefore given the fact that $\Delta U(-\mathbf{r}) = \Delta U(\mathbf{r})$, we also have that $\gamma(-\mathbf{R}) = \gamma(\mathbf{R})$. In Eq. 1.91, we can ignore the sum $\sum \alpha(\mathbf{R})e^{i\mathbf{k}\cdot\mathbf{R}}$ with respect to one since $\phi(\mathbf{r})$ is well localized so that the overlapping integral Eq. 1.88, which contains the product of two atomic wavefunctions centered on different sites, is very small. Moreover, we can make another simplifying assumption if we assume that in the sum $\sum \gamma(\mathbf{R})e^{i\mathbf{k}\cdot\mathbf{R}}$ we have to consider only the nearest neighbors (n.n.). Equation 1.91 can be written in a very simple form:

$$\mathcal{E}(\mathbf{k}) = \mathcal{E}_s - \beta - \sum_{\text{n.n.}} \gamma(\mathbf{R})e^{i\mathbf{k}\cdot\mathbf{R}}. \tag{1.92}$$

In the case of real semiconductors, we cannot limit ourself to considering a crystal with a one-atom basis and one atomic orbital. We have already observed that the outermost electrons are placed in four atomic orbitals (one s and three p orbitals, p_x, p_y, and p_z). Moreover, the corresponding lattice has a two-atom basis. In this case, the crystal states can be expanded as linear combinations of two Bloch states, corresponding to the two atoms of the basis. Therefore, by using Eqs. 1.80 and 1.81 we can write:

$$\psi(\mathbf{r}) = \sum_{\mathbf{R}} \sum_{m=1}^{4} \sum_{j=1}^{2} C_{mj}(\mathbf{k})\psi_{mj}(\mathbf{r} - \mathbf{r}_j - \mathbf{R})e^{i\mathbf{k}\cdot\mathbf{R}}, \tag{1.93}$$

where the index m refers to the different atomic orbitals (s, p_x, p_y, and p_z), the index j refers to the atom in the basis, and $\mathbf{r}_j$ gives the position of the j-th atom in the unit cell. For example, in the case of the zinc–blende lattice (which is the structure of several semiconductors), for each atom of the two-atom basis we have an s orbital and three p orbitals, thus leading to a 8×8 secular problem (eight homogeneous equations of the form 1.87). Considering also the spin degeneracy, we would end up with a 16×16 secular problem, thus increasing the complexity of the tight-binding calculations.

    In the following section, we will consider two very simple examples of lattices with a one-atom basis and a single atomic state (s-state) in order to see two simple applications of the TBM. These simple applications will allow us to discuss a few important physical properties, which can be generalized to more realistic situations.

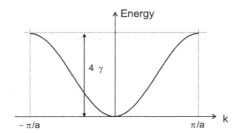

Figure 1.15 Dispersion relation, $\mathcal{E}(k)$, in the first Brillouin zone of a linear lattice.

1.6.3 Linear Lattice

In the case of a *linear lattice* with lattice constant a, the tight-binding wavefunction 1.80 can be written as:

$$\psi(x) = \sum_m e^{imka}\phi(x - ma). \tag{1.94}$$

From Eq. 1.92, we can immediately obtain the dispersion relation $\mathcal{E}(k)$ (which is shown in the first Brillouin zone in Fig. 1.15):

$$\mathcal{E}(k) = \mathcal{E}_s - \beta - \gamma(e^{ika} + e^{-ika}) = \mathcal{E}_s - \beta - 2\gamma\cos(ka). \tag{1.95}$$

The width of the band is 4γ, and it is proportional to the overlap integral. This is reasonable, since an increase of the overlap integral γ leads to a stronger interaction, and consequently a wider bandwidth. When the electron is near the bottom of the band, where $ka \ll 1$, we have:

$$\mathcal{E}(k) \simeq \mathcal{E}_s - \beta - 2\gamma\left(1 - \frac{1}{2}k^2a^2\right) = \mathcal{E}_s - \beta - 2\gamma + \gamma k^2 a^2, \tag{1.96}$$

which has the same parabolic dependence on k as the dispersion relation of a free electron. Therefore, we may conclude that an electron in that region of the k-space behaves as a free electron with an effective mass:

$$m^* = \frac{\hbar^2}{2\gamma a^2}, \tag{1.97}$$

which shows that the effective mass is inversely proportional to the overlap integral γ. This is intuitively reasonable since if the overlap between atomic wavefunctions increases, it is easier for an electron to hop from one atomic site to another, thus leading to a reduction of its effective mass (less inertia for the electronic motion in the crystal).

Upon considering the other atomic energy levels, we end up with various energy bands, as shown in Figure 1.16(a). Figure 1.16(b) shows the resulting reduced zone scheme, which we discussed as a consequence of Eq. 1.32 (band-folding procedure). Upon using Eq. 1.95, it is possible to calculate the group velocity of the electron and the effective mass in the first Brillouin zone (not only around $k = 0$):

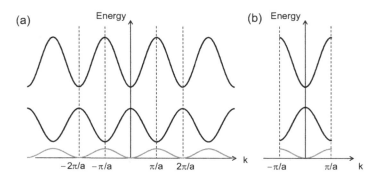

Figure 1.16 (a) Dispersion relation, $\mathcal{E}(k)$, in a linear lattice calculated by using the TBM. (b) Corresponding reduced zone scheme.

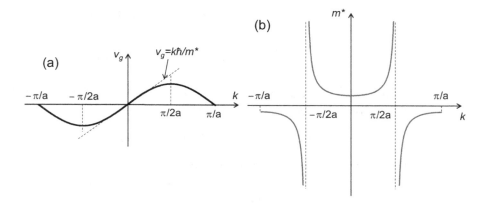

Figure 1.17 (a) Group velocity and (b) effective mass in the first Brillouin zone of a linear lattice.

$$v_g = \frac{1}{\hbar}\frac{\partial\mathcal{E}}{\partial k} = \frac{2\gamma a}{\hbar}\sin(ka), \tag{1.98}$$

$$m^* = \frac{\hbar^2}{\partial^2\mathcal{E}/\partial k^2} = \frac{\hbar^2}{2\gamma a^2\cos(ka)}. \tag{1.99}$$

Figure 1.17 shows v_g and m^* as a function of k. Note that in the central region of the first Brillouin zone, the group velocity is close to the free electron velocity, $\hbar k/m^*$. In this region, the electron can move almost freely through the lattice. This is no longer true approaching the borders of the Brillouin zone, $k = \pm\pi/a$: Here, the perturbation of the lattice is strong. This can be easily understood in terms of Bragg reflection in the linear lattice. Indeed, the Bragg condition for constructive interference of the waves scattered from each lattice point is $k = n\pi/a$. In this situation, we have a superposition between a forward ($k = \pi/a$) and a backward ($k = -\pi/a$) wave thus leading to a standing wave, which does not propagate in the crystal ($v_g = 0$).

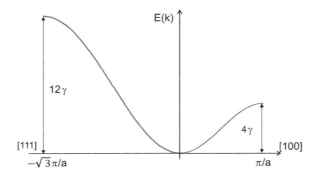

Figure 1.18 Dispersion curves along the [100] and [111] directions for a simple cubic lattice calculated by using the TBM.

1.6.4 Simple Cubic Lattice

Let us solve the problem for a simple cubic (sc) lattice. The six nearest neighbors are at:

$$a(\pm 1, 0, 0)$$
$$a(0, \pm 1, 0)$$
$$a(0, 0, \pm 1).$$

Equation 1.92 can be written as:

$$\mathcal{E}(\mathbf{k}) = \mathcal{E}_s - \beta - \gamma(e^{ik_x a} + e^{-ik_x a} + e^{ik_y a} + e^{-ik_y a} + e^{ik_z a} + e^{-ik_z a})$$
$$= \mathcal{E}_s - \beta - 2\gamma(\cos k_x a + \cos k_y a + \cos k_z a). \tag{1.100}$$

Along the [100] direction, we have ($k_x = k, k_y = k_z = 0$):

$$\mathcal{E}(\mathbf{k}) = \mathcal{E}_s - \beta - 2\gamma(\cos ka + 2). \tag{1.101}$$

The bottom of the band is at $k = 0$, and the bandwidth is 4γ. Along the [111] direction ($k_x^2 = k_y^2 = k_z^2 = k^2/3$), we have:

$$\mathcal{E}(\mathbf{k}) = \mathcal{E}_s - \beta - 6\gamma \cos \frac{ka}{\sqrt{3}}. \tag{1.102}$$

The bottom is at $k = 0$, and the bandwidth is 12γ. The top of the band is located at the extreme of the zone at $\sqrt{3}[\pi/a, \pi/a, \pi/a]$. The dispersion curves along the [100] and [111] directions are shown in Fig. 1.18. Also in this case, it is interesting to examine the band structure near the Γ point where $ka \ll 1$. Putting: $k_x^2 = k_y^2 = k_z^2 = k^2/3$, we have ($\cos x \simeq 1 - x^2/2$):

$$\mathcal{E}(\mathbf{k}) = \mathcal{E}_s - \beta - 6\gamma \cos \frac{ka}{\sqrt{3}} \simeq \mathcal{E}_s - \beta - 6\gamma \left(1 - \frac{k^2 a^2}{6}\right)$$
$$= \mathcal{E}_s - \beta - 6\gamma + \gamma k^2 a^2. \tag{1.103}$$

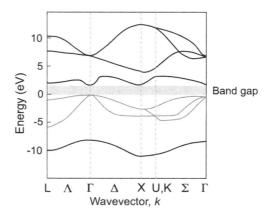

Figure 1.19   Calculated tight-binding band structure of GaAs without the effects of the spin–orbit coupling.

Comparing with the free electron problem solution, as in the previous example, we can obtain a simple expression for the effective mass (valid when $ka \ll 1$):

$$m^* = \frac{\hbar^2}{2\gamma a^2}. \tag{1.104}$$

1.6.5 Band Structure of Semiconductors Calculated by TBM

The band structure for GaAs calculated by using TBM is shown in Fig. 1.19. The top of the valence band and the bottom of the conduction band (referred to as the band edges of the semiconductor) are both at the Γ-point: GaAs is therefore a direct gap semiconductor. The states near the bottom of the conduction band have the symmetry of the s-state, whereas the states at the top of the valence band have the symmetry of the p-states. Generally, the states in the valence band can be written as linear combinations of p-orbitals. We note that the Bloch-lattice function maintains most of the symmetry properties as the original atomic orbitals. The band gap of the semiconductor and other properties in the high-symmetry points can be well fitted by using experimental inputs. Nevertheless, the calculated crystal structure shown in Fig. 1.19 does not reproduce the real structure of GaAs. This is clearly evident looking at the top of the valence band: Fig. 1.19 shows a threefold degeneracy, resulting from the degeneracy of the p_x, p_y, and p_z states. In contrast, measurements show that the top of the valence band is twofold degenerate and a third band is present below the valence band edge. This discrepancy is due to the fact that an important parameter has not been included so far in our discussion: the spin. The first result is to double the degeneracy of each state, both in the valence and conduction bands, without changing the dispersion curves. The second crucial effect is the spin–orbit interaction, which leads to changes of the energy curves in the valence band. Spin–orbit interaction is a relativistic effect: When an electron moves about the nucleus at relativistic velocity, the electric field generated by the nucleus Lorentz-transforms to a magnetic field, which interacts with the electron spin magnetic moment. In atoms, the spin–orbit interaction splits each electron energy level

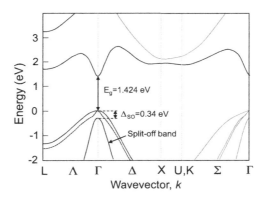

Calculated tight-binding band structure of GaAs including the effects of the spin–orbit coupling. Δ_{SO} is the spin–orbit splitting energy.

Table 1.2 Spin–orbit splitting energy for various semiconductors.	
Semiconductor	Δ_{SO} (eV)
Si	0.044
Ge	0.29
GaAs	0.34
InAs	0.41
InSb	0.85

with a nonzero orbital quantum number ℓ. Spin–orbit splitting in the band structure of crystals is due to the same effect. Spin–orbit interaction leads to modifications in the valence band, whose states are primarily p-states. Without entering into any detail, spin–orbit interaction can be included in the tight-binding calculation by adding an additional term, H_{SO}, in the Hamiltonian:

$$H_{SO} = \xi \, \mathbf{L} \cdot \mathbf{S}, \tag{1.105}$$

where $\mathbf{L}$ is the operator for orbital angular momentum, $\mathbf{S}$ is the operator for spin angular momentum, and ξ can be treated as a constant. Upon considering the spin–orbit interaction, the dispersion curves calculated for GaAs modify as shown in Fig. 1.20. The top of the valence band loses part of its degeneracy: A *split-off band* is formed, separated by an energy offset Δ_{SO} from the top of the valence band. The spin–orbit splitting for various semiconductors are reported in Table 1.2.

The spin–orbit splitting energy Δ_{SO} of semiconductors increases as the fourth power of the atomic number of the constituent elements. Intuitively this can be understood as follows. If the atomic number Z (which gives the number of protons in the nucleus) increases, the electric field seen by the valence electrons also increases and consequently also the interaction between the orbital momentum and the spin. Δ_{SO} is small for materials composed of light atoms, as Si, and may be quite large in comparison with the energy gap in

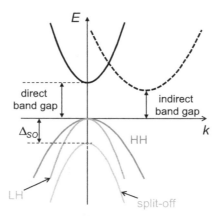

Figure 1.21 Schematic representation of the dispersion curves near the band edges of a semiconductor with direct or indirect (dashed conduction band curve) band gap. HH: heavy-hole valence band; LH: light-hole valence band.

semiconductors composed of heavy atoms, as InSb. Note that the conduction band is not affected by spin–orbit interaction (due to the *s*-symmetry) and remains doubly degenerate.

Figure 1.21 shows schematically the dispersion curves near the band edges of a semiconductor with direct or indirect (dashed conduction band curve) band gap. The doubly degenerate valence bands (fourfold degeneracy with spin) at the Γ-point are characterized by two different curvatures: These bands are called *heavy-hole* (HH) and *light-hole* (LH) bands. The band with the smaller curvature is the HH band, since the effective mass is inversely proportional to the band curvature.

So far we have considered only direct band gap semiconductors such as GaAs; there are semiconductors, like Si and Ge, where the *k*-value of the band edge of the conduction band is different from the *k*-value corresponding to the top of the valence band: These semiconductors are called *indirect*. For example, in Si the conduction band edge is close to the *X*-point and in Ge the conduction band edge is at the *L*-point, while for both semiconductors, the valence band edge is at the Γ-point. In the case of indirect semiconductors, the states at the conduction band edge present a strong anisotropy and can be described by combinations of *s*- and *p*-type states.

As will be clear in the following chapters, for most practical situations, only the dispersion relations close to the top of the valence band and to the bottom of the conduction bands have to be considered. For this reason, in most cases, parabolic dispersion relations can be considered. In this case, we can write:

$$\mathcal{E}(k) = \mathcal{E}_g + \frac{\hbar^2 k^2}{2m_c}, \tag{1.106}$$

in the conduction band, where $\mathcal{E}_g$ is the energy gap and m_c is the effective mass in the conduction band, and

$$\mathcal{E}(k) = -\frac{\hbar^2 k^2}{2m_{lh}}, \qquad \mathcal{E}(k) = -\frac{\hbar^2 k^2}{2m_{hh}}, \tag{1.107}$$

Table 1.3 Energy gap (in eV) and effective masses (in unit of the free electron mass m_0) of typical III–V semiconductors.

Material	$\mathcal{E}_g$ (eV) (0 K)	$\mathcal{E}_g$ (eV) (300 K)	m_c/m_0	m_{lh}/m_0	m_{hh}/m_0
GaAs	1.52	1.424	0.067	0.082	0.45
InP	1.42	1.27	0.08	0.089	0.6
InAs	0.43	0.354	0.023	0.025	0.41
InSb	0.23	0.17	0.014	0.016	0.40

in the valence band, where m_{lh} and m_{hh} are the effective masses for the LH and for the HH, respectively. Table 1.3 reports the values of the energy gap and effective masses for a few typical III–V semiconductors.

1.7 The k · p Method

In this section, we will briefly discuss another method, the **k** · **p** method, frequently used to calculate the band structure of semiconductors, which is quite accurate near the band edges. Here, we will just illustrate the basic idea at the heart of this method. In the following, we will use a few concepts and mathematical procedures which will be discussed in more detail in Chapter 3.

According to the Bloch theorem, the solution of the one-electron Schrödinger equation can be written as:

$$\psi_{n\mathbf{k}} = e^{i\mathbf{k}\cdot\mathbf{r}} u_{n\mathbf{k}}(\mathbf{r}), \tag{1.108}$$

where n is the band index, $\mathbf{k}$ is the wavevector in the first Brillouin zone, and $u_{n\mathbf{k}}(\mathbf{r})$ has the periodicity of the crystal. When Eq. 1.108 is substituted in the Schrödinger equation, the following equation in $u_{n\mathbf{k}}(\mathbf{r})$ is obtained:

$$\left(-\frac{\hbar^2}{2m_0}\nabla^2 - i\frac{\hbar\mathbf{k}}{m_0}\nabla + \frac{\hbar^2 k^2}{2m_0} + V(\mathbf{r}) \right) u_{n\mathbf{k}}(\mathbf{r}) = \mathcal{E}_{n\mathbf{k}} u_{n\mathbf{k}}(\mathbf{r}). \tag{1.109}$$

Since $\mathbf{p} = -i\hbar\nabla$, the previous equation can be written as:

$$\left(\frac{p^2}{2m_0} + \frac{\hbar\mathbf{k}\cdot\mathbf{p}}{m_0} + \frac{\hbar^2 k^2}{2m_0} + V \right) u_{n\mathbf{k}}(\mathbf{r}) = \mathcal{E}_{n\mathbf{k}} u_{n\mathbf{k}}(\mathbf{r}). \tag{1.110}$$

Therefore, the Hamiltonian can be written as the sum of two terms:

$$H = H_0 + H'_{\mathbf{k}}, \tag{1.111}$$

where:

$$H_0 = \frac{p^2}{2m_0} + V \tag{1.112}$$

is the unperturbed Hamiltonian and

$$H'_{\mathbf{k}} = \frac{\hbar\mathbf{k}\cdot\mathbf{p}}{m_0} + \frac{\hbar^2 k^2}{2m_0} \tag{1.113}$$

is the perturbation Hamiltonian. In the Γ-point ($\mathbf{k} = 0$) Eq. 1.110 gives:

$$\left(\frac{p^2}{2m_0} + V\right) u_{n0} = H_0 u_{n0} = \mathcal{E}_{n0} u_{n0} \qquad n = 1, 2, 3, \dots. \qquad (1.114)$$

The solutions of Eq. 1.114 form a complete and orthonormal set of basis functions. Now, we can use the terms $\hbar^2 \mathbf{k} \cdot \mathbf{p}/m_0$ and $\hbar^2 k^2/2m_0$ of the Hamiltonian as perturbations of the first and second order in $\mathbf{k}$. For this reason, the $\mathbf{k} \cdot \mathbf{p}$ model is quite accurate for small values of k. We note that the method can be used to calculate $\mathcal{E}_n(\mathbf{k})$ around any value $\mathbf{k}_0$ by expanding Eq. 1.110 around $\mathbf{k}_0$ if the wavefunctions and the energies at $\mathbf{k}_0$ are known. Now, if we assume that the band structure presents a stationary point (either a maximum or a minimum) at the energy $\mathcal{E}_{n0}$ and that the band is nondegenerate at this energy, by using standard perturbation theory it is possible to show that the wavefunctions $u_{n\mathbf{k}}$ and the energies $\mathcal{E}_{n\mathbf{k}}$ can be written as:

$$u_{n\mathbf{k}} = u_{n0} + \frac{\hbar}{m_0} \sum_{n' \neq n} \frac{\langle u_{n0}|\mathbf{k} \cdot \mathbf{p}|u_{n'0}\rangle}{\mathcal{E}_{n0} - \mathcal{E}_{n'0}} u_{n'0} \qquad (1.115)$$

$$\mathcal{E}_{n\mathbf{k}} = \mathcal{E}_{n0} + \frac{\hbar^2 k^2}{2m_0} + \frac{\hbar^2}{m_0^2} \sum_{n' \neq n} \frac{|\langle u_{n0}|\mathbf{k} \cdot \mathbf{p}|u_{n'0}\rangle|^2}{\mathcal{E}_{n0} - \mathcal{E}_{n'0}}. \qquad (1.116)$$

Note that the linear terms in k are not present since we are assuming that $\mathcal{E}_{n0}$ represents a maximum or a minimum of the band structure. In Eqs. 1.115 and 1.116, the matrix elements $\langle u_{n0}|\mathbf{k} \cdot \mathbf{p}|u_{n'0}\rangle$ can be written as:

$$\langle u_{n0}|\mathbf{k} \cdot \mathbf{p}|u_{n'0}\rangle = \int u_{n0}^* \mathbf{k} \cdot \mathbf{p} u_{n'0} \, d\mathbf{r}. \qquad (1.117)$$

Since $\mathbf{k}$ is a vector of real numbers, these matrix elements can be rewritten as:

$$\langle u_{n0}|\mathbf{k} \cdot \mathbf{p}|u_{n'0}\rangle = \mathbf{k} \cdot \langle u_{n0}|\mathbf{p}|u_{n'0}\rangle. \qquad (1.118)$$

For small values of k, it is very useful to write $\mathcal{E}_{n\mathbf{k}}$ as:

$$\mathcal{E}_{n\mathbf{k}} = \mathcal{E}_{n0} + \frac{\hbar^2 k^2}{2m^*}, \qquad (1.119)$$

where m^* is the effective mass. By comparing Eqs. 1.119 and 1.116, we obtain the following expression for the effective mass:

$$\frac{1}{m^*} = \frac{1}{m_0} + \frac{2}{m_0^2 k^2} \sum_{n' \neq n} \frac{|\langle u_{n0}|\mathbf{k} \cdot \mathbf{p}|u_{n'0}\rangle|^2}{\mathcal{E}_{n0} - \mathcal{E}_{n'0}}. \qquad (1.120)$$

This result is quite instructive since it states that the effective mass is not equal to the mass m_0 of the free electron as a result of the coupling between electronic states in different bands via the term $\mathbf{k} \cdot \mathbf{p}$. Moreover, the electron effective mass decreases upon decreasing the width of the gap $\mathcal{E}_{n0} - \mathcal{E}_{n'0}$ if the dominant term in Eq. 1.120 is given by the coupling between the conduction and valence bands.

As an example, from Eq. 1.116, it is possible to obtain an approximated expression of the dispersion curve for the conduction band assuming that $\mathcal{E}_{n0}$ corresponds to the bottom of the conduction band, $\mathcal{E}_{c0}$. The sum over n' can include only the terms with energies close to the top of the valence band, where the energy difference at the denominator, $\mathcal{E}_{n0} - \mathcal{E}_{n'0}$

is smallest, since these terms give the most important contribution. Indeed, the relative importance of a band n' to the effective mass is controlled by the energy gap between the two bands. Therefore, we can reasonably assume:

$$\mathcal{E}_{n0} - \mathcal{E}_{n'0} = \mathcal{E}_{c0} - \mathcal{E}_{v0} = \mathcal{E}_g. \tag{1.121}$$

From Eq. 1.118, we have:

$$|\langle u_{n0}|\mathbf{k} \cdot \mathbf{p}|u_{n'0}\rangle|^2 = |\mathbf{k} \cdot \langle u_{c0}|\mathbf{p}|u_{v0}\rangle|^2 = k^2|p_{cv}|^2, \tag{1.122}$$

where p_{cv} is the momentum matrix element between the two band edge Bloch functions:

$$p_{cv} = \langle u_{c0}|\mathbf{p}|u_{v0}\rangle = \int u_{c0}^*(\mathbf{r})\mathbf{p}u_{v0}^*(\mathbf{r}) \, d\mathbf{r}. \tag{1.123}$$

Assuming the zero of the energy axis at the top of the dispersion curve, from Eq. 1.116 we can write:

$$\mathcal{E}_c(\mathbf{k}) = \mathcal{E}_{c0} + \frac{\hbar^2 k^2}{2m_0} + \frac{\hbar^2 k^2}{m_0^2}\frac{|p_{cv}|^2}{\mathcal{E}_g} = \mathcal{E}_g + \frac{\hbar^2 k^2}{2m_0} + \frac{\hbar^2 k^2}{m_0^2}\frac{|p_{cv}|^2}{\mathcal{E}_g}, \tag{1.124}$$

and from Eq. 1.120 we have:

$$\frac{1}{m^*} = \frac{1}{m_0} + \frac{2}{m_0^2}\frac{|p_{cv}|^2}{\mathcal{E}_g}. \tag{1.125}$$

A usually used parameter is the *Kane energy*:

$$\mathcal{E}_P = \frac{2}{m_0}|p_{cv}|^2, \tag{1.126}$$

so that the dispersion curve in Eq. 1.124 can be written as:

$$\mathcal{E}_c(\mathbf{k}) = \mathcal{E}_g + \frac{\hbar^2 k^2}{2m_0}\left(1 + \frac{\mathcal{E}_P}{\mathcal{E}_g}\right) \tag{1.127}$$

and the effective mass is:

$$\frac{1}{m^*} = \frac{1}{m_0}\left(1 + \frac{\mathcal{E}_P}{\mathcal{E}_g}\right). \tag{1.128}$$

The Kane energy is usually much larger than the energy gap and it is, to a good approximation, almost the same for most III–V semiconductors. Table 1.4 reports the Kane energy

Table 1.4 Kane energy values of a few semiconductors [1].	
	$\mathcal{E}_P$ (eV)
GaAs	28.8
AlAs	21.1
InAs	21.5
InP	20.7
GaN	25.0
AlN	27.1

values of a few semiconductors. For the conduction band of direct gap semiconductors, if the band gap decreases also the electron effective mass decreases. In other words, the bigger the energy gap, the smaller the effect on the effective mass. For example, in InSb the conduction band effective mass is very small, $m^* = m_0/77$ as a consequence of the very small energy gap $\mathcal{E}_g = 250$ meV.

1.8 Bandstructures of a Few Semiconductors

In this final section, we will briefly discuss relevant properties of the band structure of a few representative semiconductors. We will concentrate mainly on the characteristics of the band edges since most of the optical properties of the semiconductors are determined by regions of the band structure close to the top of the valence band and to the bottom of the conduction band. We will consider three semiconductors: Si, GaAs, and GaN.

1.8.1 Silicon

We have already mentioned that Si crystal has a diamond structure, with the first Brillouin zone in the form of a truncated octahedron, as shown in Fig. 1.11(b). Silicon is an indirect semiconductor. For this reason, it is not extensively used in optical devices. The top of the valence band is at the Γ-point, while the bottom of the conduction band is at the point $\frac{2\pi}{a}(0.85, 0, 0)$, close to the X-point (the direction [100] from the Γ-point to the X-point is called Δ-line). Since there are six equivalent Δ-lines, there are six equivalent minima in the conduction band at $k = \frac{2\pi}{a}(\pm 0.85, 0, 0)$, $\frac{2\pi}{a}(0, \pm 0.85, 0)$, and $\frac{2\pi}{a}(0, 0, \pm 0.85)$. The corresponding energy gap is 1.1 eV. The effective mass at these anisotropic minima is characterized by a longitudinal mass along the six equivalent [100] directions and two transverse masses in the plane orthogonal to the longitudinal direction. The longitudinal mass is $m_l = 0.98\, m_0$, while the transverse mass is $m_t = 0.19\, m_0$. The constant energy surfaces near the conduction band minima in Si can be written as:

$$\mathcal{E}(\mathbf{k}) = \frac{\hbar^2 k_x^2}{2m_l} + \frac{\hbar^2(k_y^2 + k_z^2)}{2m_t} \tag{1.129}$$

and are ellipsoids, as shown in Fig. 1.23. Equation 1.129 represents the parabolic band approximation. The direct energy gap is ~ 3.4 eV. There is also a second minimum in the conduction band at the point L, which is located about 1.1 eV above the band edge. The top of the valence band at the Γ-point presents the HH and LH degeneracy. The top of the split-off band is just 44 meV below the top of the valence band, as shown in Fig. 1.22(b). The LH mass is $m_{lh} = 0.16\, m_0$, the HH mass is $m_{hh} = 0.46\, m_0$, and the split-off hole mass is $m_{h,SO} = 0.29\, m_0$.

1.8.2 Gallium Arsenide

Gallium arsenide is a III–V semiconductor. The electron configuration of isolated atoms is [Ar]$3d^{10}4s^24p^1$ for Ga and [Ar]$3d^{10}4s^24p^3$ for As. When forming a GaAs crystal, the

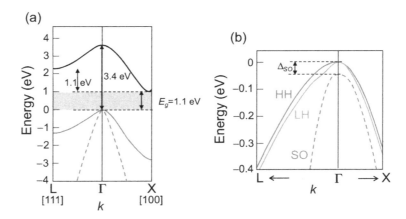

Figure 1.22 (a) Band structure of Si. (b) Zoom of a region of the valence band around the Γ-point, showing the heavy-hole (HH), the light-hole (LH), and split-off (SO) bands.

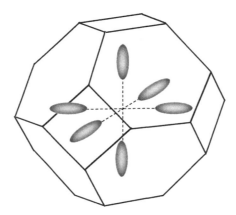

Figure 1.23 Constant energy surfaces near the conduction band minima in Si. There are six equivalent valleys in Si at the band edge.

outermost shell of each atom is hybridized in the form sp^3 so that each atom is surrounded by four neighbor atoms at the apex of a regular tetrahedron. GaAs is characterized by a zinc–blende crystal structure with a lattice constant $a = 5.65$ Å. GaAs is a direct band gap semiconductor, and it is largely used for optical devices (light-emitting diodes and lasers). The top of the valence band and the bottom of the conduction band are both at the Γ-point (see Fig. 1.20). The energy gap is 1.424 eV (at 300 K). The parabolic approximation for the conduction band is given by:

$$\mathcal{E}(k) = \frac{\hbar^2 k^2}{2m_c}, \tag{1.130}$$

where $m_c = 0.067 \, m_0$. In Eq. 1.130, we have assumed that at the band edge of the conduction band $\mathcal{E}(k) = 0$. When high fields are applied to the semiconductor, electrons can be injected in states quite far from the minima. In this situation, the parabolic approximation does not hold any more and a better approximation for the conduction band is the following:

$$\mathcal{E}(1 + \alpha\mathcal{E}) = \frac{\hbar^2 k^2}{2m_c}, \tag{1.131}$$

with $\alpha = 0.67$ eV^{-1}. The next highest minimum in the conduction band is close to the L-point. The separation between the Γ and L minima is 0.29 eV. The effective mass at the L minimum is much larger than the corresponding mass at the Γ-point: $m^* \approx 0.25\ m_0$. The next higher energy minimum in the conduction band is at the X-point; the separation between the Γ and X minima is 0.48 eV. The corresponding electron effective mass is $m_X^* \approx 0.6\ m_0$. When the semiconductor is subject to high electric fields, both the L and X valleys can be populated by electrons, in addition to the Γ valley. Therefore, these regions can be quite important to describe the electronic processes in GaAs. The valence band is characterized by the LH and HH bands and by the split-off band. The LH and HH masses are $m_{lh} = 0.082\ m_0$ and $m_{hh} = 0.45\ m_0$, respectively. The split-off energy is quite large: $\Delta = 0.34$ eV, so that, in general, this band is always completely filled by electrons and does not play any relevant role in the electronic and optical properties of GaAs.

The temperature dependence of the energy gap can be well approximated by the following expression, the *Varshni equation*:

$$\mathcal{E}_g = \mathcal{E}_g(0) - \frac{\alpha T^2}{T + \beta}, \tag{1.132}$$

where T is the temperature (in K) and $\mathcal{E}_g(0)$ is the energy gap at $T = 0$. $\mathcal{E}_g(0)$, α and β are suitable fitting parameters, which are material-specific. In the case of GaAs $\mathcal{E}_g(0) = 1.519$ eV, $\alpha = 5.405 \times 10^{-4}$ eV/K and $\beta = 204$ K. In the case of Si $\mathcal{E}_g(0) = 1.166$ eV, $\alpha = 4.73 \times 10^{-4}$ eV/K and $\beta = 636$ K. This is a quite general behavior in semiconductors: The energy gap decreases upon increasing the temperature. It is possible to show that the temperature dependence of the energy gap is due to the electron–phonon interaction in the semiconductor, which depends on the amplitude of the phonons and on the corresponding coupling constants.

1.8.3 Gallium Nitride

Gallium nitride is another III–V semiconductor with a direct band gap, largely used in optoelectronic devices like light-emitting diodes and lasers. It has a large direct energy gap $\mathcal{E}_g = 3.45$ eV, whose dependence on temperature can be taken into account by using the Varshni equation with $\mathcal{E}_g(0) = 3.51$ eV, $\alpha = 9.09 \times 10^{-4}$ eV/K, and $\beta = 830$ K (in the range 293 K $< T <$ 1237 K). It usually presents the hexagonal wurtzite crystal structure. Also in this semiconductor, each atom presents sp^3 hybridization and forms tetrahedral bonds with its nearest neighbors, with a bonding angle of 109.47° and a bond length of 19.5 nm. The large energy gap is directly related to the large bonding strength in GaN. The first Brillouin zone of a hexagonal crystal is a hexagonal prism with height $2\pi/c$, as shown in Fig. 1.11(c). The spin–orbit splitting energy is $\Delta = 15.5$ meV. The band structure

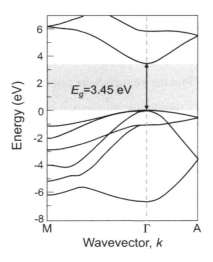

Figure 1.24 Calculated band structure of wurtzite GaN. Adapted with permission from [2].

of GaN near the band edges is shown in Fig. 1.24. The electron effective mass in the conduction band (Γ-minimum) is $m_c = 0.20\ m_0$; in the valence band, the HH effective mass along the axis a is $m_{hh,a} = 1.65\ m_0$, along the c-axis $m_{hh,c} = 1.1\ m_0$, the LH masses are: $m_{lh,a} = 0.15\ m_0$ and $m_{lh,c} = 1.1\ m_0$.

1.9 Exercises

Exercise 1.1 The atomic packing factor is defined as the fraction of the volume of a crystal which is occupied by the constituent atoms, assumed as identical hard spheres centered on the atoms and with radius such that spheres on neighboring points just touch. Determine the atomic packing factor of the following crystal structures:
a) simple cubic;
b) body-centered cubic;
c) face-centered cubic;
d) hexagonal close-packed.

Exercise 1.2 The primitive translation vectors of the face-centered cubic lattice may be taken as:

$$\mathbf{a}_1 = \frac{a}{2}(\mathbf{u}_x + \mathbf{u}_y)$$
$$\mathbf{a}_2 = \frac{a}{2}(\mathbf{u}_y + \mathbf{u}_z)$$
$$\mathbf{a}_3 = \frac{a}{2}(\mathbf{u}_z + \mathbf{u}_x)$$

a) determine the angles between these vectors;

b) determine the volume of the primitive cell;

c) determine the primitive translation vectors of the reciprocal lattice.

Exercise 1.3 Plot the Wigner–Seitz cell of a two-dimensional hexagonal lattice.

Exercise 1.4 Consider the momentum operator $\mathbf{p} = -i\hbar\nabla$ and the Bloch wavefunctions $\psi_{n\mathbf{k}}(\mathbf{r})$. Determine:

a) $\mathbf{p}\psi_{n\mathbf{k}}$;

b) $\mathbf{p}^2\psi_{n\mathbf{k}}$.

Exercise 1.5 By using the results of the previous exercise, write the Hamiltonian operator, $\mathbf{p}^2/2m + V(\mathbf{r})$, for the periodic function $u_{n\mathbf{k}}(\mathbf{r})$.

Exercise 1.6 Consider a monoatomic square lattice with only s-orbitals. Using the tight-binding method calculate and plot the dispersion curve ($\Gamma \to M \to X \to \Gamma$).

Exercise 1.7 A graphene lattice has two unit lattice vectors given by:

$$\mathbf{a}_1 = 3/2L\mathbf{u}_x - \sqrt{3}/2L\mathbf{u}_y$$
$$\mathbf{a}_2 = 3/2L\mathbf{u}_x + \sqrt{3}/2L\mathbf{u}_y$$

a) Determine the primitive vectors of the reciprocal lattice.

b) Using the Wigner–Seitz algorithm, plot the first two-dimensional Brillouin zone and indicate the position vectors of all the corners. Indicate also the Γ point.

c) Use the following dispersion relation for graphene:

$$\mathcal{E} = \pm t\sqrt{1 + 4\cos\left(\frac{3L}{2}k_x\right)\cos\left(\frac{\sqrt{3}L}{2}k_y\right) + 4\cos^2\left(\frac{\sqrt{3}L}{2}k_y\right)}$$

to plot the energy dispersion from Γ point to one of the corners ($t = 3$ eV).

d) What is the effective mass and velocity in the vicinity of the corner at $\mathcal{E} = 0$?

e) Verify that it is a good estimate to write the energy dispersion at the corner (valley) as $\mathcal{E} = \hbar vk$. What is the valley degeneracy for graphene?

Exercise 1.8 Using a particular software (QUANTUM ESPRESSO) and the generalized gradient approximation (GGA), H.Q. Yang *et al.* have calculated the effective mass of the ternary alloy $Ga_xIn_{1-x}P$ for Ga composition x varying from 0 to 1 [3]. The results are listed in Table 1.5:

a) Plot the effective mass as a function of x.

b) Write a polinomial fit for m^*/m_0 as a function of x. Note: for $0 \leq x \leq 0.625$, you can use a linear fit, while for $0.625 \leq x \leq 1$, you can use a parabolic fit. You will see that the curve $m^*(x)/m_0$ presents an inflection point at $x \approx 0.7$, where there is a change of the direct-to-indirect band gap for $Ga_xIn_{1-x}P$.

Table 1.5 Effective mass of $Ga_xIn_{1-x}P$, normalized to the free electron mass m_0, versus Ga concentration x.	
x	m^*/m_0
0	0.073060
0.125	0.089079
0.250	0.096235
0.375	0.105914
0.500	0.120593
0.625	0.133188
0.750	0.203615
0.875	0.412529
1	0.782270

Exercise 1.9 The velocity of an electron in a crystal can be written as $\mathbf{v} = (1/\hbar)\nabla_{\mathbf{k}}\mathcal{E}_n(\mathbf{k})$. Determine the velocity in the following two cases:
a) parabolic band: $\mathcal{E}(\mathbf{k}) = \hbar^2 k^2/2m^*$;
b) nonparabolic band: $\mathcal{E}(\mathbf{k})(1 + \alpha\mathcal{E}(\mathbf{k})) = \hbar^2 k^2/2m^*$.

Exercise 1.10 Assuming that $\mathcal{E} = \mathcal{E}_0 + 2\gamma\cos(ka)$, determine the electron position as a function of time (ignore scattering).

Exercise 1.11 Consider a simple cubic lattice (lattice spacing a), with one s-orbital per site. Assuming only nearest-neighbor interactions and neglecting overlap
a) show that the tight binding s-band is given by:

$$\mathcal{E}(\mathbf{k}) = \mathcal{E}_s - 2\gamma(\cos k_x a + \cos k_y a + \cos k_z a)$$

b) Plot the dispersion curve $\mathcal{E}(\mathbf{k})$ along directions $\Gamma - X - M - \Gamma - R$, where $X = \pi/a(1,0,0)$, $M = \pi/a(1,1,0)$, and $R = \pi/a(1,1,1)$.

Exercise 1.12 In a two-dimensional square lattice with lattice spacing a, the s-band can be written as:

$$\mathcal{E}(\mathbf{k}) = \mathcal{E}_s - 2\gamma(\cos k_x a + \cos k_y a).$$

a) Determine the contours of constant energy near $k = 0$.
b) Determine the contours of constant energy near the zone corners of the first Brillouin zone, showing that they are circles in the k-space. *Hint*: near the zone corners, k_x and k_y can be written as $k_x = (\pi/a) - \delta_x$ and $k_y = (\pi/a) - \delta_y$, with $\delta_{x,y}a \ll 1$.

Exercise 1.13 Write the tight-binding s-band, $\mathcal{E}(\mathbf{k})$, for a linear lattice (lattice spacing a), including a second-neighbor interaction, γ'.

Exercise 1.14 As first derived by Dresselhaus et al., the dispersion of the heavy- and light-hole bands at the top valence bands in diamond- and Zinc Blende-type semiconductors can be written as:

$$\mathcal{E}_{hh} = -Ak^2 - [B^2k^4 + C^2(k_x^2k_y^2 + k_y^2k_z^2 + k_z^2k_x^2)]^{1/2}$$

$$\mathcal{E}_{lh} = -Ak^2 + [B^2k^4 + C^2(k_x^2k_y^2 + k_y^2k_z^2 + k_z^2k_x^2)]^{1/2},$$

where the constants A, B, and C are related to the electron momentum matrix elements and to the energy gap. Determine the hole effective masses along the [100] and [111] directions.

Exercise 1.15 Plot the constant energy contour plot for the heavy-hole and light-hole dispersion reported in the previous exercise, in the plane (100) and (010).

Exercise 1.16 Calculate the density of wurtzite GaN. The lattice constants of GaN are $a = 3.186$ Å, $c = 5.186$ Å. The molar mass of GaN is 83.73 g/mol.

Exercise 1.17 Germanium has a diamond structure with a lattice constant $a = 5.64$ Å. Calculate the atomic density and the spacing between nearest-neighbor atoms.

Exercise 1.18 GaAs has a zinc–blende crystal structure. Assuming that the lattice spacing is $a = 5.65$ Å, calculate:
a) the density of GaAs in g/cm^3;
b) the density in atoms/cm^3;
c) the valence electron number per unit volume;
d) the closest spacing between adjacent As atoms.

Exercise 1.19 Consider a simple cubic semiconductor with lattice constant $a = 0.5$ nm. The conduction and valence bands have the following dispersion curve along the k_x direction:

$$\mathcal{E}_c = A\left[1 - \frac{1}{2}\sin^2\left(\frac{k_x a}{2}\right)\right]$$

$$\mathcal{E}_v = B\left[\cos\left(k_x a\right) - 1\right],$$

where $A = 1$ eV and $B = 0.5$ eV.
a) Plot the dispersion curves in the first Brillouin zone along the k_x direction and determine if the semiconductor is a direct or indirect band gap material.
b) In the ideal case of absence of scattering, evaluate the maximum velocity along the X-direction of an electron in the valence band and specify the corresponding point in the first Brillouin zone.
c) Evaluate the effective mass m_{xx} of the electron at the conduction band minimum and of the holes at the valence band maximum.

Exercise 1.20 Consider a tetragonal Bravais lattice with lattice parameters a, a and $c = 3a/4$, as shown in Fig. 1.25, where $a = 0.45$ nm. Determine the reciprocal lattice vectors $\mathbf{b}_1$, $\mathbf{b}_2$, and $\mathbf{b}_3$.

Exercise 1.21 Consider a two-dimensional hexagonal Bravais lattice, with lattice parameter $a = 0.5$ nm, as shown in Fig. 1.26. The sites of the lattice are occupied by atoms with external s orbitals. The nearest-neighbor transfer integrals $\gamma = 0.15$ eV are assigned. All

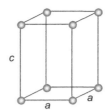

Figure 1.25   Tetragonal Bravais lattice.

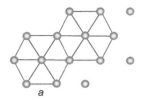

Figure 1.26   Two-dimensional hexagonal Bravais lattice, with lattice parameter $a = 0.5$ nm.

other transfer integrals and all overlap integrals are negligible. The zero of the energy is set at the atomic level, $\mathcal{E}_s = 0$.

a) Determine the dispersion $\mathcal{E}(\mathbf{k})$ of Bloch electrons within the tight-binding approximation, $\mathbf{k} = (k_x, k_y)$, being the wave vector.

b) Determine the expression and numerical values of the elements of the inverse mass tensor m_{ij}^{-1}, with $i, j = x, y$, at the Γ point of the Brillouin zone.

c) Determine the band width.

Electrons in Semiconductors

2.1 Introduction

In Chapter 1, we have seen that in a semiconductor at $T = 0$ K, the valence band is completely filled with electrons while the conduction band is empty. In this way, no current can flow in the material. When electrons are promoted in the conduction band, holes are generated in the valence band and current can now flow in the semiconductor. In this chapter, we will discuss the main properties of carriers in semiconductors. The density of states will be first calculated, and the essential concepts of carrier statistics in semiconductors will be discussed.

2.2 Periodic Boundary Conditions

We assume that the semiconductor is divided in adjacent boxes with edges $\mathbf{L}_i = N_i \mathbf{a}_i$ ($i = 1, 2, 3$), where $\mathbf{a}_i$ are primitive vectors and N_i can be considered arbitrarily large. We further assume that the crystal repeats itself exactly in the same way from one box to the other, so that:

$$\psi(\mathbf{r} + N_i \mathbf{a}_i) = \psi(\mathbf{r}). \tag{2.1}$$

The previous equation is the mathematical formulation of the so-called *Born-von Karman boundary conditions*. Since $\psi(\mathbf{r})$ is a Bloch function, Eq. 2.1 can be written as follows:

$$\psi_{n\mathbf{k}}(\mathbf{r} + N_i \mathbf{a}_i) = e^{i\mathbf{k} \cdot N_i \mathbf{a}_i} \psi_{n\mathbf{k}}(\mathbf{r}) = \psi_{n\mathbf{k}}(\mathbf{r}), \tag{2.2}$$

so that

$$e^{i N_i \mathbf{k} \cdot \mathbf{a}_i} = 1. \tag{2.3}$$

The wavevector $\mathbf{k}$ can be written as:

$$\mathbf{k} = \kappa_1 \mathbf{b}_1 + \kappa_2 \mathbf{b}_2 + \kappa_3 \mathbf{b}_3, \tag{2.4}$$

where $\mathbf{b}_j$ are the primitive vectors in the reciprocal space. Since $\mathbf{a}_i \cdot \mathbf{b}_j = 2\pi \delta_{ij}$, Eq. 2.3 can be written as follows:

$$e^{i 2\pi \kappa_i N_i} = 1, \tag{2.5}$$

so that

$$\kappa_i = \frac{m_i}{N_i}, \tag{2.6}$$

where m_i are integers. We conclude that the allowed $\mathbf{k}$ values are given by the following expression:

$$\mathbf{k} = \frac{m_1}{N_1}\mathbf{b}_1 + \frac{m_2}{N_2}\mathbf{b}_2 + \frac{m_3}{N_3}\mathbf{b}_3. \tag{2.7}$$

Therefore, each state in the $\mathbf{k}$-space occupies a volume Δk given by the volume of a parallelepiped with edges $\mathbf{b}_i/N_i$, given by:

$$\Delta k = \frac{\mathbf{b}_1}{N_1} \cdot \left(\frac{\mathbf{b}_2}{N_2} \times \frac{\mathbf{b}_3}{N_3} \right) = \frac{1}{N}\mathbf{b}_1 \cdot (\mathbf{b}_2 \times \mathbf{b}_3), \tag{2.8}$$

where $N = N_1 N_2 N_3$ is the total number of primitive cells in the crystal box and $\mathbf{b}_1 \cdot (\mathbf{b}_2 \times \mathbf{b}_3)$ represents the volume of a primitive cell in the reciprocal lattice. It is possible to demonstrate that the reciprocal lattice primitive vectors, $\mathbf{b}_i$, satisfy the following relation:

$$\mathbf{b}_1 \cdot (\mathbf{b}_2 \times \mathbf{b}_3) = \frac{(2\pi)^3}{\mathbf{a}_1 \cdot (\mathbf{a}_2 \times \mathbf{a}_3)}. \tag{2.9}$$

Since the quantity at the denominator of the previous expression represents the volume, V_u, of a Bravais lattice primitive cell in the direct lattice, so that $V_u = V/N$, the volume of the reciprocal lattice primitive cell is $(2\pi)^3 N/V$. Therefore, Eq. 2.8 can be written as:

$$\Delta k = \frac{(2\pi)^3}{V}. \tag{2.10}$$

Note that in the case of a simple cubic crystal $N_i = L/a$, $\mathbf{a}_i = a\mathbf{u}_i$, $\mathbf{b}_i = (2\pi/a)\mathbf{u}_i$ and Eq. 2.7 gives:

$$\mathbf{k} = m_1\frac{2\pi}{L}\mathbf{u}_x + m_2\frac{2\pi}{L}\mathbf{u}_y + m_3\frac{2\pi}{L}\mathbf{u}_z, \tag{2.11}$$

and the volume Δk per allowed k value is given by

$$\Delta k = \frac{(2\pi)^3}{L^3}, \tag{2.12}$$

in agreement with Eq. 2.10.

In the case of a two-dimensional lattice, the area per allowed k value is:

$$\Delta k = \frac{(2\pi)^2}{A}, \tag{2.13}$$

where, in the case of a square lattice $A = L^2$. In the case of a one-dimensional semiconductor, we have:

$$\Delta k = \frac{2\pi}{L}. \tag{2.14}$$

Exercise 2.1   Demonstrate Eq. 2.9.

We can write $\mathbf{b}_1$ by using Eq.1.6, so that:

$$\mathbf{b}_1 \cdot (\mathbf{b}_2 \times \mathbf{b}_3) = 2\pi \frac{(\mathbf{a}_2 \times \mathbf{a}_3) \cdot (\mathbf{b}_2 \times \mathbf{b}_3)}{\mathbf{a}_1 \cdot (\mathbf{a}_2 \times \mathbf{a}_3)}. \tag{2.15}$$

The previous expression can be rewritten by using the following identity related to the cross product:

$$(\mathbf{a} \times \mathbf{b}) \cdot (\mathbf{c} \times \mathbf{d}) = (\mathbf{a} \cdot \mathbf{c})(\mathbf{b} \cdot \mathbf{d}) - (\mathbf{a} \cdot \mathbf{d})(\mathbf{b} \cdot \mathbf{c}), \tag{2.16}$$

where $\mathbf{a}, \mathbf{b}, \mathbf{c}$, and $\mathbf{d}$ are generic vectors. By using Eq. 2.16 in 2.15, one obtains:

$$\mathbf{b}_1 \cdot (\mathbf{b}_2 \times \mathbf{b}_3) = 2\pi \frac{(\mathbf{a}_2 \cdot \mathbf{b}_2)(\mathbf{a}_3 \cdot \mathbf{b}_3) - (\mathbf{a}_2 \cdot \mathbf{b}_3)(\mathbf{a}_3 \cdot \mathbf{b}_2)}{\mathbf{a}_1 \cdot (\mathbf{a}_2 \times \mathbf{a}_3)}. \tag{2.17}$$

Since $\mathbf{b}_i \cdot \mathbf{a}_j = 2\pi \delta_{ij}$, the previous expression gives:

$$\mathbf{b}_1 \cdot (\mathbf{b}_2 \times \mathbf{b}_3) = \frac{(2\pi)^3}{\mathbf{a}_1 \cdot (\mathbf{a}_2 \times \mathbf{a}_3)}. \tag{2.18}$$

∎

2.3 Density of States

2.3.1 Bulk (Three-Dimensional) Semiconductors

We can now calculate the density of states, $\rho(\mathcal{E})$, in a semiconductor as a function of energy. The physical meaning of $\rho(\mathcal{E})$ is the following: $\rho(\mathcal{E}) \, d\mathcal{E}$ gives the number of states per unit volume with energy in the range between $\mathcal{E}$ and $\mathcal{E} + d\mathcal{E}$. The first step is the calculation of the density of states in the **k**-space, $\rho(k)$. We will assume a parabolic dispersion curve:

$$\mathcal{E} = \frac{\hbar^2 k^2}{2m^*} + \mathcal{E}_0. \tag{2.19}$$

In the **k**-space, the states with energy between $\mathcal{E}$ and $\mathcal{E} + d\mathcal{E}$ are the states characterized by k values contained in the region between two spherical surfaces with radii k and $k + dk$, whose volume is $4\pi k^2 \, dk$. Since the volume occupied by a single state is $(2\pi)^3/V$, the number of states with k value in the region k and $k + dk$ is:

$$\frac{4\pi k^2 dk}{8\pi^3} V = \frac{k^2}{2\pi^2} V \, dk. \tag{2.20}$$

Therefore, the density of states in the k-space is given by:

$$\rho(k) \, dk = \frac{k^2}{2\pi^2} dk, \qquad \rho(k) = \frac{k^2}{2\pi^2}. \tag{2.21}$$

Upon taking into account the electron spin degeneracy, so that each state can be occupied by two electrons with opposite spin, we obtain the following expression:

$$\rho(k) = \frac{k^2}{\pi^2}. \tag{2.22}$$

In order to calculate $\rho(\mathcal{E})$, we have to use the dispersion relation 2.19 and impose the conservation of the number of states in the **k**-space and in the $\mathcal{E}$-space:

$$\rho(\mathcal{E}) \, d\mathcal{E} = \rho(k) \, dk. \tag{2.23}$$

If we differentiate Eq. 2.19, we obtain:

$$d\mathcal{E} = \frac{\hbar^2 k}{m^*} \, dk. \tag{2.24}$$

From Eqs. 2.19, 2.23, and 2.24, we have:

$$\rho(\mathcal{E}) = \frac{1}{2\pi^2} \left(\frac{2m^*}{\hbar^2} \right)^{3/2} \sqrt{\mathcal{E} - \mathcal{E}_0}. \tag{2.25}$$

If we refer to the conduction band, in Eq. 2.25, we have $\mathcal{E}_0 = \mathcal{E}_c$, where $\mathcal{E}_c$ is the band edge (bottom energy of the conduction band).

We note here, without entering into details, that in direct band gap semiconductors, m^* is simply given by the effective mass for the conduction band: $m^* = m_c$. In the case of indirect semiconductors, m^* in the conduction band is an average effective mass of the conduction band, given by:

$$m^* = (g^2 m_1 m_2 m_3)^{1/3}, \tag{2.26}$$

where m_1, m_2, and m_3 are the effective masses along the three principal axis and g is the degeneracy factor. For example, for the X-valley of silicon $g = 6$ and Eq. 2.26 gives:

$$m_c^* = 6^{2/3} (m_l m_t^2)^{1/3}. \tag{2.27}$$

In the same way, the density of states in the valence band is given by:

$$\rho_v(\mathcal{E}) = \frac{1}{2\pi^2} \left(\frac{2m_v}{\hbar^2} \right)^{3/2} \sqrt{\mathcal{E}_v - \mathcal{E}}, \tag{2.28}$$

where $\mathcal{E}_v$ is the band edge of the valence band (i.e., the valence band maximum) and m_v is given by:

$$m_v = (m_{hh}^{3/2} + m_{lh}^{3/2})^{2/3}. \tag{2.29}$$

Figure 2.1(a) shows the density of states in the conduction band.

Note that in III–V semiconductors $m_c \ll m_v$, therefore the density of states in the conduction band is typically smaller than the density of states in the valence band. Moreover, since $m_{lh} \ll m_{hh}$, the density of states of light-holes is a small fraction of the density of states of the heavy-hole states. Therefore, light-hole states can be typically neglected in comparison with heavy-holes (i.e., true in the case of bulk semiconductors, not in the case of quantum-well semiconductors).

2.3.2 Two- and One-Dimensional Semiconductors

It is also simple to calculate the density of states for a two-dimensional semiconductor (still assuming a parabolic dispersion relation). In this case, the states with energy in the range between $\mathcal{E}$ and $\mathcal{E} + d\mathcal{E}$ in the **k**-space are represented by states with k values contained in

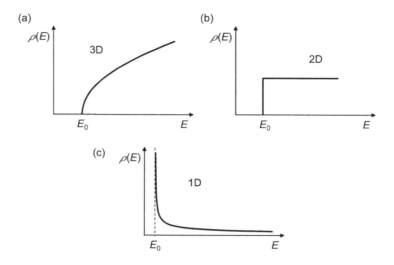

Figure 2.1   Density of states in (a) three-dimensional, (b) two-dimensional, and (c) one-dimensional systems.

the region between two circles with radii k and $k + dk$, with area $2\pi k\, dk$. Since the area occupied by each state is $(2\pi)^2/L^2$ and considering the spin degeneracy, we have:

$$\rho(k)\, dk = 2\frac{2\pi k\, dk}{4\pi^2}\frac{L^2}{L^2} = \frac{k}{\pi}\, dk. \tag{2.30}$$

Since $\rho(\mathcal{E})\, d\mathcal{E} = \rho(k)\, dk$, we obtain:

$$\rho(\mathcal{E}) = \frac{m^*}{\pi \hbar^2}. \tag{2.31}$$

Figure 2.1(b) shows the density of states in the conduction band.

In a one-dimensional system, the states with energy between $\mathcal{E}$ and $\mathcal{E}+d\mathcal{E}$ in the **k**-space are represented by states in two (linear) regions with width dk. Since each state occupies a region with length $2\pi/L$, we obtain:

$$\rho(k) = \frac{2}{\pi}, \tag{2.32}$$

and assuming a parabolic dispersion curve the corresponding expression as a function of energy is:

$$\rho_c(\mathcal{E}) = \frac{\sqrt{2m_c}}{\pi \hbar}(\mathcal{E} - \mathcal{E}_c)^{-1/2} \tag{2.33}$$

for the conduction band and

$$\rho_v(\mathcal{E}) = \frac{\sqrt{2m_v}}{\pi \hbar}(\mathcal{E}_v - \mathcal{E})^{-1/2} \tag{2.34}$$

for the valence band. Figure 2.1(c) shows the density of states in the conduction band. Zero-dimensional systems, called quantum dots, that is, structures where a three-dimensional quantum confinement is present, will be considered in Chapter 9.

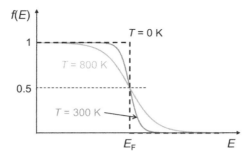

Fermi–Dirac distribution function at 0 K (dashed line), 300 K, and 800 K.

2.4 Carrier Statistics in Semiconductors

We assume that the semiconductor is in thermal equilibrium. Electrons and holes are fermions, therefore they obey the *Fermi–Dirac statistics*. The probability that a state with energy $\mathcal{E}$ is filled by an electron when the semiconductor is at thermal equilibrium at temperature T is given by the Fermi–Dirac distribution function:

$$f(\mathcal{E}) = \frac{1}{1 + \exp{(\mathcal{E} - \mathcal{E}_F)/k_B T}}, \tag{2.35}$$

where $\mathcal{E}_F$ is the *Fermi energy*, k_B is the Boltzmann constant, and T is the temperature in K. Figure 2.2 shows the Fermi–Dirac distribution function for three different temperatures. From Eq. 2.35, it is evident that $f(\mathcal{E}_F) = 1/2$. $f(\mathcal{E})$ is very small for energies larger than $\mathcal{E}_F$, meaning that only a few electrons are found in high-energy states at thermal equilibrium. At $T = 0$, the Fermi–Dirac function changes abruptly at $\mathcal{E} = \mathcal{E}_F$ from $f(\mathcal{E}) = 1$ when $\mathcal{E} < \mathcal{E}_F$ to $f(\mathcal{E}) = 0$ when $\mathcal{E} > \mathcal{E}_F$ as shown in Fig. 2.2, so that all allowed states below $\mathcal{E}_F$ are filled with electrons while all states above $\mathcal{E}_F$ are empty.

The probability that a state with energy $\mathcal{E}$ is empty is given by:

$$\bar{f}(\mathcal{E}) = 1 - f(\mathcal{E}) = \frac{1}{1 + \exp{(\mathcal{E}_F - \mathcal{E})/k_B T}}. \tag{2.36}$$

The number of electrons per unit volume in the conduction band is given by:

$$n = \int_{cb} \rho_c(\mathcal{E}) f(\mathcal{E}) \, d\mathcal{E}, \tag{2.37}$$

where the integral is calculated over the conduction band. In the same way, the density of holes in the valence band can be calculated as follows:

$$p = \int_{vb} \rho_v(\mathcal{E}) \bar{f}(\mathcal{E}) \, d\mathcal{E}. \tag{2.38}$$

By using Eq. 2.25 in Eq. 2.37, one obtains:

$$n = \frac{1}{2\pi^2} \left(\frac{2m_c}{\hbar^2} \right)^{3/2} \int_{\mathcal{E}_c}^{\infty} \sqrt{\mathcal{E} - \mathcal{E}_c} \frac{1}{1 + \exp{(\mathcal{E} - \mathcal{E}_F)/k_B T}} \, d\mathcal{E}. \tag{2.39}$$

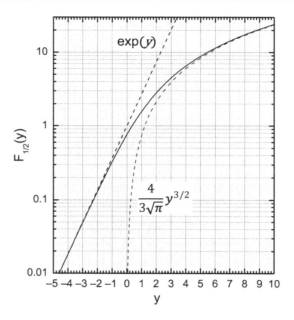

Figure 2.3 Fermi function $F_{1/2}(y)$ (solid line) and approximations (dashed lines): exp (y), $\frac{4}{3\sqrt{\pi}}y^{3/2}$.

If we introduce a new variable x defined as $x = (\mathcal{E} - \mathcal{E}_c)/k_BT$, Eq. 2.39 can be written as follows:

$$n = \frac{1}{4}\left(\frac{2m_ck_BT}{\pi\hbar^2}\right)^{3/2}\frac{2}{\sqrt{\pi}}\int_0^\infty\frac{\sqrt{x}}{1+e^{x-y}}\,dx, \qquad (2.40)$$

where:

$$y = \frac{\mathcal{E}_F - \mathcal{E}_c}{k_BT}. \qquad (2.41)$$

Equation 2.40 can be rewritten as follows:

$$n = N_cF_{1/2}(y) = N_cF_{1/2}\left(\frac{\mathcal{E}_F - \mathcal{E}_c}{k_BT}\right), \qquad (2.42)$$

where:

$$N_c = \frac{1}{4}\left(\frac{2m_ck_BT}{\pi\hbar^2}\right)^{3/2} \qquad (2.43)$$

is called *effective density of states* in the conduction band and $F_{1/2}(y)$ is the *Fermi function* of order $n = 1/2$, shown in Fig. 2.3. In the case of large negative arguments (i.e., $y < 0$ and $|y| \gg 1$) $F_{1/2}(y) \approx \exp(y)$, which is the Boltzmann distribution function. When $y \gg 1$ we have:

$$F_{1/2}(y) \approx \frac{2}{\sqrt{\pi}}\frac{2}{3}y^{3/2}. \qquad (2.44)$$

These two approximations are shown by dashed curves in Fig. 2.3. A few values of the Fermi function $F_{1/2}(y)$ are reported in Table 2.1, for y values between -5 and 10.

y	$F_{1/2}(y)$
-5	6.722×10^{-3}
-4	1.820×10^{-2}
-3	4.893×10^{-2}
-2	1.293×10^{-1}
-1	3.278×10^{-1}
0	7.652×10^{-1}
1	1.576
2	2.824
3	4.488
4	6.512
5	8.844
6	11.447
7	14.291
8	17.355
9	20.624
10	24.085

Table 2.1 A few values of the Fermi function $F_{1/2}(y)$, for y values between -5 and 10.

2.4.1 Nondegenerate Semiconductors

In the case of nondegenerate semiconductors, the Fermi energy is well below the bottom of the conduction band and well above the top of the valence band; we assume that the Fermi energy is at least $3k_BT$ away from either band edge. In this case, we can assume that $(\mathcal{E} - \mathcal{E}_F) \gg k_BT$, and the Fermi–Dirac distribution function can be approximated by the simpler Maxwell–Boltzmann distribution function (nonquantum regime):

$$f_c(\mathcal{E}) \approx \exp\left(-\frac{\mathcal{E} - \mathcal{E}_F}{k_BT}\right), \qquad (2.45)$$

which is valid when we remove the limitations imposed by Pauli exclusion principle. This is reasonable since at energies well above Fermi energy the probability to find an electron in an allowed state is so small that the Pauli exclusion principle does not have practical effects and the classical Maxwell–Boltzmann statistics can be used. As already mentioned, since in this case y is negative and $|y| \gg 1$, one obtains:

$$F_{1/2} \approx e^y \frac{2}{\sqrt{\pi}} \int_0^\infty \frac{\sqrt{x}}{e^x}\, dx = e^y, \qquad (2.46)$$

so that the electron density in the conduction band is given by (see Eq. 2.42):

$$n = N_c \exp\left(-\frac{\mathcal{E}_c - \mathcal{E}_F}{k_BT}\right). \qquad (2.47)$$

From this expression, it is clear the origin of the name "effective density of states" for N_c. Indeed, it is as all the energy states in the conduction band were effectively concentrated into a single energy level, $\mathcal{E}_c$, which can allocate N_c electrons per unit volume. Therefore,

Table 2.2 Values of N_c and N_v for Ge, Si, GaAs, and InP at $T = 300$ K.				
	Ge	Si	GaAs	InP
N_c (cm^{-3})	1.04×10^{19}	3.2×10^{19}	4.4×10^{17}	5.7×10^{17}
N_v (cm^{-3})	5.0×10^{18}	1.83×10^{19}	8.2×10^{18}	1.2×10^{19}

the electron density is given by the product of N_c and the probability that the state with energy $\mathcal{E}_c$ is occupied.

In the same way, it is possible to calculate the hole density in the valence band:

$$p = N_v \exp\left(-\frac{\mathcal{E}_F - \mathcal{E}_v}{k_B T}\right), \tag{2.48}$$

where:

$$N_v = \frac{1}{4}\left(\frac{2m_v k_B T}{\pi \hbar^2}\right)^{3/2}. \tag{2.49}$$

Using the numerical values in Eqs. 2.47 and 2.48, we can write:

$$N_c = 2.5 \times 10^{19} \left(\frac{m_c}{m_0}\right)^{3/2} \left(\frac{T}{300}\right)^{3/2} \text{cm}^{-3}$$

$$N_v = 2.5 \times 10^{19} \left(\frac{m_v}{m_0}\right)^{3/2} \left(\frac{T}{300}\right)^{3/2} \text{cm}^{-3}. \tag{2.50}$$

The expressions of N_c and N_v differ only because there are differences in the values of the effective masses m_c and m_v. N_c and N_v are of the order of $10^{18} - 10^{19}$ cm^{-3}, as reported in Table 2.2 in the case of Ge, Si, GaAs, and InP. Equations 2.47 and 2.48 allow one to calculate the position of the Fermi level once the electron or hole densities are known. Indeed:

$$\mathcal{E}_F = \mathcal{E}_c - k_B T \ln \frac{N_c}{n}$$

$$\mathcal{E}_F = \mathcal{E}_v + k_B T \ln \frac{N_v}{p} \tag{2.51}$$

We note that the product np is independent of the Fermi level and it is given by the following expression:

$$np = N_c N_v \exp\left(-\frac{\mathcal{E}_g}{k_B T}\right), \tag{2.52}$$

where $\mathcal{E}_g = \mathcal{E}_c - \mathcal{E}_v$ is the width of the band gap. This is an important formula since it gives the relationship between electron and hole densities in a semiconductor at thermal equilibrium. Expression 2.52 is called *mass-action law*.

2.4.2 Intrinsic and Extrinsic Semiconductors

When the electrons in the conduction band are the result of thermal excitation from the valence band, the semiconductor is called *intrinsic*. An intrinsic semiconductor is also called undoped semiconductor, since it is a pure semiconductor without any dopant species. In an intrinsic semiconductor, the number of electrons in the conduction band is equal to the number of holes in the valence band, since the holes result from electrons that have been thermally excited to the conduction band. Therefore, we have:

$$n = p = n_i = \sqrt{N_c N_v} \, \exp\left(-\frac{\mathcal{E}_g}{2k_B T}\right), \tag{2.53}$$

where n_i is called intrinsic carrier density. The intrinsic carrier density can be written as follows:

$$n_i = N_c \exp\left(-\frac{\mathcal{E}_c - \mathcal{E}_{Fi}}{k_B T}\right) = N_v \exp\left(-\frac{\mathcal{E}_{Fi} - \mathcal{E}_v}{k_B T}\right), \tag{2.54}$$

where $\mathcal{E}_{Fi}$ is the Fermi energy of the intrinsic semiconductor (intrinsic Fermi energy). In an intrinsic semiconductor, since $n = p = n_i$ and $N_c \approx N_v$, we have $(\mathcal{E}_c - \mathcal{E}_{Fi}) \approx (\mathcal{E}_{Fi} - \mathcal{E}_v)$. Therefore, the Fermi level $\mathcal{E}_{Fi}$ is close to the middle of the energy gap $\mathcal{E}_{Fi} \approx (\mathcal{E}_c + \mathcal{E}_v)/2$.

Also in the case of extrinsic semiconductors, the intrinsic Fermi level is frequently used as a reference level. Indeed, the expressions for the carrier densities n and p in an extrinsic semiconductor can be written in terms of the intrinsic carrier density and the intrinsic Fermi level:

$$n = n_i \exp\left(\frac{\mathcal{E}_F - \mathcal{E}_{Fi}}{k_B T}\right) \tag{2.55}$$

$$p = n_i \exp\left(\frac{\mathcal{E}_{Fi} - \mathcal{E}_F}{k_B T}\right). \tag{2.56}$$

Therefore, the energy separation between the Fermi level and the intrinsic Fermi level is a measure of the departure of the semiconductor from an intrinsic material.

2.5 Mass-Action Law

The mass-action law can be deduced in a more general way by using thermodynamic considerations. We will assume, as we have done so far, that the semiconductor is in thermal equilibrium. This is a dynamical condition where every process is balanced by its inverse process. At thermal equilibrium, some electrons are continuously excited by thermal energy from the valence band to the conduction band, while other electrons lose energy and from the conduction band return to the valence band. The excitation of an electron from the valence to the conduction band leads to the generation of a hole and an electron, while when an electron returns to the valence band, an electron–hole recombination event is produced. It is reasonable to assume that the generation rate, G, depends on temperature T while, to first order, it does not depend on the number of carriers already present in the conduction and valence bands. This last assumption is related to the extremely high number of bonds which can be broken by thermal energy, thus generating electrons in the conduction band and holes in the valence band. Therefore, we can write:

$$G = f_1(T), \tag{2.57}$$

where $f_1(T)$ is a function, which depends on the semiconductor characteristics and on temperature. On the other hand, the recombination rate, R, depends not only on temperature but also on the density of electrons in the conduction band and of holes in the valence band, since each recombination process requires that an electron is present in the conduction band and a hole is present in the valence band. We can write:

$$R = npf_2(T), \tag{2.58}$$

where $f_2(T)$, as $f_1(T)$, is a function which depends on the semiconductor characteristics and on temperature. At thermal equilibrium, the generation rate is equal to the recombination rate so that:

$$npf_2(T) = f_1(T) \tag{2.59}$$

or

$$np = \frac{f_1(T)}{f_2(T)} = f_3(T). \tag{2.60}$$

This is an important result: In a given semiconductor at thermal equilibrium, the product of the hole and electron densities is a function only of temperature.

In an intrinsic semiconductor, we have:

$$np = n_i^2 = f_3(T). \tag{2.61}$$

The intrinsic carrier density is a function of temperature since the thermal energy is the origin of the excitation of the electrons from the valence band to the conduction band. By using Eq. 2.53, we can write:

$$n_i^2 = N_c N_v \exp\left(-\frac{\mathcal{E}_g}{k_B T}\right). \tag{2.62}$$

The intrinsic carrier density is obviously a function of the energy gap since upon increasing the gap, fewer electrons can be excited into the conduction band by thermal energy. For silicon ($\mathcal{E}_g = 1.1$ eV) around room temperature n_i doubles for every 8-K increase in temperature. Equation 2.60 can be rewritten as:

$$np = n_i^2, \tag{2.63}$$

which is the mass-action law. This law holds for both intrinsic and extrinsic (i.e., doped, see next section) semiconductors.

2.6 Doped Semiconductors

An intrinsic semiconductor is of limited use in applications. The situation is completely different when the material is doped, that is when particular chemical impurities are incorporated substitutionally in the semiconductor. This means that the impurities occupy lattice sites in place of the atoms of the pure semiconductor. For example, if an atom of Si, with four valence electrons, is substituted by an atom from group V, with five valence electrons, such as arsenic, four of the valence electrons of the impurity atom fill bonds between the

a) *n*-type semiconductor   b) *p*-type semiconductor

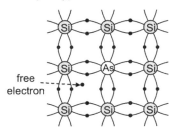 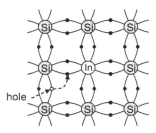

Figure 2.4 (a) *n*-type silicon with arsenic donor impurity; (b) *p*-type silicon with indium acceptor impurity.

impurity atom and four adjacent Si atoms. The remaining fifth electron is not covalently bound to adjacent atoms, and it is only weakly bound to the impurity atom by the excess positive charge of the nucleus. Therefore, only a small amount of energy is required to break this weak bond so that the fifth electron is free to move in the material, as shown in Fig. 2.4(a), thus contributing to electrical conduction. Since in this process the impurity atoms donate electrons to the semiconductor, they are called *donors* and the resulting semiconductor is called an *n*-type semiconductor.

The energy required to break the bond to a donor atom can be estimated by using a simple hydrogen-like model. Indeed, the fifth electron experiences the attractive potential, $U(r)$, of the single net positive charge of the nucleus of the impurity atom, weakened by the polarization effects of the semiconductor material:

$$U(r) = -\frac{e^2}{4\pi\epsilon_o\epsilon_r r}, \tag{2.64}$$

where ϵ_r is the relative dielectric constant of the semiconductor. Therefore, we have essentially the same situation of the electron in the hydrogen atom, with two important differences: The electron mass is m^* (not the free electron mass) and $\epsilon_r \neq 1$, that is the Coulomb potential experienced by the electron is reduced by the factor ϵ_r. The Bohr theory of hydrogen can be easily modified to take into account these two differences. The ionization energy of atomic hydrogen is:

$$\mathcal{E}_H = \frac{e^4 m_0}{2(4\pi\epsilon_0\hbar)^2} = 13.6\,\text{eV}. \tag{2.65}$$

The ionization energy of a donor impurity in the semiconductor, that is, the binding energy, $\mathcal{E}_b$, of the electron to the core, can be obtained from Eq. 2.65 by using the dielectric constant of the semiconductor, $\epsilon = \epsilon_0\epsilon_r$ and by replacing m_0 with the effective mass m^*:

$$\mathcal{E}_b = \frac{e^4 m^*}{2(4\pi\epsilon_r\epsilon_0\hbar)^2} = 13.6\left(\frac{1}{\epsilon_r^2}\frac{m^*}{m_0}\right)\,\text{eV}. \tag{2.66}$$

The situation is schematically illustrated in Fig. 2.5. The effective mass which must be used in this case is the *conductivity effective mass*, m_σ^*, whose value depends on how electrons respond to an external potential. For direct band gap semiconductors, m_σ^* is simply

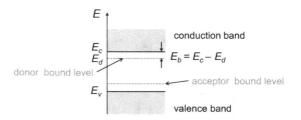

Figure 2.5 Schematic representation of the energy level of a donor impurity in a semiconductor. The binding energy is given by $\mathcal{E}_b = \mathcal{E}_c - \mathcal{E}_d$. Figure also shows an acceptor bound level.

the effective mass. For semiconductors such as Si, the conductivity mass is given by the following equation:

$$m_\sigma^* = 3 \left(\frac{2}{m_t} + \frac{1}{m_l} \right)^{-1}. \tag{2.67}$$

Exercise 2.2 Calculate the binding energy of the donor states in GaAs and Si. Numerical data for Si: conduction band effective mass $m_l = 0.98\, m_0$, $m_t = 0.2\, m_0$, $\epsilon_r = 11.9$. Numerical data for GaAs: conduction band effective mass $m_c = 0.067\, m_0$, $\epsilon_r = 13.2$.

In the case of GaAs, we have:

$$\mathcal{E}_b = 13.6 \left(\frac{1}{\epsilon_r^2} \frac{m^*}{m_0} \right) = 5.2 \text{ meV}.$$

In the case of Si, we have first to calculate the conductivity effective mass:

$$m_\sigma^* = 3 \left(\frac{2}{m_t} + \frac{1}{m_l} \right)^{-1} = 0.26\, m_0.$$

The ionization energy of the donor is given by $\mathcal{E}_b = 25$ meV, which is only about 3% of the energy gap.   ∎

We can also obtain a simple estimation of the mean distance of the electron from the donor atom. The Bohr radius of the ground state of hydrogen is related to the ionization energy by the following formula:

$$\mathcal{E}_H = \frac{e^2}{2(4\pi\epsilon_0 R_H)}, \tag{2.68}$$

so that

$$R_H = \frac{4\pi\epsilon_0\hbar^2}{m_0 e^2} = 0.53 \text{ Å}. \tag{2.69}$$

The Bohr radius of the donor can be obtained from the previous expression by considering the relative dielectric constant of the semiconductor and the effective mass:

$$R_d = \frac{4\pi\epsilon_r\epsilon_0\hbar^2}{m^* e^2} = 0.53 \left(\epsilon_r \frac{m_0}{m^*} \right) \text{ Å}. \tag{2.70}$$

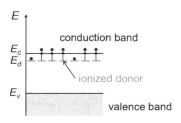

Figure 2.6   Donor states are neutral when occupied by an electron and positively charged when empty.

The simple hydrogen-like model we have used so far predicts that the binding energy of a donor state depends only on the semiconductor crystal (through ϵ_r and m^*) and not on the particular dopant atom. In a real situation, the ionization energy depends also on the donor atom as a result of a small distortion of the atomic potential. For example, measured values of the donor ionization energies in Si give $\mathcal{E}_b = 45$ meV for P, $\mathcal{E}_b = 49$ meV for As, and $\mathcal{E}_b = 39$ meV for Sb. In Ge $\mathcal{E}_b = 12$ meV for P, $\mathcal{E}_b = 12.7$ meV for As, and $\mathcal{E}_b = 9.6$ meV for Sb. The Bohr radii of the donor states obtained from Eq. 2.70 are $(132)(0.53)$ Å $\simeq 70$ Å for Ge, $(45)(0.53)$ Å $\simeq 24$ Å for Si, and $(193)(0.53)$ Å $\simeq 102$ Å for GaAs, which are quite large values, which lead to overlap of the dopant orbitals even at relatively low donor concentrations compared to the number of atoms of the host semiconductor. The orbital overlap leads to the generation of an impurity band formed from the donor states, as it will be discussed in Section 2.6.2.

Due to the small ionization energy of the donor states, thermal energy is large enough to excite electrons from the bound donor states to semiconductor conduction band already at temperatures >150 K. When an electron is excited in the conduction band, a fixed and positively charged atom core is left in the semiconductor in the dopant site. Therefore, the allowed energy states introduced by donors are neutral when occupied by an electron and positively charged when they are empty, as schematically shown in Fig. 2.6.

When a semiconductor is doped with donor atoms, the number of electrons in the conduction band is much larger than the number of holes in the valence band. Therefore, in a n-type semiconductor, the electrons are majority carriers and the holes are minority carriers. Figure 2.7 shows the electron density in the conduction band of Si as a function of temperature in the case of Si doped with $N_d = 1 \times 10^{16}$ arsenic atoms cm^{-3}. At very low temperature, the donor electrons are tied to the donor sites: This effect is called *carrier freeze-out*. At $T > 150$ K, all donors are ionized and $n \simeq N_d$. This region is called the saturation region. Upon increasing the temperature up to a few hundreds of K (~ 600 K in Si), the thermal energy is able to ionize many electrons from silicon–silicon bonds so that the intrinsic carrier population starts to increase and becomes the dominant source of electrons in the conduction band. Semiconductor devices usually work in the saturation region, where the free carrier density is almost independent of temperature and it is approximately equal to the density of the dopant atoms. Semiconductor devices cannot operate in the high-temperature intrinsic region since it is not possible to control the intrinsic carrier density by applying external bias. For this reason, the devices which must work at high

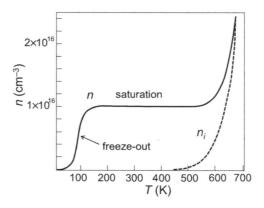

Electron concentration versus temperature for silicon doped with 1×10^{16} arsenic atoms cm^{-3} (solid line). The dashed line shows the intrinsic carrier density, n_i.

temperature are based on the use of large band gap semiconductors in order to increase the temperature required to ionize the bonds between the semiconductor atoms.

If a semiconductor with four valence electrons as Si is doped with impurity atoms with three valence electrons (such as B, Al, Ga, and In), the three electrons fill three of the four covalent bonds of Si and one bond remains vacant. This bond can be filled by an electron coming from a nearby covalent bond, so that the vacant bond is moved carrying with it a positive charge, as shown in Fig. 2.4(b). In this case, the trivalent dopant atoms are called *acceptors* since they accept electrons from the valence band of the host semiconductor to complete the four covalent bonds with neighbor atoms, thus leaving holes in the valence band. The material is called *p*-type semiconductor, since most of the conduction is carried by positive charges (holes). The energy required to excite an electron from the valence band to the vacant bond is typically of the same order of magnitude of the binding energy of a donor state. For example, the acceptor energies $\mathcal{E}_a$ of trivalent atoms in Si are $\mathcal{E}_a = 45$ meV for B, $\mathcal{E}_a = 57$ meV for Al, and $\mathcal{E}_a = 65$ meV for Ga. The corresponding acceptor state is represented by an energy level slightly above the top of the valence band, as shown in Fig. 2.5. An acceptor level is neutral when empty and negatively charged when it is occupied by an electron. The hydrogen-like model used in the case of donor impurities can be applied also in the case of acceptor atoms, but its application is complicated by the degeneracy at the valence band edge and it does not give accurate results.

In the case of compound semiconductors like GaAs, tetravalent impurities may substitutionally replace either Ga or As. For example, Si acts as a donor impurity when it replaces Ga, whereas it acts as an acceptor impurity when it substitutes for As. A different situation is encountered for example when tin is used in GaAs since it replaces almost exclusively Ga in GaAs thus leading to an effective *n*-type doping. *n*-type GaAs can be obtained by using impurities from group VI of the periodic table, such as tellurium, selenium, or sulfur. *p*-type doping of GaAs can be obtained by using impurities from group II, like zinc or cadmium.

2.6.1 Carrier Density and Fermi Energy in Doped Semiconductors

We assume to have a semiconductor doped with N_d donor atoms per unit volume and N_a acceptors per unit volume. Moreover, we assume that all dopants are ionized and that $N_d > N_a$. Due to the charge neutrality of the material, we have:

$$n + N_a = p + N_d. \tag{2.71}$$

Using the mass-action law ($np = n_i^2$) in the previous equation, we obtain

$$n - \frac{n_i^2}{n} = N_d - N_a, \tag{2.72}$$

which may be solved for the electron concentration n:

$$n = \frac{N_d - N_a}{2} + \left[\left(\frac{N_d - N_a}{2} \right)^2 + n_i^2 \right]^{1/2}. \tag{2.73}$$

If we consider a n-type semiconductor $N_d - N_a > 0$ and $N_d - N_a \gg n_i$. Therefore, from Eq. 2.73, we have:

$$n \approx N_d - N_a \tag{2.74}$$

$$p = \frac{n_i^2}{n} \approx \frac{n_i^2}{N_d - N_a}. \tag{2.75}$$

From the last equation, we see that the electron density depends on the net excess of ionized donors over acceptors. Thus, a p-type material containing N_a acceptors can be converted into an n-type material by adding an excess of donors so that $N_d > N_a$. For silicon at room temperature, $n_i = 1.45 \times 10^{10}$ cm^{-3} while the net donor density in n-type silicon is typically about 10^{15} cm^{-3} or greater, so that $(N_d - N_a) \gg n_i$. Thus, for $N_d - N_a = 10^{15}$ cm^{-3}, we have $n \approx N_d - N_a$ and $p = 2 \times 10^5$ cm^{-3}, so that the minority-carrier concentration is nearly 10 orders of magnitude below the majority-carrier population. In general, the concentration of one type of carrier is many orders of magnitude greater than that of the other in extrinsic semiconductors.

By using Eq. 2.51, it is possible to calculate the position of the Fermi level:

$$\mathcal{E}_F = \mathcal{E}_c - k_B T \log \frac{N_c}{N_d - N_a}. \tag{2.76}$$

In the same way, assuming a p−type doping ($N_a > N_d$), we have:

$$\mathcal{E}_F = \mathcal{E}_v + k_B T \log \frac{N_v}{N_a - N_d}. \tag{2.77}$$

2.6.2 Degenerate Semiconductors

The previous equations show that, if the doping level is less than the effective density of states (N_c and N_v), the semiconductor remains nondegenerate, that is, the Fermi level is located in the energy gap. When the semiconductor is heavily doped, that is, when N_d or N_a approaches the effective density of states in the conduction or valence band (therefore $\sim 10^{18}$–10^{19} cm^{-3} for typical semiconductors), we cannot employ the approximation used

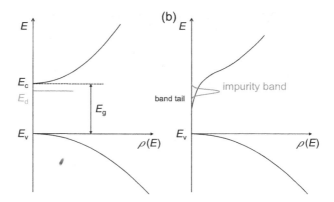

Figure 2.8 Schematic energy band diagram for (a) a nondegenerate semiconductor and for (b) a heavily doped semiconductor. $\rho(\mathcal{E})$ is the density of states.

in the case of nondegenerate semiconductors. Since we cannot approximate the Fermi–Dirac distribution function by the Maxwell–Boltzmann distribution function because it is not possible to neglect the effects of Pauli exclusion principle, Eqs. 2.47 and 2.48 are no longer valid. A semiconductor with very high doping levels ($N_d \gtrsim N_c$ or $N_a \gtrsim N_v$) is called *degenerate semiconductor* because the corresponding Fermi level is located within the conduction or the valence band. In the case of heavy doping, other effects must be taken into account, which are completely negligible for nondegenerate semiconductors. In the case of small dopant concentration, the mean distance between dopant atoms is typically much larger than the extension of their wavefunction, and the overlapping between the wavefunctions of the dopant atoms is very small. For this reason, the energy levels of the dopants can be considered as discrete levels, as shown in Fig. 2.8(a). In heavily doped semiconductors, the separation between dopant atoms reduces to a few angstroms, so that impurity bands are generated, as shown in Fig. 2.8(b).

Another important effect is related to electron–electron interaction, when a large number of electrons are introduced in the conduction band, or when a large number of holes is introduced in the valence band. Electron–electron interaction produces a narrowing of the semiconductor energy gap. For example, in the case of bulk GaAs, the energy gap narrowing is given by [4]:

$$\Delta \mathcal{E}_g = \mathcal{E}_g - \mathcal{E}_{g0} = -1.6 \times 10^{-8}(p^{1/3} + n^{1/3}) \, \text{eV}, \tag{2.78}$$

where $\mathcal{E}_{g0}$ is the energy gap at zero doping, and n and p are the electron and hole densities measured in cm^{-3}. Assuming a donor density of $\sim 10^{18}$ cm^{-3}, an energy gap reduction of $\Delta \mathcal{E}_g \simeq 16$ meV is obtained. In the case of silicon, the band gap narrowing can be well reproduced by the following expression, where T is the temperature in K:

$$\Delta \mathcal{E}_g \simeq -22.5 \left(\frac{N_d}{10^{18}} \frac{300}{T} \right)^{1/2} \, \text{meV}. \tag{2.79}$$

Another effect it is worth to mention is related to the fact that a very high dopant concentration disrupts the perfect periodicity of the crystal lattice, thus leading to the formation of a band tail, as shown in Fig. 2.8(b).

2.7 Quasi-Fermi Levels in Nonequilibrium Systems

So far we have always assumed semiconductors in thermal equilibrium. There are situations in which the material is in a nonequilibrium condition. Indeed, all semiconductor devices work under nonequilibrium conditions caused, for example, by the application of an external voltage or by exposing the semiconductor to light with photon energy greater than the energy gap. In all these cases, electron and hole populations are no longer in thermodynamic equilibrium with each other. Let us assume that electrons have been excited into the conduction band by an external excitation. Typically, the intraband relaxation processes, with a characteristic relaxation time of the order of ~ 1 ps (1 ps = 10^{-12} s), due to electron–phonon scattering processes, is much faster than the interband relaxation processes, with a characteristic relaxation time of ~ 1 ns (1 ns = 10^{-9} s), related to electron–hole recombination processes. Due to the very different temporal scales of the intraband and interband relaxation processes, a thermal equilibrium condition can be reached in a few picoseconds both in the conduction and in the valence band. The semiconductor is still in a nonequilibrium situation as a whole: Conduction and valence bands are independently in thermal equilibrium. For this reason, we can introduce two Fermi–Dirac distribution functions, $f_c(\mathcal{E})$ and $f_v(\mathcal{E})$, for the conduction and valence bands, respectively, given by:

$$f_c(\mathcal{E}) = \frac{1}{1 + \exp{(\mathcal{E} - \mathcal{E}_{Fc})/k_B T}} \tag{2.80}$$

$$f_v(\mathcal{E}) = \frac{1}{1 + \exp{(\mathcal{E} - \mathcal{E}_{Fv})/k_B T}}, \tag{2.81}$$

where $\mathcal{E}_{Fc}$ and $\mathcal{E}_{Fv}$ are the energies of the so-called *quasi-Fermi levels* for the conduction and valence bands, respectively. If Maxwell–Boltzmann statistics can be applied ($n \ll N_c$, $p \ll N_v$), the electron and hole densities can be calculated by using Eqs. 2.47–2.48 and 2.55–2.56, replacing $\mathcal{E}_F$ with $\mathcal{E}_{Fc}$ and $\mathcal{E}_{Fv}$, respectively, so that:

$$n = N_c \exp{\left(-\frac{\mathcal{E}_c - \mathcal{E}_{Fc}}{k_B T}\right)} = n_i \exp{\left(\frac{\mathcal{E}_{Fc} - \mathcal{E}_{Fi}}{k_B T}\right)}, \tag{2.82}$$

$$p = N_v \exp{\left(-\frac{\mathcal{E}_{Fv} - \mathcal{E}_v}{k_B T}\right)} = n_i \exp{\left(\frac{\mathcal{E}_{Fi} - \mathcal{E}_{Fv}}{k_B T}\right)}, \tag{2.83}$$

The position of the quasi-Fermi levels can be obtained in terms of the intrinsic Fermi level, $\mathcal{E}_i$, and the intrinsic carrier density:

$$\mathcal{E}_{Fc} = \mathcal{E}_{Fi} + k_B T \log{\left(\frac{n}{n_i}\right)} \tag{2.84}$$

$$\mathcal{E}_{Fv} = \mathcal{E}_{Fi} - k_B T \log{\left(\frac{p}{n_i}\right)}. \tag{2.85}$$

It is not surprising that under nonequilibrium conditions, the product np is not equal to the thermal equilibrium value n_i^2. Indeed, by using Eqs. 2.82 and 2.83, we obtain:

$$np = n_i^2 \exp\left(\frac{\mathcal{E}_{Fc} - \mathcal{E}_{Fv}}{k_B T}\right). \tag{2.86}$$

Therefore, the energy separation between the two quasi-Fermi levels can be considered as a measure of the departure of the semiconductor from the thermal equilibrium situation. At thermal equilibrium ($\mathcal{E}_{Fc} = \mathcal{E}_{Fv}$), the previous equation reduces to the usual expression: $np = n_i^2$.

Exercise 2.3 Calculate the energy of the quasi-Fermi levels in GaAs at $T = 0$ K and at $T = 300$ K for an injected electron density $n = 2.86 \times 10^{18}$ cm^{-3} and for an injected hole density $p = 2.86 \times 10^{18}$ cm^{-3}. Numerical values: $m_c = 0.067\, m_0$, $m_{hh} = 0.45\, m_0$, $m_{lh} = 0.082\, m_0$.

At $T = 0$ K, the calculation of the quasi-Fermi energy is particularly simple since, in this case, all allowed states with $\mathcal{E} < \mathcal{E}_{Fc}$ are filled by electrons, while all allowed states with $\mathcal{E} > \mathcal{E}_{Fc}$ are empty, so that the carrier density can be calculated analytically:

$$n = \frac{1}{2\pi^2}\left(\frac{2m_c}{\hbar^2}\right)^{3/2} \int_{\mathcal{E}_c}^{\mathcal{E}_{Fc}} \sqrt{\mathcal{E} - \mathcal{E}_c}\, d\mathcal{E} = \frac{1}{3\pi^2}\left(\frac{2m_c}{\hbar^2}\right)^{3/2}(\mathcal{E}_{Fc} - \mathcal{E}_c)^{3/2},$$

so that:

$$\mathcal{E}_{Fc} - \mathcal{E}_c = \frac{\hbar^2}{2m_c}(3\pi^2 n)^{2/3} = 109 \text{ meV}.$$

We can already anticipate that, since in this case $\mathcal{E}_{Fc} - \mathcal{E}_c \gg k_B T$ at room temperature, the previous result can be considered a good approximation of the position of the quasi-Fermi level also at room temperature.
In the same way for the valence band, one obtains:

$$\mathcal{E}_v - \mathcal{E}_{Fv} = \frac{\hbar^2}{2m_v}(3\pi^2 p)^{2/3} = 16 \text{ meV},$$

where the effective mass m_v is given by Eq. 2.29: $m_v = (m_{hh}^{3/2} + m_{lh}^{3/2})^{2/3} = 0.47\, m_0$. Since in this case $\mathcal{E}_v - \mathcal{E}_{Fv} < k_B T$ (at room temperature), we expect that the result of Eq. 2.87 is valid only at very low temperatures.

At $T = 300$ K, the electron density in the conduction band can be calculated by using Eq. 2.42. We have first to calculate the effective density of states in the conduction band:

$$N_c = \frac{1}{4}\left(\frac{2m_c k_B T}{\pi \hbar^2}\right)^{3/2} = 4.4 \times 10^{17} \text{ cm}^{-3}.$$

From Eq. 2.42, one obtains:

$$F_{1/2}(y) = \frac{n}{N_c} = 6.5.$$

From Fig. 2.3, it is possible to obtain the corresponding y value: $y \simeq 4$, so that:

$$\mathcal{E}_{Fc} - \mathcal{E}_c \simeq 4k_BT = 103.5 \text{ meV}.$$

Thus, confirming that the result obtained at $T = 0$ K was a good approximation of the correct value of the quasi-Fermi energy in the conduction band. Since $y \gg 1$, we can use the approximation of the Fermi function given by Eq. 2.44:

$$y = \left[\frac{3}{4}\sqrt{\pi}F_{1/2}(y)\right]^{2/3} \simeq 4.21,$$

thus giving $\mathcal{E}_{Fc} - \mathcal{E}_c \simeq 4.21k_BT = 109$ meV, which represents a good approximation of the correct result.

The same procedure can be adopted in the case of hole injection. In this case:

$$N_v = \frac{1}{4}\left(\frac{2m_vk_BT}{\pi\hbar^2}\right)^{3/2} = 8.16 \times 10^{18} \text{ cm}^{-3},$$

$$F_{1/2}(y) = \frac{p}{N_v} = 0.35,$$

where $y = \mathcal{E}_v - \mathcal{E}_{Fv}/k_BT$. From Fig. 2.3, it is possible to obtain the corresponding y value: $y \simeq -0.95$, so that:

$$\mathcal{E}_v - \mathcal{E}_{Fv} = yk_BT \simeq -24.6 \text{ meV},$$

thus showing that, as already anticipated, the result obtained at $T = 0$ does not represent a reasonable approximation of the correct one. Since y is negative, we can use the exponential approximation of the Fermi function: $F_{1/2}(y) \approx e^y$, so that $y = \log[F_{1/2}(y)] = -1.05$ and $\mathcal{E}_v - \mathcal{E}_{Fv} = -27$ meV. ∎

2.8 Charge Transport in Semiconductors

In this section, we will briefly discuss the charge transport processes in semiconductors. When the material is in thermal equilibrium, the electrons in the conduction band move in random motion as almost free particles: The influence of the crystal lattice is contained in the effective mass. According to the Boltzmann's law of equi-distribution of kinetic energy per degrees of freedom of particles, the average kinetic energy of a particle is $(1/2)k_BT$ per each degree of freedom. As a first order approximation, the average thermal kinetic energy of the electrons can be written as:

$$\frac{1}{2}m^*v_{th}^2 = \frac{3}{2}k_BT, \tag{2.87}$$

where m^* is the effective mass of electrons in the conduction band. For example, in the case of silicon at $T = 300$ K, the thermal velocity v_{th} turns out to be of the order of $\sim 10^7$ cm/s. In a perfect lattice, no scattering events are present and the effects related to the lattice are taken into account in the effective mass. In a real lattice, various scattering mechanisms are present. Indeed, at $T > 0$ K, the lattice atoms vibrate about their equilibrium position. In quantum mechanics, this process is described in terms of electron–phonon scattering. In a real crystal, we can also have scattering at ionized impurities, which is relevant at high dopant densities; electron–electron scattering, important at high carrier density; scattering at crystal defects. At thermal equilibrium, the motion of electrons is completely random so that the net current in any direction is zero. The collisions with lattice determine a transfer of energy between the electrons and the atoms, which form the lattice. The average time interval between consecutive collisions, τ_n, is called mean scattering time, and it is of the order of 1 ps. Therefore, the average distance between consecutive collisions, called mean free path, is roughly given by $\ell_n = v_{th}\tau_n \approx 0.1\ \mu$m.

If an electric field is applied to the crystal, the electrons are accelerated in the opposite direction and an overall motion of carriers in the direction of the electric field is produced, superposed to the random thermal motion. If the applied field is not too high, we can assume that the mean scattering time is not notably changed by the application of the field. The average net velocity in the direction of the field is called drift velocity, v_d. In a steady-state condition, the momentum acquired by the electron between consecutive collisions is lost to the lattice in the collision. The momentum gained by an electron between collisions is equal to the impulse of the force acting on the electron:

$$m_n^* v_d = (-eE)\tau_n, \tag{2.88}$$

so that the drift velocity can be written as:

$$v_d = -\frac{e\tau_n}{m_n^*}E = -\mu_n E, \tag{2.89}$$

where

$$\mu_n = \frac{e\tau_n}{m_n^*} \tag{2.90}$$

is the electron mobility and describes how an electron responds to an applied electric field. The drift of electrons induced by the applied field gives rise to a current flow, with a current density given by:

$$J_n = -nev_d = ne\mu_n E. \tag{2.91}$$

We can repeat the same discussion for holes in the valence band. The hole mobility can be written as:

$$\mu_p = \frac{e\tau_p}{m_p^*} \tag{2.92}$$

and the total current density can be written as:

$$J = J_n + J_p = (ne\mu_n + pe\mu_p)E = \sigma E, \tag{2.93}$$

where

$$\sigma = ne\mu_n + pe\mu_p \tag{2.94}$$

is the conductivity of the semiconductor. In an n-type semiconductor $J \simeq J_n$, while in a p-type semiconductor $J \simeq J_p$.

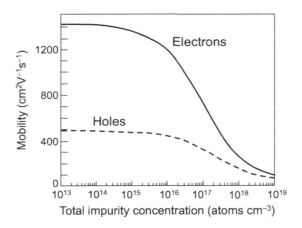

Figure 2.9   Electron and hole mobility versus doping density for silicon at room temperature.

Various scattering processes are present in a semiconductor, each characterized by a given mean scattering time τ_i and each giving rise to a different mobility μ_i. If we assume that the probability that a carrier is scattered in a time interval dt by the scattering process i is dt/τ_i, the total scattering probability in the same time interval can be written as:

$$\frac{dt}{\tau_c} = \sum_i \frac{dt}{\tau_i}. \tag{2.95}$$

Therefore, the mobility $\mu = e\tau_c/m^*$ can be written as (*Matthiessen's rule*):

$$\frac{1}{\mu} = \sum_i \frac{1}{\mu_i}. \tag{2.96}$$

It is therefore evident that the value of the mobility is determined mainly by the scattering process characterized by the smallest scattering time. The mobility depends on total dopant concentration as shown in Fig. 2.9, which displays electron and hole mobility for Si at 300 K as a function of the total dopant density. In the case of low dopant density (of the order of 10^{15} cm^{-3}), the mobility resulting from lattice phonon scattering is smaller than that due to ionized impurity scattering, therefore for dopant densities smaller than $\sim 10^{15}$ cm^{-3} (in Si) the electron and hole mobility is almost constant, as clearly shown in Fig. 2.9. In other words, at low doping lattice (phonon) scattering dominates. Upon increasing the dopant density, the mobility related to impurity scattering becomes comparable to that related to lattice scattering and the total mobility decreases. It is evident that in a compensated semiconductor (i.e., when both donor and acceptor atoms are introduced) while the carrier density depends on the difference between the dopant concentrations, $N_d - N_a$, the scattering depends on the total dopant density, $N_d + N_a$. Therefore, compensated semiconductors have a lower carrier mobility for the same net dopant density than that of an un-compensated semiconductor (i.e., with only a single dopant type). A very good approximation of measured dependence of μ versus dopant density in Si is given by the following expression:

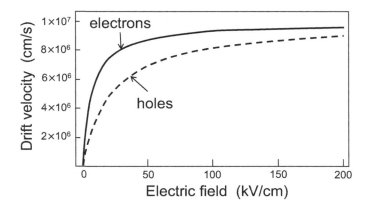

Figure 2.10 Drift velocity of electrons and holes as a function of the electric field amplitude in silicon.

$$\mu = \mu_{min} + \frac{\mu_{max} - \mu_{min}}{1 + (N/N_{ref})^{\alpha}}, \qquad (2.97)$$

where N is the total dopant density and the parameters μ_{min}, μ_{max}, N_{ref}, and α depend on the dopant atom.

The different scattering processes show different temperature dependences, thus leading to a complex dependence of the total mobility versus temperature. Without entering into details, upon increasing the temperature, lattice vibrations increase, thus increasing the lattice scattering component and reducing the corresponding mobility:

$$\mu_{lattice} \propto T^{-n}, \qquad n \sim 1.5. \qquad (2.98)$$

In contrast, the impurity scattering component is reduced at high temperature since the carriers move faster thus remaining near the impurities for shorter time:

$$\mu_{impurity} \propto T^{n}, \qquad n \sim 1.5. \qquad (2.99)$$

Therefore, at higher temperature, lattice scattering tends to dominate over impurity scattering and the mobility decreases. At lower temperature, the reverse situation is found and the mobility increases with temperature. This behavior gives rise to a maximum of the curve $\mu(T)$.

Before closing this section, we note that the linear dependence of the drift velocity versus applied electric field, $v_d = \mu E$, is valid only for small values of the applied field, as shown in Fig. 2.10, which displays the drift velocity of electrons and holes as a function of the electric field amplitude in silicon. In the derivation of the linear relation between v_d and E, we have assumed that the average collision time τ does not depend on the applied electric field. Note that for low electric field amplitudes, the drift velocity is much smaller than the thermal velocity. In the case of Si $\mu_n \approx 1400\,\text{cm}^2\text{V}^{-1}\text{s}^{-1}$, assuming a field amplitude $E = 100\,\text{Vcm}^{-1}$, the drift velocity is $v_d \approx 1.4 \times 10^5$ cm/s, while $v_{th} \approx 10^7$ cm/s, so that the applied electric field does not significantly alter the total electron velocity. Upon increasing the applied electric field, the electron kinetic energy increases and overcomes the thermal energy. In this regime, the high-energy electrons are called *hot electrons*. The energy transferred from the hot electrons to the lattice is high. When the hot electron density reaches a critical value, a new scattering process comes into play: the scattering with high-energy or *optical phonons*. In this situation, the drift velocity cannot increase linearly with E and reaches a saturation value at high E. A very good approximation of the dependence of the drift velocity on the applied electric field in Si is given by the following empirical expression:

$$v_d = v_\ell \frac{E}{E_c} \left[\frac{1}{1 + (E/E_c)^\beta} \right]^{1/\beta}, \tag{2.100}$$

where v_ℓ, E_c and β are suitable parameters (e.g., for electrons in Si at 300 K $v_\ell = 1.07 \times 10^7$ cm/s, $E_c = 6.91 \times 10^3$ V/cm and $\beta = 1.11$).

2.9 Diffusion Current

In a semiconductor, there is a second component of current, which is generated by a nonuniform spatial distribution of carrier density: This current component, which is particularly important in semiconductors, is called diffusion current. Diffusion of carriers occurs from regions of high density toward regions of low carrier density. In other words, gradients of the carrier densities, ∇n or ∇p, generate electron or hole currents, according to the Fick's law:

$$\mathbf{J}_n = eD_n\nabla n \tag{2.101}$$

$$\mathbf{J}_p = -eD_p\nabla p, \tag{2.102}$$

where D_n and D_p are the diffusion coefficients for electrons and holes, respectively. The current densities $\mathbf{J}_n$ and $\mathbf{J}_p$ in a semiconductor with an applied electric field are given by the sum of the drift current and of the diffusion current:

$$\mathbf{J}_n = e\mu_n n\mathbf{E} + eD_n\nabla n \tag{2.103}$$

$$\mathbf{J}_p = e\mu_p p\mathbf{E} - eD_p\nabla p, \tag{2.104}$$

and the total current density is given by $\mathbf{J} = \mathbf{J}_n + \mathbf{J}_p$. D_n and D_p are given by the Einstein relations:

$$\frac{D_n}{\mu_n} = \frac{D_p}{\mu_p} = \frac{k_B T}{e}. \tag{2.105}$$

Table 2.3 shows the mobility and diffusion coefficients at $T = 300$ K for Si, GaAs, and Ge corresponding to the typical mobility values at the same temperature and low dopant density.

Equations 2.103 and 2.104 can be written in a different way in terms of the quasi-Fermi levels. Quasi-Fermi levels have to be used since when current is flowing the semiconductor is not in thermal equilibrium. The potential, ϕ, is related to the applied electric field by the following relation:

$$\mathbf{E} = -\nabla\phi, \tag{2.106}$$

Table 2.3 Typical mobility and diffusion coefficient at $T = 300$ K and low dopant concentration for a few semiconductors.			
	μ_n (cm²/Vs)	D_n (cm²/s)	μ_p (cm²/Vs)	D_p (cm²/s)
Si	≤ 1400	≤ 36	≤ 450	≤ 12
Ge	≤ 3900	≤ 100	≤ 1900	≤ 50
GaAs	≤ 8500	≤ 200	≤ 400	≤ 10
InP	≤ 5400	≤ 130	≤ 200	≤ 5
InAs	$\leq 4 \times 10^4$	≤ 1000	≤ 500	≤ 13
InSb	$\leq 7.7 \times 10^4$	$\leq 2 \times 10^3$	≤ 850	≤ 22

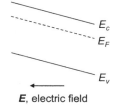

E, electric field

Figure 2.11   Schematic energy-band of an n-type semiconductor with an applied electric field **E**.

since the electron energy, $\mathcal{E}_c - \mathcal{E}_0$ (where $\mathcal{E}_0$ is an arbitrary fixed energy level) is given by:

$$\mathcal{E}_c - \mathcal{E}_0 = -e\phi, \tag{2.107}$$

the electric field can be written in terms of the gradient of the energy $\mathcal{E}_c$:

$$\mathbf{E} = \frac{1}{e}\nabla\mathcal{E}_c. \tag{2.108}$$

In the same way, we can write:

$$\mathbf{E} = \frac{1}{e}\nabla\mathcal{E}_c = \frac{1}{e}\nabla\mathcal{E}_v = \frac{1}{e}\nabla\mathcal{E}_i. \tag{2.109}$$

Therefore, when an electric field is present in the semiconductor, the conduction, valence, and intrinsic energies vary with position, as schematically shown in Fig. 2.11. If Maxwell–Boltzmann statistics can be used ($n \ll N_c$, $p \ll N_v$), the electron and hole densities are given by Eqs. 2.82 and 2.83. The electron current density is given by Eq. 2.103. Δn can be calculated by using Eq. 2.82:

$$\nabla n = \frac{n}{k_B T}(\nabla\mathcal{E}_{Fc} - \nabla\mathcal{E}_c) = \frac{n}{k_B T}\nabla\mathcal{E}_{Fc} - \frac{ne}{k_B T}\mathbf{E}, \tag{2.110}$$

where we have used Eq. 2.108. Therefore, by using Eq. 2.110 and the Einstein relation, J_n can be written as:

$$\mathbf{J}_n = \mu_n n\nabla\mathcal{E}_{Fc}. \tag{2.111}$$

In the same way for holes, we obtain:

$$\mathbf{J}_p = \mu_p p\nabla\mathcal{E}_{Fv}. \tag{2.112}$$

Therefore, the total current density for electrons and holes is proportional to the gradient of the quasi-Fermi level of the respective carrier type.

2.10 Exercises

Exercise 2.1 A GaAs wafer at room temperature ($T = 300$ K) is doped with 5×10^{14} cm^{-3} donor atoms.

a) Calculate the electron and hole densities and the position of the Fermi level with respect to the intrinsic Fermi level.

b) Upon irradiation with light, a steady-state density of electron and holes equal to 5×10^{12} cm^{-3} is photogenerated. Assuming that the carriers are uniformly photogenerated in the GaAs wafer, determine the total electron and hole densities and the positions of the quasi-Fermi levels with respect to the intrinsic Fermi level.

c) Repeat point b), assuming that the density of the photogenerated carriers is 4×10^{18} cm^{-3}.

Exercise 2.2 Calculate the position of the Fermi level with respect to the top of the valence band in intrinsic GaAs at room temperature and the density of the intrinsic carriers. Numerical data: energy gap $\mathcal{E}_g = 1.424$ eV, $m_c = 0.067\, m_0$, $m_{hh} = 0.45\, m_0$, $m_{lh} = 0.082\, m_0$.

Exercise 2.3 Consider GaAs at $T = 350$ K with an injected electron density $n = 3.88 \times 10^{18}$ cm^{-3}. Calculate the position of the quasi-Fermi level in the conduction band with respect to the bottom of the conduction band (write the result in eV). Repeat the same calculation at $T = 0$ K.

Exercise 2.4 Consider a semiconductor with two valence bands (the heavy-hole band and the light-hole band) that are degenerate at the Γ-point. Demonstrate that the density of states in the valence band can be written as

$$\rho_v(\mathcal{E}) = \frac{1}{2\pi^2} \left(\frac{2m_v}{\hbar^2} \right)^{3/2} \sqrt{\mathcal{E}_v - \mathcal{E}},$$

where:

$$m_v = (m_{hh}^{3/2} + m_{lh}^{3/2})^{2/3}.$$

Exercise 2.5 The effective mass matrix of a semiconductor is:

$$m/m_0 = \begin{pmatrix} 0.3750 & 0.0217 & 0.0375 \\ 0.0217 & 0.2313 & 0.0541 \\ 0.0375 & 0.0541 & 0.2938 \end{pmatrix}$$

determine the density of states for the conduction band.

Exercise 2.6 On the basis of the hydrogen-like model used for the calculation of the impurity levels in doped semiconductors, calculate the ionization energy for donors in germanium, assuming that the refractive index of Ge is 4 and that $m^* = 0.12\, m_0$.

Exercise 2.7   Determine the amount of boron atoms (in grams) that has to be incorporated into 1 kg of germanium to obtain a doping level of 4×10^{16} cm^{-3}.

Exercise 2.8   Referring to Exercise 2.3, estimate the electron temperature at $T = 0$ K. *Hint*: the electron temperature is defined by assuming an equivalent thermal energy $\langle \mathcal{E} \rangle = 3k_B T/2$, where $\langle \mathcal{E} \rangle$ is the average electron energy.

Exercise 2.9 n-type silicon is doped with $N_d = 10^{15}$ atoms cm^{-3} of arsenic. Determine the temperature at which half the arsenic atoms are ionized.

Exercise 2.10   In the case of high applied fields, a good approximation of the dispersion curve in the conduction band is (Kane dispersion):

$$\mathcal{E}(1 + \alpha \mathcal{E}) = \frac{\hbar^2 k^2}{2m^*}.$$

Calculate the density of states.

Exercise 2.11   Consider GaAs at room temperature. Assuming that the Fermi level is located 40 meV inside the conduction band, determine the corresponding electron density.

Exercise 2.12   In a linear lattice, the dispersion relation can be written as

$$\mathcal{E}(k) = \mathcal{E}_s - 2\gamma \cos(ka),$$

calculate the corresponding density of states and plot it as a function of energy.

Exercise 2.13   In silicon, there are six equivalent minima in the conduction band at $\mathbf{k}_0 = (2\pi/a)(\pm 0.85, 0, 0)$, $(2\pi/a)(0, \pm 0.85, 0)$, $(2\pi/a)(0, 0, \pm 0.85)$ and the constant energy surfaces near the conduction band minima, $\mathcal{E}_c$, can be written as:

$$\mathcal{E}_c(k) = \mathcal{E}_c + \frac{\hbar^2 (k_x - k_{0x})^2}{2m_l} + \frac{\hbar^2 (k_y - k_{0y})^2}{2m_t} + \frac{\hbar^2 (k_z - k_{0z})^2}{2m_t}.$$

If we define

$$q_x = \sqrt{\frac{m}{m_l}}(k_x - k_{0x}), \qquad q_y = \sqrt{\frac{m}{m_t}}(k_y - k_{0y}), \qquad q_z = \sqrt{\frac{m}{m_t}}(k_z - k_{0z}),$$

a) Demonstrate that the dispersion is isotropic in the q-space.
b) Demonstrate that the total electron density in the conduction band (considering the six equivalent valleys) can be written as:

$$n = 6\sqrt{\frac{m_l m_t^2}{m^3}} \int_0^\infty \frac{q^2}{\pi^2} f_c(\mathcal{E}_c(q) - \mathcal{E}_F) dq,$$

where $f_c(\mathcal{E})$ is the Fermi–Dirac distribution function.

Exercise 2.14   By using the results of Exercise 2.13, demonstrate that the effective mass m_c^* for the X-valley of silicon is given by

$$m_c^* = 6^{2/3} (m_l m_t^2)^{1/3}.$$

Exercise 2.15   Calculate the position of the intrinsic Fermi level in Si at $T = 300$ K.

Exercise 2.16   Determine the position of the intrinsic Fermi level with respect to the center of the band gap in GaAs at $T = 300$ K.

Exercise 2.17   Plot the carrier density in the conduction band of a semiconductor in thermal equilibrium as a function of the Fermi energy by using the following expression:
a) $n = N_c F_{1/2}(y)$;
b) $n = N_c \exp\left[-(\mathcal{E}_c - \mathcal{E}_F)/(k_B T)\right]$;
c) the expression of n versus $\mathcal{E}_F$ when $T = 0$ K;
d) the Joyce-Dixon approximation

$$\frac{\mathcal{E}_F - \mathcal{E}_c}{k_B T} \simeq \ln\frac{n}{N_c} + \frac{1}{\sqrt{8}}\frac{n}{N_c} - \left(\frac{3}{16} - \frac{\sqrt{3}}{9}\right)\left(\frac{n}{N_c}\right)^2 + \cdots$$

and discuss the quality of the different approximations.

Exercise 2.18   Demonstrate that the density of electrons occupying a donor level is given by the following expression:

$$n_d = \frac{N_d}{1 + \frac{1}{2}\exp\left(\frac{\mathcal{E}_d - \mathcal{E}_F}{k_B T}\right)},$$

where $\mathcal{E}_d$ is the energy of the donor states in the gap and N_d is the density of donor atoms.

Exercise 2.19   A silicon sample is doped with $N_d = 5 \times 10^{16}$ cm^{-3} atoms of phosphorous. Calculate:
a) the effective density of states, N_c;
b) the ionization energy of the donor, assuming a hydrogen-like model ($m_l = 0.98\ m_0$, $m_t = 0.2\ m_0$, $\epsilon_r = 11.9$);
c) the ratio between the electron density in the donor state at room temperature and the total density of electrons;
d) calculate the same ratio for a doping level of $N_d = 5 \times 10^{18}$ cm^{-3}.

Exercise 2.20   Calculate the electron and hole density in a silicon sample for each of the following conditions:
a) intrinsic sample;
b) $N_d = 10^{14}$ cm^{-3}, $N_a = 0$;
c) $N_d = 10^{16}$ cm^{-3}, $N_a = 4 \times 10^{16}$ cm^{-3}.

Exercise 2.21   Consider p-type silicon doped with $N_a = 10^{16}$ cm^{-3} of boron. Determine the temperature at which 80% of the acceptor atoms are ionized.

Exercise 2.22   A device based on n-type silicon operates at $T = 450$ K. At this temperature, the intrinsic carrier density must contribute no more than 4% of the total electron density. Determine the minimum doping density to meet this condition. As a first-order approximation, neglect the variation of the energy gap with temperature.

Exercise 2.23   Derive the density of states and zero temperature electron density as a function of energy for only a single valley for the following materials:

a) graphene: $\mathcal{E} = \hbar v k$;

b) parabolic band semiconductor: $\mathcal{E} = \hbar^2 k^2 / 2m$.

Comment on their differences. Do you think the electron density is temperature dependent? Invoke the Fermi–Dirac function in your argument. Note: there is a spin-degeneracy factor of 2.

Exercise 2.24   In a Si wafer with thickness $L = 2$ μm, the excess carrier density at $x = 0$ and $x = L$ are: $\delta n(0) = 10^{16}$ cm^{-3} and $\delta n(L) = 0$. Assuming $D_n = 20$ cm^{2}s^{-1} and $\tau_n = 100$ ns, calculate the electron diffusion current at $x = 0$ and at $x = L/2$.

3 Basic Concepts of Quantum Mechanics

3.1 Quantum Mechanics Fundamentals

In quantum mechanics, it is assumed that a particle can be described by a wavefunction $\psi(\mathbf{r}, t)$, which satisfies the Schrödinger equation:

$$H\psi(\mathbf{r}, t) = i\hbar \frac{\partial \psi(\mathbf{r}, t)}{\partial t}, \tag{3.1}$$

where H is the Hamiltonian operator, which contains the information about the system of interest. It is assumed that $\psi(\mathbf{r}, t)$ is normalized:

$$\int \psi^* \psi \, d\tau = 1, \tag{3.2}$$

where the asterisk indicates that complex conjugate of ψ is taken and the integral is calculated over all space. We will generally use the Dirac notation, so that the integral $\int \phi^* \psi \, d\tau$, where $\phi(\mathbf{r}, t)$ and $\psi(\mathbf{r}, t)$ are two wavefunctions, can be represented as:

$$\int \phi^* \psi \, d\tau = \langle \phi | \psi \rangle. \tag{3.3}$$

Each wavefunction, ψ, can be represented by a *ket* vector $|\psi\rangle$. There is a one-to-one correspondence between a ket vector and another vector, $\langle \psi |$, called *bra* vector, such that the product $\langle \psi | \psi \rangle$ represents the integral $\int \psi^* \psi \, d\tau$. Equation 3.3 defines the scalar product (or inner product) between the two wavevectors $|\phi\rangle$ and $|\psi\rangle$. The scalar product has the following properties:

$$\langle \phi | \psi \rangle = \langle \psi | \phi \rangle^*, \tag{3.4}$$

$$\langle \phi | \phi \rangle \text{ is real and positive (it is zero only if } |\phi\rangle = 0). \tag{3.5}$$

Indeed:

$$\langle \phi | \psi \rangle = \int \phi^* \psi \, d\tau = \left(\int \psi^* \phi \, d\tau \right)^* = \langle \psi | \phi \rangle^*, \tag{3.6}$$

$$\langle \phi | \phi \rangle = \int \phi^* \phi \, d\tau = \int |\phi|^2 \, d\tau \geq 0. \tag{3.7}$$

Moreover:

$$\langle \phi | \alpha \psi_1 + \beta \psi_2 \rangle = \alpha \langle \phi | \psi_1 \rangle + \beta \langle \phi | \psi_2 \rangle, \tag{3.8}$$

$$\langle \alpha \phi_1 + \beta \phi_2 | \psi \rangle = \alpha^* \langle \phi_1 | \psi \rangle + \beta^* \langle \phi_2 | \psi \rangle. \tag{3.9}$$

By using the Dirac notation, the Schrödinger equation (Eq. 3.1) can be written as:

$$H|\psi\rangle = i\hbar \frac{\partial|\psi\rangle}{\partial t}, \tag{3.10}$$

with the normalization condition:

$$\langle\psi|\psi\rangle = 1. \tag{3.11}$$

Let us first consider an isolated system, without any perturbation or interaction, described by the unperturbed Hamiltonian H_0, which is assumed time-independent. For this system, there is a set of stationary states $|\phi_k\rangle$, which can be calculated by solving the Schrödinger equation:

$$H_0|\phi_k\rangle = i\hbar \frac{\partial|\phi_k\rangle}{\partial t}. \tag{3.12}$$

A solution of Eq. 3.12 can be written as:

$$|\phi_k(\mathbf{r}, t)\rangle = e^{-i(\mathcal{E}_k/\hbar)t}|u_k\rangle = e^{-i\omega_k t}|u_k\rangle, \tag{3.13}$$

where $\hbar\omega_k = \mathcal{E}_k$ and $|u_k\rangle$ is a solution of the eigenvalue equation:

$$H_0|u_k\rangle = \mathcal{E}_k|u_k\rangle. \tag{3.14}$$

The wavefunctions $\{|u_k\rangle\}$ form a complete set of orthonormal functions, called *representation* (in particular, energy representation).

If the system is in a given stationary state $|\phi_k\rangle$ and the energy is measured, the result of the measurement is given by the expectation value of H_0

$$\langle H_0\rangle = \langle\phi_k|H_0|\phi_k\rangle = \langle u_k|H_0|u_k\rangle = \langle u_k|\mathcal{E}_k|u_k\rangle = \mathcal{E}_k\langle u_k|u_k\rangle = \mathcal{E}_k. \tag{3.15}$$

which does not change with time: This is why the states $|\phi_k\rangle$ are called stationary states.

A very important general result is that an operator A which corresponds to a physically observable property is Hermitian, that is:

$$\langle\phi|A|\psi\rangle = \langle\psi|A|\phi\rangle^*, \tag{3.16}$$

where $|\phi\rangle$ and $|\psi\rangle$ are two given wavevectors. Therefore, a Hermitian operator is a self-adjoint operator, since the adjoint operator of A, which is indicated by $A^\dagger$, is given by the following formula:

$$\langle\phi|A|\psi\rangle = \langle\psi|A^\dagger|\phi\rangle^*, \tag{3.17}$$

if A is Hermitian $A^\dagger = A$.

A more accurate definition of adjoint operator is the following:

$$\int_a^b f^*(x)[Ag(x)]dx = \int_a^b [A^\dagger f(x)]^* g(x)dx + \text{boundary terms}, \tag{3.18}$$

where the boundary terms are calculated at the end points of the integration interval (a, b). An operator is called self-adjoint if $A^\dagger = A$. In addition, if particular boundary conditions are met by the functions f and g or by the operator itself, so that the boundary terms in Eq. 3.18 vanish, the operator is called Hermitian over the interval $a \leq x \leq b$, so that, in this case:

$$\int_a^b f^*(x)[Ag(x)]dx = \int_a^b [A^\dagger f(x)]^* g(x)dx, \tag{3.19}$$

It is possible to demonstrate that any two eigenfunctions $|u_n\rangle$ and $|u_m\rangle$ of a Hermitian operator with different eigenvalues, a_n and a_m, are orthogonal:

$$\langle u_n | u_m \rangle = 0 \qquad (m \neq n). \tag{3.20}$$

We will also assume that the eigenfunctions are normalized, so that:

$$\langle u_n | u_m \rangle = \delta_{nm}, \tag{3.21}$$

where δ_{nm} is the Kronecker delta symbol. These eigenstates form a complete, orthonormal set so that any state, $|\psi\rangle$, of the system can be written as:

$$|\psi\rangle = \sum_n c_n |u_n\rangle. \tag{3.22}$$

The coefficients c_n can be calculated as:

$$c_n = \langle u_n | \psi \rangle. \tag{3.23}$$

Indeed:

$$\langle u_n | \psi \rangle = \sum_k c_k \langle u_n | u_k \rangle = \sum_k c_k \delta_{nk} \overset{!}{=} c_n,$$

where we have used the orthonormality condition 3.21. Given that c_n can be calculated as the scalar product between the generic state $|\psi\rangle$ and the eigenstate $|u_n\rangle$, it is useful to think of c_n as the projection of $|\psi\rangle$ on $|u_n\rangle$. Therefore, $|\psi\rangle$ can be represented as a vector in an infinite-dimension abstract space, and Eq. 3.22 can be considered as the expansion of a given $|\psi\rangle$ vector in terms of the unit vectors $|u_k\rangle$ in this space.

The physical meaning of the coefficients c_k can be understood by calculating the expectation value of an operator A corresponding to a physical observable, with a discrete set of values $\{a_n\}$. The eigenstates $\{|u_n\rangle\}$ associated with the operator A satisfy the eigenvalue equation

$$A|u_n\rangle = a_n |u_n\rangle, \tag{3.24}$$

and any state $|\psi\rangle$ of the system can be written as in Eq. 3.22. The corresponding expectation value, $\langle A \rangle$ is given by:

$$\langle A \rangle = \langle \psi | A | \psi \rangle = \sum_{kn} c_k^* c_n \langle u_k | A | u_n \rangle = \sum_{kn} a_n c_k^* c_n \langle u_k | u_n \rangle = \sum_n a_n |c_n|^2, \tag{3.25}$$

which represents the average of all possible outcomes, a_n, of a measurement of the observable associated to the operator A, weighted by the probability to find the system in a particular eigenstate n, given by $|c_n|^2$.

Exercise 3.1   Calculate the average value of the momentum of an electron in a one-dimensional periodic potential.

The wavefunction describing the system is the Bloch wavefunction:

$$|\psi\rangle = e^{ikx}|u(x)\rangle.$$

The momentum operator is:

$$p = -i\hbar\frac{d}{dx},$$

and its average value can be calculated as follows:

$$\langle p \rangle = \langle\psi| - i\hbar\frac{d}{dx}|\psi\rangle,$$

where:

$$\langle\psi| = e^{-ikx}\langle u|$$
$$\frac{d|\psi\rangle}{dx} = ike^{ikx}|u\rangle + e^{ikx}\frac{d|u\rangle}{dx},$$

so that:

$$\langle p \rangle = \hbar k\langle u|u\rangle + \langle u| - i\hbar\frac{d}{dx}|u\rangle = \hbar k + \langle p_u \rangle, \tag{3.26}$$

where $\langle p_u \rangle$ represents the average momentum of the electron resulting from the interaction with the lattice, since this is the physical meaning of the periodic function $|u(x)\rangle$. Equation 3.26 clearly shows that the electron momentum is composed by two components: the crystal momentum, $\hbar k$, and the momentum due to the interaction with the lattice.   ∎

3.1.1   Quantum Analogy of the Classical Motion of a Particle in Confined Structures

Let us consider the simplest superposition of states: We will assume a linear combination of only two states:

$$|\psi_1\rangle = e^{-i\omega_1 t}|u_1(x)\rangle \qquad \text{and} \qquad |\psi_2\rangle = e^{-i\omega_2 t}|u_2(x)\rangle, \tag{3.27}$$

so that:

$$|\psi\rangle = c_1 e^{-i\omega_1 t}|u_1\rangle + c_2 e^{-i\omega_2 t}|u_2\rangle. \tag{3.28}$$

For example, we can consider a particle put in a superposition of its ground state $|\psi_1\rangle$, with energy $\mathcal{E}_1 = \hbar\omega_1$, and the first excited state $|\psi_2\rangle$, with energy $\mathcal{E}_2 = \hbar\omega_2$. This superposition state is called wave packet, and the variation of the position of the center of mass of this wave packet is the closest quantum mechanical analogy of the classical motion of the particle. The probability density is given by $|\psi(x,t)|^2$, where:

$$|\psi(x,t)|^2 = \psi(x,t)\psi^*(x,t) = \qquad\qquad\qquad (3.29)$$
$$= |c_1|^2|u_1(x)|^2 + |c_2|^2|u_2(x)|^2 + 2\text{Re}\{c_1^*c_2u_1^*(x)u_2(x)e^{-i(\omega_2-\omega_1)t}\}.$$

The first two terms do not depend on time, while the third one, generated by the quantum interference between the two states, oscillates in time with a period given by:

$$T = \frac{2\pi}{|\omega_2 - \omega_1|} = \frac{h}{|\mathcal{E}_2 - \mathcal{E}_1|} = \frac{h}{\Delta\mathcal{E}}, \qquad\qquad (3.30)$$

where $\Delta\mathcal{E} = \mathcal{E}_2 - \mathcal{E}_1$ is the energy separation between the two eigenstates. Therefore, the larger the energy difference $\Delta\mathcal{E}$, the faster is the motion of the particle in the superposition state. On the other hand, the energy separation $\Delta\mathcal{E}$ is related to the mass of the particle and to the spatial extent of the potential well, which confines the particle motion. In this way, a link is created between the temporal evolution of the particle dynamics and the spatial extension of the structure where the motion is developing.

From this general introduction, it is simple to understand which are the relevant timescales associated to various processes in atoms, molecules, and nanostructures. Let's start from a single molecule. A molecule can rotate and vibrate. By using the quantum terminology, we can say that molecular rotations are related to the generation of a coherent superposition of rotational states. Since typical energy difference between these states is in the range between 10^{-4} eV and 2×10^{-3} eV, molecular rotations evolve on the picosecond times scale; indeed, if we take $\Delta\mathcal{E} = 10^{-4}$ eV, we obtain $T \simeq 40$ ps ($h = 4.13 \times 10^{-15}$ eV·s), while if we take $\Delta\mathcal{E} = 2 \times 10^{-3}$ eV, we obtain $T \simeq 2$ ps. Molecular vibrations are faster, of the order of tens of femtoseconds, being related to the excitation of vibrational levels separated by a few tens of milli-eV. Indeed, if we take $\Delta\mathcal{E} = 100$ meV, we obtain $T \simeq 40$ fs. Therefore, the elementary dynamical processes in all chemical transformations, which are associated to molecular rotations and vibrations, typically occur on an ultrafast timescale.

At a more fundamental level, the timescale relevant for processes at the atomic or molecular level is set by the motion of electrons, which occurs on shorter timescale, ranging from a few femtoseconds to tens of attoseconds (1 as=10^{-18} s). In this case, the elementary dynamical steps are related to the excitation of electronic states separated by a few eVs. In the case of valence electrons in atoms, the energy separation can be of the order of a few eV; if we assume $\Delta\mathcal{E} = 10$ eV, we obtain $T \simeq 400$ as. In the case of electrons in core shells $\Delta\mathcal{E} \simeq 1$ keV so that $T \simeq 4$ as. Motion within nuclei is much faster, typically in the zeptosecond (1 zs=10^{-21} s) temporal range. ∎

If a perturbation is applied to the system, the Hamiltonian can be written as the sum of the unperturbed Hamiltonian and a perturbation Hamiltonian H':

$$H = H_0 + H'. \qquad\qquad (3.31)$$

This Hamiltonian must be used in the Schrödinger equation (Eq. 3.10). Also in this case, it is useful to expand the state function $|\psi\rangle$ in terms of the eigenfunctions of the unperturbed

Hamiltonian and write an expression similar to Eq. 3.22, with the important difference that the coefficients c_k are now time-dependent:

$$|\psi\rangle = \sum_k c_k(t)|\phi_k\rangle = \sum_k c_k(t)e^{-i\omega_k t}|u_k\rangle. \tag{3.32}$$

This expansion can be inserted in the Schrödinger equation (Eq. 3.10) and a set of differential equations in $c_k(t)$ can be obtained, which can be solved by using various techniques. In the following section, we will discuss a particular technique, which can be used to calculate approximated solutions of the Schrödinger equation when a small perturbation is applied to the system. Another technique to study the temporal evolution of a system is based on the density matrix formalism, which we will introduce in Section 3.4.

3.2 Time-Dependent Perturbation Theory

Let us consider a system, which at time $t = 0$ is perturbed. The wavefunction $|\psi\rangle$ is the solution of the Schrödinger equation:

$$(H_0 + H')|\psi\rangle = i\hbar\frac{\partial|\psi\rangle}{\partial t}, \tag{3.33}$$

where H' is called interaction Hamiltonian. By using 3.32 in 3.33, we have:

$$\sum_n (H_0 + H')c_n e^{-i\omega_n t}|u_n\rangle = \sum_n \left(i\hbar\frac{dc_n}{dt} + \mathcal{E}_n c_n\right)e^{-i\omega_n t}|u_n\rangle. \tag{3.34}$$

Since $H_0|u_n\rangle = \mathcal{E}_n|u_n\rangle$, Eq. 3.34 can be written as:

$$\sum_n c_n e^{-i\omega_n t}H'|u_n\rangle = i\hbar\sum_n \frac{dc_n}{dt}e^{-i\omega_n t}|u_n\rangle. \tag{3.35}$$

If we now perform the scalar product with $|u_k\rangle$, we obtain:

$$i\hbar\sum_n \frac{dc_n}{dt}e^{-i\omega_n t}\langle u_k|u_n\rangle = \sum_n c_n e^{-i\omega_n t}\langle u_k|H'|u_n\rangle, \tag{3.36}$$

where $\langle u_k|H'|u_n\rangle$ is the matrix element H'_{kn} of H'. By using the orthonormality condition $\langle u_k|u_n\rangle = \delta_{kn}$, the previous equation gives:

$$\frac{dc_k}{dt} = -\frac{i}{\hbar}\sum_n c_n H'_{kn}e^{i\omega_{kn}t}, \tag{3.37}$$

where $\omega_{kn} = \omega_k - \omega_n = (\mathcal{E}_k - \mathcal{E}_n)/\hbar$. So far no approximations have been performed. We will now assume that the Hamiltonian describing the perturbation can be written as $\lambda H'$, where λ is a perturbation parameter, which we will assume very small($\lambda \ll 1$) since we are considering a small perturbation of the system. λ is a very useful parameter since it allows one to distinguish among the various approximation orders in the perturbation calculations. In terms of λ, the power-series expansion for c_n can be written as:

$$c_n = c_n^{(0)} + \lambda c_n^{(1)} + \lambda^2 c_n^{(2)} + \dots \tag{3.38}$$

By using 3.38 in 3.37, we obtain:

$$\frac{dc_k^{(0)}}{dt} + \lambda \frac{dc_k^{(1)}}{dt} + \lambda^2 \frac{dc_k^{(2)}}{dt} + \ldots =$$

$$= -\frac{i}{\hbar} \sum_n [c_n^{(0)} + \lambda c_n^{(1)} + \lambda^2 c_n^{(2)} + \ldots] \lambda H'_{kn} e^{i\omega_{kn} t}. \tag{3.39}$$

Equating the same powers of λ, the following set of equations is obtained:

$$\frac{dc_k^{(0)}}{dt} = 0$$

$$\frac{dc_k^{(1)}}{dt} = -\frac{i}{\hbar} \sum_n c_n^{(0)} H'_{kn}(t) e^{i\omega_{kn} t}$$

$$\frac{dc_k^{(2)}}{dt} = -\frac{i}{\hbar} \sum_n c_n^{(1)} H'_{kn}(t) e^{i\omega_{kn} t}. \tag{3.40}$$

The solution of the zero-order equation is $c_k^{(0)} = $ constant. Therefore, $c_k^{(0)}$ are given by the initial conditions and no transition is occurring. If we assume that at $t = 0$ the system is in the mth state, we can write:

$$c_m^{(0)} = 1$$

$$c_n^{(0)} = 0 \qquad n \neq m, \tag{3.41}$$

so that the second equation in the set 3.40 can be written as:

$$\frac{dc_k^{(1)}}{dt} = -\frac{i}{\hbar} H'_{km} e^{i\omega_{km} t}, \tag{3.42}$$

which gives:

$$c_k^{(1)}(t) = -\frac{i}{\hbar} \int_0^t H'_{km} e^{i\omega_{km} t} \, dt. \tag{3.43}$$

Since at $t = 0$, the system occupies the mth state, $|c_k^{(1)}(t)|^2$ gives the probability (to first order) to find the system in the kth state at time t; therefore, it gives the transition probability, $P_{mk}(t)$, from the m state to the k state in the temporal interval t under the action of the external perturbation:

$$P_{mk}(t) = \frac{1}{\hbar^2} \left| \int_0^t H'_{km}(t) e^{i\omega_{km} t} \, dt \right|^2. \tag{3.44}$$

A particularly interesting situation is represented by a harmonic perturbation.

3.2.1 Harmonic Perturbation

In the case of a harmonic perturbation with frequency ω, the interaction Hamiltonian can be written as ($t \geq 0$):

$$H'(\mathbf{r}, t) = 2H'(\mathbf{r}) \cos(\omega t) = H'(\mathbf{r}) e^{-i\omega t} + H'^*(\mathbf{r}) e^{i\omega t}, \tag{3.45}$$

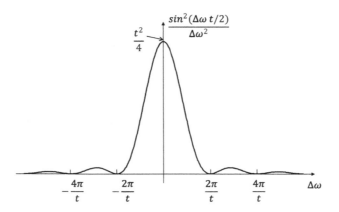

Figure 3.1 Plot of the function $\sin^2 (\Delta\omega t/2)/\Delta\omega^2$ versus $\Delta\omega$.

and the corresponding matrix element km can be written as:

$$H'_{km}e^{-i\omega t} + H'^*_{km}e^{i\omega t}. \tag{3.46}$$

Equation 3.43 can be written in the following way:

$$c_k^{(1)} = -\frac{i}{\hbar} \int_0^t [H'_{km}e^{i(\omega_{km}-\omega)t} + H'^*_{km}e^{i(\omega_{km}+\omega)t}] \, dt \tag{3.47}$$

$$c_k^{(1)} = -\frac{1}{\hbar} \left[H'_{km}\frac{e^{i(\omega_{km}-\omega)t}-1}{\omega_{km}-\omega} + H'^*_{km}\frac{e^{i(\omega_{km}+\omega)t}-1}{\omega_{km}+\omega} \right]. \tag{3.48}$$

In the following, we will assume that $\omega \approx |\omega_{km}|$: This approximation is called *quasi-resonance approximation* or *rotating phase approximation*. When $\omega \simeq \omega_{km}$ the first term at the right side of Eq. 3.48 is the dominant one, therefore:

$$c_k^{(1)} = -\frac{1}{\hbar}H'_{km}\frac{e^{i(\omega_{km}-\omega)t}-1}{\omega_{km}-\omega} \tag{3.49}$$

and the transition probability is given by:

$$P_{mk}(t) = |c_k^{(1)}(t)|^2 = \frac{4|H'_{km}|^2}{\hbar^2} \frac{\sin^2\left[\frac{1}{2}(\omega_{km}-\omega)t\right]}{(\omega_{km}-\omega)^2}. \tag{3.50}$$

The first term at the right side of Eq. 3.48 is dominant when $\mathcal{E}_k > \mathcal{E}_m$ and we have an absorption transition: The system, initially in the state m, absorbs a photon with energy $\hbar\omega \approx \mathcal{E}_k - \mathcal{E}_m$ thus producing a transition into a state with higher energy. When $\mathcal{E}_k < \mathcal{E}_m$ and the photon energy of the harmonic perturbation is $\hbar\omega \simeq \mathcal{E}_m - \mathcal{E}_k$, we have a stimulated emission transition from a state with energy $\mathcal{E}_m$ to a state with energy $\mathcal{E}_k$. In this case, the transition probability is given by:

$$P_{mk}(t) = \frac{4|H'_{km}|^2}{\hbar^2} \frac{\sin^2\left[\frac{1}{2}(\omega_{km}+\omega)t\right]}{(\omega_{km}+\omega)^2}. \tag{3.51}$$

Figure 3.1 shows the function $f(\Delta\omega) = \sin^2 (\Delta\omega t/2)/\Delta\omega^2$, plotted versus the frequency detuning from resonance, $\Delta\omega = \omega - \omega_{km}$.

When the observation time t is sufficiently long $f(\Delta\omega)$ can be approximated as follows:[1]

$$\frac{\sin^2\left[\frac{1}{2}(\omega_{km} - \omega)t\right]}{(\omega_{km} - \omega)^2} \simeq \frac{\pi t}{2}\delta(\omega_{km} - \omega), \tag{3.52}$$

where $\delta(\omega_{km} - \omega)$ is the Dirac-δ function and the transition probability can be written as:

$$P_{mk} = \frac{4|H'_{km}|^2}{\hbar^2}\frac{\pi t}{2}\delta(\omega_{km} - \omega) = \frac{2\pi t}{\hbar^2}|H'_{km}|^2\delta(\omega_{km} - \omega). \tag{3.53}$$

The corresponding transition rate is:

$$W_{mk} = \frac{d}{dt}P_{mk}(t) = \frac{2\pi}{\hbar^2}|H'_{km}|^2\delta(\omega_{km} - \omega). \tag{3.54}$$

Equation 3.54 is called *Fermi's Golden Rule*, and it is a particularly important formula. Fermi's golden rule can be written in terms of frequency, ν, and energy $\mathcal{E}$:

$$W_{mk} = \frac{1}{\hbar^2}|H'_{km}|^2\delta(\nu_{km} - \nu) \tag{3.55}$$

$$W_{mk} = \frac{2\pi}{\hbar}|H'_{km}|^2\delta(\mathcal{E}_k - \mathcal{E}_m - \hbar\omega). \tag{3.56}$$

3.2.2 Transitions from a State to a Band of States

In several cases of practical importance, we have to consider a transition from a state m to a group of states (a band of states) around state k. Let the density of these states be $g(\mathcal{E}_k)$. The transition probability from state m to a final state within this band can be written as:

$$\bar{P}_{mk} = \int_{\text{band},k} P_{mk}(\mathcal{E}_k)\, g(\mathcal{E}_k)\, d\mathcal{E}_k, \tag{3.57}$$

where the integral is calculated over the band around energy $\mathcal{E}_k$ and $P_{mk}(\mathcal{E}_k)$ can be obtained from Eq. 3.50:

$$P_{mk}(\mathcal{E}_k) = \frac{|H'_{mk}|^2}{\hbar^2}\frac{\sin^2\left[t(\mathcal{E}_k - \mathcal{E}_m - \hbar\omega)/2\hbar\right]}{[(\mathcal{E}_k - \mathcal{E}_m - \hbar\omega)/2\hbar]^2}. \tag{3.58}$$

If we now introduce the variable β defined as:

$$\beta = \frac{t}{2\hbar}(\mathcal{E}_k - \mathcal{E}_m - \hbar\omega) \tag{3.59}$$

so that

$$d\beta = \frac{t}{2\hbar}\, d\mathcal{E}_k \tag{3.60}$$

Equation 3.57 can be written in the following way:

$$\bar{P}_{mk} = \frac{2t}{\hbar}\int |H'_{km}|^2 g(\mathcal{E}_k)\frac{\sin^2\beta}{\beta^2}\, d\beta. \tag{3.61}$$

The function $\sin^2\beta/\beta^2$ is very small outside a region around $\beta = 0$ ($\omega = \omega_{km}$) with width $2\pi/t$ (see Fig. 3.1), which can be made arbitrarily small by increasing the observation

---

[1] $\int_{-\infty}^{\infty} \frac{\sin^2 x}{x^2}dx = \pi$, so that $\int_{-\infty}^{\infty} \frac{\sin^2 (xt/2)}{x^2}\, dx = \frac{\pi t}{2}$.

time t. If $2\pi/t$ is much smaller than the width of the band $g(\omega)$ and we assume that $|H'_{km}|^2$ does not strongly depends on energy, the integral 3.61 becomes:

$$\bar{P}_{mk} = \frac{2t}{\hbar} g(\mathcal{E}_k)|H'_{km}|^2 \int_{-\infty}^{\infty} \frac{\sin^2 \beta}{\beta^2} d\beta = \frac{2\pi}{\hbar} t\, g(\mathcal{E}_k)|H'_{km}|^2. \tag{3.62}$$

The corresponding transition rate is:

$$W_{mk} = \frac{2\pi}{\hbar} |H'_{km}|^2 g(\mathcal{E}_k), \tag{3.63}$$

which can be also written as a function of frequency ν. Indeed, $g(\mathcal{E}_k)\, d\mathcal{E}_k = g(\nu_k)\, d\nu_k$, where $\mathcal{E}_k = h\nu_k$, so that:

$$g(\mathcal{E}_k) = \frac{1}{h} g(\nu_k). \tag{3.64}$$

From 3.63 and 3.64, we find:

$$W_{mk} = \frac{1}{\hbar^2} |H'_{km}|^2 g(\nu_k), \tag{3.65}$$

which represents another form of the Fermi's Golden Rule, which will be used extensively in the next chapters.

As already mentioned, another technique to study a quantum system is based on the use of the density matrix formalism, which is the subject of the final section of this chapter. To introduce the density matrix formalism, it is useful to summarize a few general properties of operators, which we have already partially used in this chapter.

3.3 Properties of Operators: A Short Summary

The application of a linear operator A on a ket results in general in another ket, which describes a different state:

$$A|\psi\rangle = |\psi_1\rangle. \tag{3.66}$$

Therefore, an expression such as $\langle\phi|A|\psi\rangle$, which we have already used in this chapter, has the following meaning:

$$\langle\phi|A|\psi\rangle = \langle\phi|\psi_1\rangle, \tag{3.67}$$

that is, it is the inner product of $|\psi_1\rangle$ and $|\phi\rangle$. We have already mentioned the definition of the adjoint operator $A^\dagger$:

$$\langle\phi|A|\psi\rangle = \langle\psi|A^\dagger|\phi\rangle^*, \tag{3.68}$$

whose matrix elements are:

$$A^\dagger_{ij} = \langle u_i|A^\dagger|u_j\rangle = \langle u_j|A|u_i\rangle^* = A^*_{ji}. \tag{3.69}$$

If $A = A^\dagger$, the operator A is Hermitian. In this case:

$$A_{ij} = A^*_{ji}, \tag{3.70}$$

therefore the diagonal elements of a Hermitian operator are real ($A_{ii} = A^*_{ii}$).

It is interesting to observe that, while the product between a bra and a ket is a scalar (it is the inner product of the two vectors), the product between a ket and a bra, $|\phi\rangle\langle\psi|$, is an operator. Indeed:

$$\{|\phi\rangle\langle\psi|\}\,|\xi\rangle = |\phi\rangle\langle\psi|\xi\rangle = \langle\psi|\xi\rangle|\phi\rangle. \tag{3.71}$$

Since $|\phi\rangle\langle\psi|$ transforms the vector $|\xi\rangle$ in another vector, $\langle\psi|\xi\rangle|\phi\rangle$, it is an operator. This is useful for the definition of the *identity operator*, I, that is, and operator such that $IA = AI = A$, where A is any operator. By using a set of orthonormal basis vectors, the identity operator can be written as:

$$I = \sum_k |u_k\rangle\langle u_k|. \tag{3.72}$$

Indeed, a generic wavefunction $|\psi\rangle = \sum_k c_k|\phi_k\rangle$, where $c_k = \langle\phi_k|\psi\rangle$, can be written as follows:

$$|\psi\rangle = \sum_k \langle\phi_k|\psi\rangle|\phi_k\rangle = \sum_k |\phi_k\rangle\langle\phi_k|\psi\rangle = \left(\sum_k |u_k\rangle\langle u_k|\right)|\psi\rangle = I|\psi\rangle. \tag{3.73}$$

The matrix elements of I are:

$$I_{ij} = \langle u_i|I|u_j\rangle = \langle u_i|u_j\rangle = \delta_{ij}, \tag{3.74}$$

where we have used the fact that $I|u_j\rangle = |u_j\rangle$. The diagonal elements of I are equal to 1, and the off-diagonal ones are equal to 0.

We note that, given an Hermitian operator A on a linear space $\mathcal{H}$, there exists an orthonormal basis of $\mathcal{H}$ consisting of eigenvectors of A. In the same way, we can state that the operator A can be written as a linear combination of orthogonal projections, formed from its eigenvectors. This representation of A is called its spectral decomposition. Indeed:

$$\begin{aligned} A = IAI &= \sum_{ij} |u_i\rangle\langle u_i|A|u_j\rangle\langle u_j| = \sum_{ij} a_j|u_i\rangle\langle u_i|u_j\rangle\langle u_j| = \\ &= \sum_i a_i|u_i\rangle\langle u_i|, \end{aligned} \tag{3.75}$$

where we have used the eigenvalue equation:

$$A|u_j\rangle = a_j|u_j\rangle. \tag{3.76}$$

In this basis, the matrix representation of A is diagonal. The spectral decomposition allows one to write a given operator in terms of its eigenvalues.

3.4 The Density Matrix

So far we have considered a system in a *pure state*, described by the wavefunction $|\psi\rangle$. We have seen that the state $|\psi\rangle$ can be written as a superposition of stationary states $|\phi_k\rangle$ and quantum mechanics tells us that the squared modulus of the projection of $|\psi\rangle$ on a given stationary state $|\phi_k\rangle$ gives the probability to find, by a measurement, the system in that particular stationary state. We could confirm this result by preparing the system in a known initial state, perform the measurement (e.g., one can measure the energy), and repeat several times the same procedure, paying attention to always prepare the system in the same identical initial state. As we have seen, in this case the expectation value of a given operator A can be written as $\langle A \rangle = \sum_n a_n |c_n|^2$, which is nothing but the weighted average of all possible outcomes of the measurement of the physical observable corresponding to the operator A.

Generally we have to deal with statistical (incoherent) mixtures, and we cannot know the exact initial quantum state of a system. For example, we know only the temperature T of the system, so that the state of the system cannot be described by a wavevector $|\psi\rangle$. What it is possible to know, using for example the laws of statistical mechanics, is the probability, p_n, that the system is in the pure state $|\psi_n\rangle$. Note that the probability p_n is completely different from the probability $|c_n|^2$ discussed before in the case of a pure state $|\psi\rangle$. The probability $|c_n|^2$ does not arise from any lack of knowledge, but it is related to the process of quantum mechanical measurement: Unless the system is in a particular eigenstate, the measurement places the system in one of its eigenstates with a probability $|c_n|^2$. On the contrary, the probability p_n results from our lack of information about the exact preparation of the system. In this case, the quantum state must be described by a weighted mixture (statistical mixture) of states and it is called a *mixed state*.

A pure state is a state which can be described by a single ket vector, while a mixed state is a statistical ensemble of pure states and cannot be described by a single ket vector. It is useful to point out the difference between coherent and incoherent superposition of states. Let us first consider two states, $|\psi_1\rangle$ and $|\psi_2\rangle$, their coherent superposition can be written as:

$$|\psi\rangle = c_1|\psi_1\rangle + c_2|\psi_2\rangle, \tag{3.77}$$

where c_1 and c_2 are, in general, complex numbers. If the states are assumed normalized, we have:

$$|c_1|^2 + |c_2|^2 = 1. \tag{3.78}$$

As discussed in 3.1, the corresponding probability density, $|\psi|^2$, contains a term which originates from quantum interference between the states $|\psi_1\rangle$ and $|\psi_2\rangle$ (see Eq. 3.29), related to the information about the relative phase of the states $|\psi_1\rangle$ and $|\psi_2\rangle$, contained in the complex numbers c_1 and c_2. The presence of interference terms is the crucial characteristic of a pure state.

On the contrary, the incoherent superposition of two states, forming a mixed state, can be written as:

$$p_1|\psi_1\rangle + p_2|\psi_2\rangle, \tag{3.79}$$

where p_1 and p_2 are necessarily real numbers, since they are statistical weights giving the fractional population of the state and

$$p_1 + p_2 = 1 \tag{3.80}$$

p_1 and p_2 should never be interpreted as components in a given two-dimensional space. Moreover, we remark the complete lack of information about the relative phase between the states, which is implied by the fact that p_1 and p_2 are real. In the case of incoherent superposition of states, no interference terms can be present. In this case, the measurement of an observable associated to a given operator A does not correspond to the quantum expectation value of A, but to the ensemble average of A between the mixed state:

$$\langle A \rangle = p_1 \langle \psi_1 | A | \psi_1 \rangle + p_2 \langle \psi_2 | A | \psi_2 \rangle, \tag{3.81}$$

where interference terms are absent. We can say that the incoherence (the mixed state nature) implies the absence of interference terms.

For example, it is known that a given macrostate can be described in terms of macroscopic parameters, such as temperature, pressure, and volume. A microstate specifies a system in terms of the properties of each of the constituent particles, such as the position and the momentum of each molecule in a gas. From statistical mechanics, it is known that many different microstates can correspond to the same macrostate. The probability to find the system in a particular microstate with energy $\mathcal{E}_n$ is:

$$p_n = \frac{\exp\left(-\mathcal{E}_n/k_B T\right)}{Z} = \frac{\exp\left(-\mathcal{E}_n/k_B T\right)}{\sum_k \exp\left(-\mathcal{E}_k/k_B T\right)}, \tag{3.82}$$

where Z is the partition function.

We are interested in calculating the average value of an observable corresponding to a given operator A. As pointed out before, in the case of a mixed state, it is reasonable to write:

$$\langle A \rangle = \sum_n p_n \langle \psi_n | A | \psi_n \rangle, \tag{3.83}$$

indeed, p_n represents the probability that the system is in the pure state $|\psi_n\rangle$ and $\langle \psi_n | A | \psi_n \rangle$ is the average value of A when the system is in this state. Therefore, Eq. 3.83 is the probabilistic weighted average of the expectation values of the pure states $|\psi_n\rangle$. It is evident that $\sum_n p_n = 1$. Equation 3.83 can be written in a different way:

$$\langle A \rangle = \sum_n p_n \langle \psi_n | A I | \psi_n \rangle = \sum_{n,k} p_n \langle \psi_n | A | u_k \rangle \langle u_k | \psi_n \rangle, \tag{3.84}$$

where we have used the identity operator. By changing the position of the two scalar products in the previous expression, we have:

$$\langle A \rangle = \sum_{n,k} p_n \langle u_k | \psi_n \rangle \langle \psi_n | A | u_k \rangle. \tag{3.85}$$

If we now introduce the *density operator* ρ as:

$$\rho \equiv \sum_n p_n | \psi_n \rangle \langle \psi_n |, \tag{3.86}$$

Equation 3.85 can be rewritten as:

$$\langle A \rangle = \sum_k \langle u_k | \rho A | u_k \rangle = \text{Tr}(\rho A), \tag{3.87}$$

where $\mathrm{Tr}(\rho A)$ is the trace of the matrix (ρA). Therefore, the expectation value of any operator A is the sum of the diagonal matrix elements of the product ρA. It is evident that, since we can always obtain the average value of an observable as $\mathrm{Tr}(\rho A)$, the density operator ρ must contain all the physically relevant information that we can know about a system. Note that in the definition of ρ (Eq. 3.86), the state functions $|\psi_n\rangle$ are normalized but are not necessarily orthogonal.

An important result is that ρ is Hermitian. Indeed:

$$
\begin{aligned}
\langle\psi|\rho|\phi\rangle &= \sum_n p_n\langle\psi|\psi_n\rangle\langle\psi_n|\phi\rangle = \sum_n p_n\langle\psi_n|\psi\rangle^*\langle\phi|\psi_n\rangle^* = \\
&= \sum_n p_n(\langle\phi|\psi_n\rangle\langle\psi_n|\psi\rangle)^* = \langle\phi|\rho|\psi\rangle^*
\end{aligned}
\tag{3.88}
$$

Moreover, $\mathrm{Tr}(\rho) = 1$:

$$
\begin{aligned}
\mathrm{Tr}(\rho) &= \sum_k \langle u_k|\rho|u_k\rangle = \sum_{k,n} p_n\langle u_k|\psi_n\rangle\langle\psi_n|u_k\rangle = \\
&= \sum_{k,n} p_n\langle\psi_n|u_k\rangle\langle u_k|\psi_n\rangle = \sum_n p_n\langle\psi_n|\psi_n\rangle = 1.
\end{aligned}
\tag{3.89}
$$

Indeed, $\langle\psi_n|\psi_n\rangle = 1$ and $\sum_n p_n = 1$. By writing the pure states $|\psi_n\rangle$ in 3.86 as:

$$
|\psi_n\rangle = \sum_k c_{nk}e^{-i\omega_k t}|u_k\rangle,
$$

the density operator can be written as:

$$
\rho = \sum_{n,k,\ell} p_n c_{nk}c_{n\ell}^* e^{-i(\omega_k-\omega_\ell)t}|u_k\rangle\langle u_\ell|.
\tag{3.90}
$$

The matrix elements ρ_{ij} are given by:

$$
\begin{aligned}
\rho_{ij} &= \langle u_i|\rho|u_j\rangle = \sum_{n,k,\ell} p_n c_{nk}c_{n\ell}^* e^{-i(\omega_k-\omega_\ell)t}\langle u_i|u_k\rangle\langle u_\ell|u_j\rangle = \\
&= \sum_n p_n c_{ni}c_{nj}^* e^{-i(\omega_i-\omega_j)t},
\end{aligned}
\tag{3.91}
$$

the diagonal elements are:

$$
\rho_{jj} = \sum_n p_n|c_{nj}|^2.
\tag{3.92}
$$

so that ρ_{jj} gives the probability to find the system in the state $|u_j\rangle$. The off-diagonal elements are called *coherence elements*, and their physical meaning will be discussed in Section 3.4.1.

Exercise 3.2 Consider a pure state $|\psi\rangle$. Calculate $\mathrm{Tr}(\rho)$, $\mathrm{Tr}(\rho^2)$, the expectation value of the observable associated to the operator A, the diagonal and the off-diagonal elements of ρ.

The density operator is:

$$\rho = |\psi\rangle\langle\psi|. \tag{3.93}$$

Its trace can be easily calculated:

$$\mathrm{Tr}(\rho) = \sum_k \langle u_k|\rho|u_k\rangle = \sum_k \langle u_k|\psi\rangle\langle\psi|u_k\rangle = \sum_k \langle\psi|u_k\rangle\langle u_k|\psi\rangle = \langle\psi|\psi\rangle = 1. \tag{3.94}$$

Now we calculate ρ^2:

$$\rho^2 = (|\psi\rangle\langle\psi|)(|\psi\rangle\langle\psi|) = |\psi\rangle\langle\psi|\psi\rangle\langle\psi| = |\psi\rangle\langle\psi| = \rho. \tag{3.95}$$

Therefore

$$\mathrm{Tr}(\rho^2) = \mathrm{Tr}(\rho) = 1. \tag{3.96}$$

The expectation value of A is:

$$\begin{aligned}
\langle A\rangle &= \mathrm{Tr}(\rho A) = \sum_k \langle u_k|\rho A|u_k\rangle = \sum_k \langle u_k|\psi\rangle\langle\psi|A|u_k\rangle = \\
&= \sum_k a_k \langle u_k|\psi\rangle\langle\psi|u_k\rangle = \sum_k a_k |c_k|^2,
\end{aligned} \tag{3.97}$$

where we have used the eigenvalue equation $A|u_k\rangle = a_k|u_k\rangle$.
The diagonal elements of ρ are:

$$\begin{aligned}
\rho_{jj} &= \langle u_j|\rho|u_j\rangle = \langle u_j|\psi\rangle\langle\psi|u_j\rangle = \langle u_j|\psi\rangle\langle u_j|\psi\rangle^* = \\
&= c_j e^{-i\omega_j t} c_j^* e^{i\omega_j t} = |c_j|^2.
\end{aligned} \tag{3.98}$$

The off-diagonal elements of ρ are:

$$\begin{aligned}
\rho_{ij} &= \langle u_i|\rho|u_j\rangle = \langle u_i|\psi\rangle\langle\psi|u_j\rangle = \langle u_i|\psi\rangle\langle u_j|\psi\rangle^* = \\
&= c_i c_j^* e^{-i(\omega_i - \omega_j)t}.
\end{aligned} \tag{3.99}$$

In the absence of relaxation processes the populations of the different energy levels, given by the diagonal elements of ρ, are time invariant so that the time dependence is present only in the off-diagonal matrix elements.   ∎

3.4.1 Density Matrix of a Canonical Ensemble.
A *canonical ensemble* is a statistical ensemble of systems which are all in contact (i.e., in thermal equilibrium) with a heat reservoir at temperature T, so that the overall system is in thermal equilibrium with its environment. The probability to find the system in a particular energy eigenstate $|u_n\rangle$ with energy $\mathcal{E}_n$ is given by the classical Boltzmann distribution 3.82:

$$p_n = \frac{1}{Z}\exp(-\beta\mathcal{E}_n), \tag{3.100}$$

where $\beta = 1/k_B T$. The thermal equilibrium density operator can be obtained by using 3.100 into 3.86:

$$\rho = \frac{1}{Z} \sum_n \exp(-\beta \mathcal{E}_n) |u_n\rangle\langle u_n|, \tag{3.101}$$

which represents the spectral decomposition of ρ in the energy representation. It is possible to write the previous expression in a more compact form by introducing the exponential Hamiltonian operator $\exp(-\beta H)$, where:

$$e^{-\beta H} = \sum_{m=0}^{\infty} \frac{(-\beta H)^m}{m!}. \tag{3.102}$$

If we apply the exponential Hamiltonian to vector $|u_k\rangle$, we obtain:

$$e^{-\beta H}|u_k\rangle = \sum_{m=0}^{\infty} \frac{(-\beta H)^m}{m!} |u_k\rangle = \sum_{m=0}^{\infty} \frac{(-\beta)^m}{m!} H^m |u_k\rangle =$$

$$= \sum_{m=0}^{\infty} \frac{(-\beta \mathcal{E}_k)^m}{m!} |u_k\rangle = e^{-\beta \mathcal{E}_k} |u_k\rangle, \tag{3.103}$$

where we have used the equation:

$$H^m |u_k\rangle = \mathcal{E}_k^m |u_k\rangle. \tag{3.104}$$

We now apply the operator ρ as defined by Eq. 3.101 to an arbitrary wavefunction $|\xi\rangle$

$$\rho|\xi\rangle = \frac{1}{Z} \sum_n e^{-\beta \mathcal{E}_n} |u_n\rangle\langle u_n|\xi\rangle = \frac{1}{Z} \sum_n e^{-\beta H} |u_n\rangle\langle u_n|\xi\rangle = \frac{1}{Z} e^{-\beta H} |\xi\rangle,$$

where we have used the identity operator and Eq. 3.103. Since $|\xi\rangle$ is an arbitrary wavefunction, it may be cancelled out of the previous equation to yield:

$$\rho = \frac{e^{-\beta H}}{Z}. \tag{3.105}$$

The partition function can be written as:

$$Z = \text{Tr}(e^{-\beta H}) = \sum_n e^{-\beta \mathcal{E}_n}, \tag{3.106}$$

and the matrix element of the density matrix can be calculated as:

$$\rho_{ij} = \langle u_i | \frac{e^{-\beta H}}{Z} | u_j \rangle = \frac{e^{-\beta \mathcal{E}_j}}{Z} \delta_{ij}, \tag{3.107}$$

thus meaning that the density matrix is diagonal with diagonal elements given by $e^{-\beta \mathcal{E}_j}/Z$. Therefore, if the system is initially out of thermal equilibrium, it must evolve so that the diagonal elements redistribute to a Boltzmann distribution and the off-diagonal elements must go to zero. In the first case, we have a population relaxation, and in the second case, we have a dephasing process. The relaxation processes will be discussed in Section 3.4.2.

For a **mixed state**, the property 3.95 does not hold. Indeed in general:

$$\rho^2 = \sum_n \sum_m p_n p_m |\psi_n\rangle\langle\psi_n|\psi_m\rangle\langle\psi_m| \neq \rho, \tag{3.108}$$

moreover, it is possible to demonstrate that:

$$\text{Tr}(\rho^2) \leq 1. \tag{3.109}$$

In Eq. 3.109 $\text{Tr}(\rho^2) = 1$ if and only if the system can be in only one physical state, that is, all but one of the probabilities p_n corresponding to independent states must be zero. It is possible to conclude that a state of a physical system is a pure state if $\text{Tr}(\rho^2) = 1$, whereas if $\text{Tr}(\rho^2) < 1$ the system is in a mixed state. $\text{Tr}(\rho^2)$ can be interpreted as a measure of the *mixed nature* of a given system.

Exercise 3.3   Demonstrate that in general $\text{Tr}(\rho^2) \leq 1$.

$$\text{Tr}(\rho^2) = \sum_k \langle u_k| \left\{ \sum_{nm} p_n p_m |\psi_n\rangle\langle\psi_n|\psi_m\rangle\langle\psi_m| \right\} |u_k\rangle =$$

$$= \sum_k \sum_{nm} p_n p_m \langle u_k|\psi_n\rangle\langle\psi_n|\psi_m\rangle\langle\psi_m|u_k\rangle =$$

$$= \sum_k \sum_{nm} p_n p_m \langle\psi_m|u_k\rangle\langle u_k|\psi_n\rangle\langle\psi_n|\psi_m\rangle, \tag{3.110}$$

in the last expression we recognize the identity operator I, so that:

$$\text{Tr}(\rho^2) = \sum_{nm} p_n p_m \langle\psi_m|\psi_n\rangle\langle\psi_n|\psi_m\rangle = \sum_{nm} p_n p_m |\langle\psi_n|\psi_m\rangle|^2, \tag{3.111}$$

From the Schwarz inequality, one has:

$$|\langle\psi_n|\psi_m\rangle|^2 \leq \langle\psi_n|\psi_n\rangle\langle\psi_m|\psi_m\rangle, \tag{3.112}$$

so that:

$$\text{Tr}(\rho^2) \leq \sum_{nm} p_n p_m \langle\psi_n|\psi_n\rangle\langle\psi_m|\psi_m\rangle = \sum_n p_n \sum_m p_m = 1. \tag{3.113}$$

∎

3.4.2 Temporal Dependence of ρ

We have seen that by using the density matrix formalism, it is possible to obtain the average value of a given observable. What we need to know in addition is the temporal evolution of ρ. We start from the definition of ρ:

$$\rho = \sum_n p_n |\psi_n\rangle\langle\psi_n|,$$

by calculating the temporal derivative we obtain:

$$\frac{\partial\rho}{\partial t} = \sum_n p_n \left[\frac{\partial|\psi_n\rangle}{\partial t}\langle\psi_n| + |\psi_n\rangle\frac{\partial\langle\psi_n|}{\partial t} \right]. \tag{3.114}$$

The Schrödinger equation for $|\psi_n\rangle$ is:

$$H|\psi_n\rangle = i\hbar\frac{\partial|\psi_n\rangle}{\partial t}, \tag{3.115}$$

the corresponding equation for $\langle\psi_n|$ is:

$$\langle\psi_n|H = -i\hbar\frac{\partial\langle\psi_n|}{\partial t}. \tag{3.116}$$

By using Eqs. 3.115 and 3.116 in 3.114, we obtain:

$$i\hbar\frac{\partial\rho}{\partial t} = \sum_n p_n[H|\psi_n\rangle\langle\psi_n| - |\psi_n\rangle\langle\psi_n|H] = [H\rho - \rho H]. \tag{3.117}$$

The operator $H\rho - \rho H$ is called *commutator* and can be written as $[H, \rho]$. Therefore, we have:

$$i\hbar\frac{\partial\rho}{\partial t} = [H, \rho], \tag{3.118}$$

or, equivalently:

$$\frac{\partial\rho}{\partial t} = -\frac{i}{\hbar}[H, \rho] = \frac{i}{\hbar}[\rho, H] \tag{3.119}$$

called the *von Neumann equation*, which is the quantum analogue of the classical Liouville equation. In Section 4.5, this equation will be used to calculate the susceptibility of a two-level system, which we will subsequently extend to the case of a semiconductor.

Exercise 3.4   By using the von Neumann equation derive the Ehrenfest theorem in the formulation of Heisenberg:

$$\frac{d}{dt}\langle A\rangle = \left\langle\frac{\partial A}{\partial t}\right\rangle + \frac{1}{i\hbar}\langle[A, H]\rangle, \tag{3.120}$$

where A is a given quantum mechanical operator.

By using the density operator formalism, the expectation value of an operator can be written as in Eq. 3.87:

$$\langle A\rangle = \text{Tr}(\rho A),$$

therefore, upon calculating the time derivative of $\langle A\rangle$, we obtain:

$$\frac{d}{dt}\langle A\rangle = \frac{d}{dt}\text{Tr}(\rho A) = \text{Tr}\left(\frac{\partial\rho}{\partial t}A\right) + \text{Tr}\left(\rho\frac{\partial A}{\partial t}\right). \tag{3.121}$$

By using the von Neumann equation:

$$\begin{aligned}
\frac{d}{dt}\langle A\rangle &= \text{Tr}\left(\frac{1}{i\hbar}[H, \rho]A\right) + \text{Tr}\left(\rho\frac{\partial A}{\partial t}\right) = \\
&= \frac{1}{i\hbar}\text{Tr}(H\rho A - \rho HA) + \text{Tr}\left(\rho\frac{\partial A}{\partial t}\right).
\end{aligned} \tag{3.122}$$

Since trace is ciclic $\text{Tr}(ABC) = \text{Tr}(BCA) = \text{Tr}(CAB)$:

$$\text{Tr}(H\rho A) = \text{Tr}(\rho AH) = \langle AH\rangle, \tag{3.123}$$

so that Eq. 3.122 can be written as:

$$\frac{d}{dt}\langle A\rangle = \frac{1}{i\hbar}\left(\langle AH\rangle - \langle HA\rangle\right) + \left\langle\frac{\partial A}{\partial t}\right\rangle = \frac{1}{i\hbar}\langle[A,H]\rangle + \left\langle\frac{\partial A}{\partial t}\right\rangle, \tag{3.124}$$

which is the formulation of the Ehrenfest theorem.

It is interesting to note that Ehrenfest theorem as given by Eq. 3.124 closely follows the expression of the time derivative of a function $f(q,p,t)$, where q and p are the generalized coordinate and momentum, in terms of the Hamilton's equations:

$$\dot{q} = \frac{\partial H}{\partial p} \tag{3.125}$$

$$\dot{p} = -\frac{\partial H}{\partial q}. \tag{3.126}$$

Applying the chain rule, we have:

$$\frac{d}{dt}f(q,p,t) = \frac{\partial f}{\partial q}\dot{q} + \frac{\partial f}{\partial p}\dot{p} + \frac{\partial f}{\partial t}$$

$$= \frac{\partial f}{\partial q}\frac{\partial H}{\partial p} - \frac{\partial f}{\partial p}\frac{\partial H}{\partial q} + \frac{\partial f}{\partial t} = \{f,H\} + \frac{\partial f}{\partial t}, \tag{3.127}$$

where $\{f,H\}$ is defined as the Poisson bracket of f and H. Previous classical equation elegantly compares with Eq. 3.120 upon replacing the Poisson bracket of classical mechanics with the commutator of quantum mechanics. ∎

3.4.3 Temporal Evolution of the Matrix Elements of ρ

The temporal evolution of the matrix elements of the density operator can be evaluated by using the von Neumann equation:

$$\frac{\partial\rho_{ij}}{\partial t} = -\frac{i}{\hbar}\langle u_i|H\rho - \rho H|u_j\rangle. \tag{3.128}$$

By using the identity operator, the previous equation can be rewritten as:

$$\frac{\partial\rho_{ij}}{\partial t} = -\frac{i}{\hbar}\sum_k\left\{\langle u_i|H|u_k\rangle\langle u_k|\rho|u_j\rangle - \langle u_i|\rho|u_k\rangle\langle u_k|H|u_j\rangle\right\} =$$

$$= -\frac{i}{\hbar}\sum_k\left\{H_{ik}\rho_{kj} - \rho_{ik}H_{kj}\right\}. \tag{3.129}$$

We can also investigate the evolution of the system under the action of an external perturbation described by the interaction Hamiltonian H'. In order to take into account the presence of interactions internal to the system, due for example to interactions between an atom and lattice or, in the case of gases or liquids, to collisions between molecules, we will introduce the Hamiltonian H^r. H^r is generally called randomizing Hamiltonian, since the internal interactions that are taken into account by H^r determine a variation of the matrix elements of the density operator without any external perturbation. The total Hamiltonian can be written as:

$$H = H_0 + H' + H^r, \tag{3.130}$$

and the matrix elements of the commutator $[H, \rho]$ are given by:

$$
\begin{aligned}
[H, \rho]_{ij} &= [H_0, \rho]_{ij} + [H', \rho]_{ij} + [H^r, \rho]_{ij} = \\
&= (\mathcal{E}_i - \mathcal{E}_j)\rho_{ij} + [H', \rho]_{ij} + [H^r, \rho]_{ij}.
\end{aligned} \tag{3.131}
$$

From the von Neumann equation, we obtain:

$$
\frac{\partial \rho_{ij}}{\partial t} = -\frac{i}{\hbar}(\mathcal{E}_i - \mathcal{E}_j)\rho_{ij} - \frac{i}{\hbar}[H', \rho]_{ij} - \frac{i}{\hbar}[H^r, \rho]_{ij}. \tag{3.132}
$$

Without going into details, we can assume that in the case of small coupling with environment, the effect of the randomizing Hamiltonian on the evolution of the system can be well described by replacing in Eq. 3.132 the term $[H^r, \rho]_{ij}$ with a simple relaxation term. Indeed, when the external perturbation is turned off, the system evolves to equilibrium with its environment, due to internal interactions. Since the off-diagonal matrix element of ρ is null at equilibrium, Eq. 3.132 can be written as:

$$
\frac{\partial \rho_{ij}}{\partial t} = -i\omega_{ij}\rho_{ij} - \frac{i}{\hbar}[H', \rho]_{ij} - \frac{\rho_{ij}}{\tau_{ij}}, \tag{3.133}
$$

where $\omega_{ij} = (\mathcal{E}_i - \mathcal{E}_j)/\hbar$ and we have replaced the term related to the internal interactions with a relaxation term, which, when $H' = 0$, determines an exponential relaxation of the off-diagonal elements ρ_{ij} with a time constant τ_{ij}, which is assumed real and positive. Moreover, since ρ is Hermitian, $\tau_{ji} = \tau_{ij}$.

Referring to the diagonal matrix elements of ρ, it is evident that when an external perturbation applied to a system is turned off, the populations of the states have to evolve toward their thermal equilibrium values, $N\rho_{jj}^e$, where N is the number of atoms or molecules per unit volume, with a given time constant, which depends on the particular nature of the system under investigation. From Eq. 3.132, we obtain:

$$
\frac{\partial \rho_{jj}}{\partial t} = -\frac{i}{\hbar}[H', \rho]_{jj} - \frac{i}{\hbar}[H^r, \rho]_{jj}. \tag{3.134}
$$

The last term, $[H^r, \rho]_{jj}$, takes into account the relaxation processes leading to the evolution of the population of state $|j\rangle$ toward its equilibrium value, when the external perturbation is switched off. In this case, we have to consider the transitions from state $|k\rangle$ to state $|j\rangle$ and the transitions from state $|j\rangle$ to state $|k\rangle$. At equilibrium, when $H' = 0$, the occupation probability of a given state (e.g., the state $|j\rangle$) must be time-independent, thus meaning that at equilibrium the number of transitions per unit time from state $|j\rangle$ to state $|k\rangle$ must equal the number of transitions per unit time from state $|k\rangle$ to state $|j\rangle$. A simple and effective way to express the number of transitions per unit time can be obtained by introducing a transition probability, W_{ij}, defined as the probability per unit time of a transition from state $|i\rangle$ to state $|j\rangle$, when $H' = 0$. In this way, the number of transitions per unit time from state $|j\rangle$ to state $|k\rangle$ (thus leading to a decrease of the population of state $|j\rangle$) can be written as $\rho_{jj}W_{jk}$, while the number of transitions per unit time from state $|k\rangle$ to state $|j\rangle$ (thus leading to a increase of the population of state $|j\rangle$) can be written as $\rho_{kk}W_{kj}$. At equilibrium, we must have:

$$
\rho_{jj}^e W_{jk} = \rho_{kk}^e W_{kj}. \tag{3.135}
$$

With these definitions, the term $[H^r, \rho]_{jj}$ in Eq. 3.134 associated with the randomizing Hamiltonian can be written as follows:

$$[H^r, \rho]_{jj} = i\hbar \sum_k (\rho_{kk} W_{kj} - \rho_{jj} W_{jk}). \tag{3.136}$$

The previous expression takes into account the processes leading to both an increase and a decrease of the population of the generic state $|j\rangle$. Equation 3.136 can be written in a different and useful way by introducing the relaxation times T_{jk} defined as:

$$T_{jk} = \frac{\rho_{kk}^e}{W_{jk}}. \tag{3.137}$$

On the basis of Eq. 3.135, it is evident that $T_{jk} = T_{kj}$. Therefore, Eq. 3.134 can be written as:

$$\frac{\partial \rho_{jj}}{\partial t} = -\frac{i}{\hbar}[H', \rho]_{jj} + \sum_k \left(\rho_{kk} \frac{\rho_{jj}^e}{T_{jk}} - \rho_{jj} \frac{\rho_{kk}^e}{T_{jk}} \right). \tag{3.138}$$

Assuming that all the relaxation times T_{jk} are equal ($T_{jk} = T_1$), the previous equation can be rewritten:

$$\frac{\partial \rho_{jj}}{\partial t} = -\frac{i}{\hbar}[H', \rho]_{jj} + \frac{1}{T_1} \sum_k \left(\rho_{kk} \rho_{jj}^e - \rho_{jj} \rho_{kk}^e \right). \tag{3.139}$$

Since $\mathrm{Tr}(\rho) = \mathrm{Tr}(\rho^e) = 1$, the previous equation can be written in a simpler form ($\sum_k \rho_{kk} = \sum_k \rho_{kk}^e = 1$):

$$\frac{\partial \rho_{jj}}{\partial t} = -\frac{i}{\hbar}[H', \rho]_{jj} - \frac{1}{T_1} \left(\rho_{jj} - \rho_{jj}^e \right). \tag{3.140}$$

It is evident that when the external perturbation is turned off ($H' = 0$), the probability of occupation of state $|j\rangle$ decays to the equilibrium value, ρ_{jj}^e, with a time constant T_1, which is generally called *longitudinal relaxation time* or population lifetime. The longitudinal relaxation time is a measure of the time required for a system to reach the thermodynamic equilibrium with its environment. The population relaxation can be due to interactions between an atom and the lattice, or, in the case of gases or liquids to collisions between molecules or to spontaneous emission.

   The time constant for the off-diagonal matrix elements, τ_{ij}, is generally written as T_2, in particular when the analysis of the system can be reduced to two states. T_2 is called *transverse relaxation time*, and it is related to dephasing of the dipoles generated by the perturbation. We note that any process contributing to population relaxation also contributes to dephasing. Moreover, there are processes which contribute to dephasing but do not alter the population of the states: For example, an elastic collision, in which one molecule is excited from the state $|1\rangle$ to the state $|2\rangle$ and another makes the inverse transition, from state $|2\rangle$ to state $|1\rangle$, does not change the population of a given state but can lead to a dephasing of the off-diagonal elements. Therefore, if all molecules are initially oscillating in phase, after a given time they will oscillate with arbitrary phases relative to each other. We will see that T_2 is associated with the linewidth of a transition. Since all processes contributing to T_1 also contribute to T_2 and there are dephasing processes, which contribute only to T_2, in general $T_2 < T_1$.

3.4.4   Effect of H' on ρ_{11}

Let as assume that the system is in the state $|1\rangle$ at $t = 0$ and that the external perturbation is turned off at $t = 0$, so that $H' = 0$ for $t > 0$. In this case at $t = 0$, we have $\rho_{ij}(0) = \delta_{i1}\delta_{j1}$. For small values of t, ρ_{11} is the largest element of the density matrix, so that Eq. 3.132 can be written as ($H' = 0$):

$$i\hbar\frac{\partial \rho_{1j}}{\partial t} = \hbar\omega_{1j}\rho_{1j} - \rho_{11}H'_{1j} \qquad j \neq i, \tag{3.141}$$

while Eq. 3.134 can be written as:

$$i\hbar\frac{\partial \rho_{11}}{\partial t} = \sum_k \left[H'_{1k}\rho_{k1} - \rho_{1k}H'_{k1} \right]. \tag{3.142}$$

Using the method of Laplace transform, we can write:

$$i\hbar s\rho_{1j} = \hbar\omega_{1j}\rho_{1j} - \rho_{11}H'_{1j} \tag{3.143}$$

$$i\hbar(s\rho_{11} - 1) = \sum_k \left[H'_{1k}\rho_{k1} - \rho_{1k}H'_{k1} \right]. \tag{3.144}$$

From the first equation:

$$\rho_{1j} = \frac{\rho_{11}H'_{1j}}{\hbar(\omega_{1j} - is)}. \tag{3.145}$$

By using this result in Eq. 3.144 (where $\rho_{j1} = \rho^*_{1j}$), we obtain:

$$\rho_{11}(s) = \left[s + \frac{i}{\hbar^2} \sum_k \left(\frac{|H'_{1k}|^2}{\omega_{1k} + is} - \frac{|H'_{1k}|^2}{\omega_{1k} - is} \right) \right]^{-1}. \tag{3.146}$$

If the coupling with the environment is completely negligible, Eq. 3.146 presents a pole at $s = 0$, so that $\rho_{11}(t)$ is a constant, thus meaning that the system remains in the state $|1\rangle$. If the coupling is small, the pole of Eq. 3.146 will be slightly shifted due to the small interaction of the medium with its environment. Therefore, we have to evaluate the sum over k in Eq. 3.146 for $s \to 0$. It is convenient to write this sum in a different way:

$$\frac{i}{\hbar^2} \sum_k \left(\frac{|H'_{1k}|^2}{\omega_{1k} + is} - \frac{|H'_{1k}|^2}{\omega_{1k} - is} \right) = \sum_k \frac{2s|H'_{1k}|^2}{s^2 + \omega^2_{1k}}. \tag{3.147}$$

We can now assume that the coupling between the states of the system is determined by interactions with the vibrational modes of the lattice, but the conclusions we will derive are more general, thus including all the possible internal interactions. Since the vibrational modes are closely spaced, the sum in Eq. 3.147 can be replaced by an integral:

$$\sum_k \frac{2s|H'_{1k}|^2}{s^2 + \omega^2_{1k}} \to \int_{-\infty}^{\infty} g(\omega_{1k})\frac{2s|H'_{1k}|^2}{s^2 + \omega^2_{1k}} \, d\omega_{1k}. \tag{3.148}$$

Now, we have to evaluate the following limit:

$$\lim_{s\to 0^+} \int_{-\infty}^{\infty} g(\omega_{1k}) \frac{2s|H_{1k}^r|^2}{s^2 + \omega_{1k}^2} \, d\omega_{1k}. \tag{3.149}$$

We observe that:

$$\lim_{s\to 0^+} \frac{s}{\pi(s^2 + \omega_{1k}^2)} = \begin{cases} 0 & \omega_{1k} \neq 0 \\ \infty & \omega_{1k} = 0 \end{cases} \tag{3.150}$$

and:

$$\int_{-\infty}^{\infty} \frac{s}{\pi(s^2 + \omega_{1k}^2)} \, d\omega_{1k} = 1. \tag{3.151}$$

These are the characteristics of a Dirac δ-function, so that the limit of Eq. 3.149 can be easily calculated:

$$\lim_{s\to 0^+} \int 2\pi g(\omega_{1k})|H_{1k}^r|^2 \frac{s}{\pi(s^2 + \omega_{1k}^2)} \, d\omega_{1k} = 2\pi g(\omega_{1k})|H_{1k}^r|^2 \Big|_{\omega_{1k}=0}. \tag{3.152}$$

If we now introduce a time constant τ defined as follows (with $\omega_{1k} = 0$):

$$\frac{1}{\tau} = \frac{2\pi}{\hbar^2} g(\omega_{1k})|H_{1k}^r|^2. \tag{3.153}$$

Eq. 3.146 can be written as:

$$\rho_{11}(s) = \frac{1}{s + 1/\tau}, \tag{3.154}$$

which gives the following time evolution of $\rho_{11}(t)$:

$$\rho_{11}(t) = \exp(-t/\tau). \tag{3.155}$$

This result is more general: In the regime of small coupling, the interaction of the medium with its environment, described by the term $[H^r, \rho]$ in the von Neumann equation, causes a relaxation of the matrix elements of the density operator with a relaxation time constant which depends on the particular coupling mechanisms. ∎

3.4.5 Density Matrix and Entropy

In terms of the density operator, the entropy S is given by the following expression:

$$S = -k_B \text{Tr}(\rho \log \rho), \tag{3.156}$$

which is generally called von Neumann entropy. Considering some complete orthonormal set, Eq. 3.156 can be written as:

$$S = -k_B \sum_n \langle u_n|\rho \log \rho|u_n\rangle = -k_B \sum_{n,m} \langle u_n|\rho|u_m\rangle \langle u_m|\log \rho|u_n\rangle. \tag{3.157}$$

Since the trace is basis independent, the definition of entropy given by Eq. 3.156 implies that it is independent of the basis in which the quantum states are expressed. Therefore, we can use a basis in which ρ is diagonal, so that:

$$S = -k_B \sum_k \rho_{kk} \log(\rho_{kk}). \tag{3.158}$$

For a pure state only one eigenvalue of ρ is 1 and all others are zero, thus implying that $S(\rho) = 0$ for a pure state. Since $S(\rho) = 0$ if and only if the system is in a pure state, $S(\rho)$ can be considered a measure of the mixed nature of a given state.

If we maximize the entropy subject to the constraint that

$$\mathrm{Tr}(\rho) = 1, \tag{3.159}$$

we have:

$$\delta S = -k_B \delta \mathrm{Tr}(\rho \log \rho) = 0, \tag{3.160}$$

which gives:

$$\sum_{k=1}^{N} (\log \rho_{kk} + 1) \delta \rho_{kk} = 0, \tag{3.161}$$

due to the constraint (Eq. 3.159), $\delta \rho_{kk}$ are not independent. Indeed, from 3.159:

$$\delta \mathrm{Tr}(\rho) = 0, \tag{3.162}$$

so that

$$\sum_k \delta \rho_{kk} = 0. \tag{3.163}$$

We will use the method of Lagrange undetermined multipliers. We multiply Eq. 3.163 by an as yet unknown number λ and we add it to Eq. 3.161, thus obtaining:

$$\sum_{k=1}^{N} (\log \rho_{kk} + 1 + \lambda) \delta \rho_{kk} = 0, \tag{3.164}$$

where λ is called a Lagrange undetermined multiplier. In Eq. 3.164, $\delta \rho_{kk}$ are all independent, so that from Eq. 3.164 we obtain that each term in parenthesis must be zero. Therefore:

$$\log \rho_{kk} = -(1 + \lambda)$$
$$\rho_{kk} = e^{-(1+\lambda)}. \tag{3.165}$$

Since $\mathrm{Tr}(\rho) = 1$, we have:

$$\sum_{k=1}^{N} \rho_{kk} = \sum_{k=1}^{N} e^{-(1+\lambda)} = N e^{-(1+\lambda)} = 1. \tag{3.166}$$

From Eqs. 3.165 and 3.166, we have:

$$\rho_{kk} = \frac{1}{N}, \tag{3.167}$$

thus meaning that the entropy is maximized when the probability to find the system in any of its possible states is the same. In this case, the entropy can be written as follows:

$$S = -k_B \sum_{k=1}^{N} \rho_{kk} \log \rho_{kk} = k_B \log N. \tag{3.168}$$

In thermal equilibrium, we may have the additional constraint that the average energy is fixed. The average energy is given by the following expression:

$$\langle \mathcal{E} \rangle = \text{Tr}(\rho H), \tag{3.169}$$

where H is the Hamiltonian operator. We want to write the density operator in order to maximize the entropy of the system subject to the constraints 3.159 and 3.169. Then, upon varying ρ, we have:

$$\text{Tr}(1 + \log \rho)\delta\rho = 0 \tag{3.170}$$

$$\text{Tr}\,\delta\rho = 0 \tag{3.171}$$

$$\text{Tr}\,H\delta\rho = 0. \tag{3.172}$$

We will now apply the method of Lagrange undetermined multipliers. We multiply Eq. 3.171 by the undetermined multiplier λ and Eq. 3.172 by β and we add to Eq. 3.170, thus obtaining:

$$\text{Tr}(1 + \lambda + \log \rho + \beta H)\delta\rho = 0, \tag{3.173}$$

where $\delta\rho$ is arbitrary and all variations are now independent. Therefore, from the previous equation, we obtain:

$$\log \rho = -1 - \lambda - \beta H, \tag{3.174}$$

which can be written as:

$$\rho = e^{-(1+\lambda)} e^{-\beta H}. \tag{3.175}$$

Since $\text{Tr}(\rho) = 1$, we have:

$$e^{1+\lambda} = \text{Tr}\left(e^{-\beta H}\right). \tag{3.176}$$

From Eqs. 3.175 and 3.176, we can obtain the expression of the density operator:

$$\rho = \frac{e^{-\beta H}}{\text{Tr}\left(e^{-\beta H}\right)}. \tag{3.177}$$

If we now define the partition function Z as:

$$Z \equiv \text{Tr}\left(e^{-\beta H}\right), \tag{3.178}$$

we obtain:

$$\rho = \frac{1}{Z} e^{-\beta H}, \tag{3.179}$$

which is the expression of the density operator of a statistical ensemble of systems in thermodynamic equilibrium with a heat bath at temperature T, as we have already written in Eq. 3.105.

In the energy representation, in which $H|\psi_n\rangle = \mathcal{E}_n|\psi_n\rangle$, the matrix elements of the density operator in Eq. 3.179 are given by Eq. 3.107. Therefore, the probability of finding the system in the state $|\psi_m\rangle$ is given by:

$$p_m = \rho_{mm} = \frac{e^{-\beta\mathcal{E}_m}}{Z}, \tag{3.180}$$

which corresponds to the Maxwell–Boltzmann probability distribution. In this energy representation, the density operator can be written as:

$$\rho = \sum_m p_m|\psi_m\rangle\langle\psi_m|, \tag{3.181}$$

where p_m is given by Eq. 3.180.

3.5 Exercises

Exercise 3.1   Consider the operators $\hat{A} = \frac{d}{dx} + x$ and $\hat{B} = \frac{d}{dx} - x$. Determine the operator $\hat{A}\hat{B}$.

Exercise 3.2   Evaluate the following commutators:
a) $[x, p_x]$
b) $[x, p_x^2]$
c) $[x, H]$
d) $[p_x, H]$
where $H = p^2/(2m) + V(x)$.

Exercise 3.3   Show that $[x^n, p_x] = i\hbar n x^{n-1}$, where n is a positive integer.

Exercise 3.4   Show that $[f(x), p_x] = i\hbar \frac{df(x)}{dx}$.

Exercise 3.5   If $|\psi\rangle$ is an eigenfunction of H with eigenvalue $\mathcal{E}$, show that for any operator $\hat{A}$, the expectation value of $[H, \hat{A}]$ vanishes, that is, $\langle\psi|[H, \hat{A}]|\psi\rangle = 0$.

Exercise 3.6   A Hermitian operator $\hat{A}$ has only three normalized eigenfunctions, $|\psi_1\rangle$, $|\psi_2\rangle$, and $|\psi_3\rangle$, with corresponding eigenvalues $a_1 = 1$, $a_2 = 2$, and $a_3 = 3$, respectively. For a particular state $|\phi\rangle$ of the system, there is a 50% chance that a measure of A produces a_1 and equal chances for either a_2 or a_3.
a) Calculate $\langle A\rangle$;
b) Express the normalized wave function $|\phi\rangle$ of the system in terms of the eigenfunctions of $\hat{A}$.

Exercise 3.7   Demonstrate that the propagator operator $\hat{U} = \exp(-iHt/\hbar)$ gives the quantum state $|\psi(t)\rangle = \hat{U}(t)|\psi(0)\rangle$ that satisfies the Schrödinger equation:

$$i\hbar\frac{d}{dt}|\psi(t)\rangle = H|\psi(t)\rangle.$$

Exercise 3.8   The eigenstates of a particle in an infinite potential well are:

$$\phi_n(x) = \sqrt{\frac{2}{L}} \sin\left(n\frac{\pi}{L}x\right).$$

a) Determine the matrix elements of the position operator, $\langle\phi_n|x|\phi_m\rangle$.
b) Determine the expectation value of the position x.

Exercise 3.9   As Exercise 3.8.
a) Determine the matrix elements of the momentum operator $\hat{p} = -i\hbar d/dx$.
b) Determine the expectation value of the momentum of the particle.

Exercise 3.10   Determine the density matrices associated to the following mixtures of subsystems:
a) $|\psi_1\rangle$, $|\psi_2\rangle$, where:

$$|\psi_1\rangle = \begin{pmatrix} 1 \\ 0 \end{pmatrix} \qquad |\psi_2\rangle = \begin{pmatrix} 0 \\ 1 \end{pmatrix},$$

with probabilities $p_1 = p_2 = 1/2$.
b) $|\psi_1\rangle$, $|\psi_2\rangle$, $|\psi_3\rangle$ where:

$$|\psi_1\rangle = \frac{1}{\sqrt{281}}\begin{pmatrix} 9 \\ -i10\sqrt{2} \end{pmatrix} \qquad |\psi_2\rangle = \frac{1}{\sqrt{194}}\begin{pmatrix} 12 \\ -i5\sqrt{2} \end{pmatrix} \qquad |\psi_3\rangle = \frac{1}{\sqrt{17}}\begin{pmatrix} -i3 \\ 2\sqrt{2} \end{pmatrix},$$

with probabilities $p_1 = 281/900$, $p_2 = 97/450$, $p_3 = 17/36$.

Exercise 3.11   At time $t = 0$, a one-dimensional system is described by the wave function:

$$\phi(x) = \begin{cases} Ax/a & 0 < x < a \\ A(b-x)/b - a & a \le x < b \\ 0 & \text{otherwise} \end{cases}$$

with A, a and b constants.
a) Find A in terms of a and b so that ϕ is normalized.
b) Sketch $\phi(x)$ as a function of x.
c) Determine the probability of finding the system to the left of a. Check your result in the limiting cases $b = a$ and $b = 2a$.
d) Determine the expectation value of x.

Exercise 3.12   Consider the one-dimensional wave function in the domain $-\infty < x < \infty$, $\psi(x,t) = Ae^{-\lambda|x|}e^{-i\omega t}$, where A, λ, and ω are real and positive constants.
a) Normalize $\psi(x,t)$.
b) Determine the expectation values of x and x^2. Useful integrals are:

$$\int x^n e^{-x}dx = -x^n e^{-x} + n\int x^{n-1}e^{-x}dx \qquad n \ge 1$$

$$\int_0^\infty x^n e^{-x}dx = n!$$

Exercise 3.13 A wave function of a particle with mass m is given by:

$$\psi(x) = \begin{cases} A \cos \alpha x & -\pi/2\alpha \le x \le \pi/2\alpha \\ 0 & \text{otherwise} \end{cases}$$

where $\alpha = 10^{10}$ m^{-1}.
a) Find the normalization constant.
b) Find the probability that the particle can be found on the interval $0 \le x \le 0.5 \times 10^{-10}$ m.
c) Find the particle's average position.
d) Finds its average momentum.
e) Find its average kinetic energy when -0.5×10^{-10} m $\le x \le 0.5 \times 10^{-10}$ m.

Exercise 3.14 Demonstrate that the density operator $\rho(t)$ can be written as $\rho(t) = e^{-iHt}\rho(0)e^{iHt}$.
Note: the time-evolution operator, or propagator, can be defined as:

$$U(t - t_0) = e^{-iH\cdot(t-t_0)}.$$

By using the propagator, the wavefunction $|\psi(t)\rangle$ can be written as:

$$|\psi(t)\rangle = U(t, t_0)|\psi(t_0)\rangle$$

so that $\rho(t) = U(t, t_0)\rho(t_0)U^\dagger(t, t_0)$.

Exercise 3.15 Use the result of Exercise 3.14 to deduce the von Neumann's equation.

Exercise 3.16 Determine the density matrix associated to the mixed state composed by the following wavefunctions: $|\psi_1\rangle = |0\rangle + |1\rangle$, with probability $p_1 = 1/2$ and $|\psi_2\rangle = |0\rangle - |1\rangle$, with probability $p_2 = 1/2$, where:

$$|0\rangle = \begin{pmatrix} 1 \\ 0 \end{pmatrix} \qquad |1\rangle = \begin{pmatrix} 0 \\ 1 \end{pmatrix}.$$

Exercise 3.17 Show that $\rho = (1/2)[|0\rangle + |1\rangle][\langle 0| + \langle 1|]$ is the density matrix of a pure state.

Exercise 3.18 The density operator of a canonical ensemble is:

$$\rho = \frac{e^{-\beta H}}{\text{Tr}\left(e^{-\beta H}\right)},$$

where $\beta = 1/k_B T$. Derive the energy representation of ρ.

Exercise 3.19 A pure qubit can be written as (Bloch sphere representation):

$$|\psi\rangle = \cos\left(\frac{\theta}{2}\right)|0\rangle + e^{i\phi}\sin\left(\frac{\theta}{2}\right)|1\rangle.$$

Demonstrate that the corresponding density matrix is given by:

$$\rho = \frac{1}{2}\begin{pmatrix} 1 + \cos\theta & \sin\theta(\cos\phi - i\sin\phi) \\ \sin\theta(\cos\phi + i\sin\phi) & 1 - \cos\theta \end{pmatrix}.$$

Exercise 3.20   Prove that $\text{Tr}(\rho^2)$ is time-independent.
Hint: write $\rho(t)$ as demonstrated in Exercise 3.14: $\rho(t) = U(t, t_0)\rho(t_0)U^\dagger(t, t_0)$.

Exercise 3.21   The partition function can be written as $Z = \text{Tr}\left(e^{-\beta H}\right)$. Show that the ground state energy, $\mathcal{E}_0$, is given by:

$$\mathcal{E}_0 = \lim_{\beta \to \infty} \left(-\frac{1}{Z}\frac{\partial Z}{\partial \beta}\right).$$

Exercise 3.22   The density operator of a two-level system is:
a)

$$\rho = \frac{1}{4}|a\rangle\langle a| + \frac{3}{4}|b\rangle\langle b| = \begin{pmatrix} 1/4 & 0 \\ 0 & 3/4 \end{pmatrix}$$

b)

$$\rho = \frac{1}{4}|a\rangle\langle a| + \frac{3}{4}|b\rangle\langle b| + \frac{\sqrt{3}}{4}|a\rangle\langle b| + \frac{\sqrt{3}}{4}|b\rangle\langle a| = \begin{pmatrix} 1/4 & \sqrt{3}/4 \\ \sqrt{3}/4 & 3/4 \end{pmatrix}$$

determine if the system is in a pure state or in a mixed state.

4 Electron–Photon Interaction

4.1 Introduction

In this section, we will study the interaction of a quantized medium with an electromagnetic field, which will be treated classically. This approach is called *semiclassical approximation*. After an introduction on general aspects of electromagnetic theory, which will be used in this chapter, the Hamiltonian describing the interaction of electrons with an electromagnetic field will be introduced. Finally, by using the density matrix formalism, the linear optical susceptibility of a two-level system will be calculated.

4.2 Classical Electromagnetic Theory

Electric and magnetic fields satisfy Maxwell's equations:

$$\nabla \cdot \mathbf{D} = \rho$$
$$\nabla \times \mathbf{E} = -\frac{\partial \mathbf{B}}{\partial t}$$
$$\nabla \cdot \mathbf{B} = 0$$
$$\nabla \times \mathbf{H} = \mathbf{J} + \frac{\partial \mathbf{D}}{\partial t}, \tag{4.1}$$

where $\mathbf{E}$ and $\mathbf{B}$ are the electric and magnetic fields, $\mathbf{D} = \epsilon \mathbf{E}$ is the electric displacement vector, ϵ is the dielectric constant given by $\epsilon = \epsilon_r \epsilon_0$, $\mathbf{H} = \mathbf{B}/\mu$, where μ is the permeability (we will assume $\mu = \mu_0$), ρ is the carrier density, and $\mathbf{J}$ is the current density. Since $\nabla \cdot \mathbf{B} = 0$, the magnetic field can be written as the curl of a vector potential:

$$\mathbf{B} = \nabla \times \mathbf{A}. \tag{4.2}$$

This equation does not define $\mathbf{A}$ in a unique way; indeed, we can define a new vector potential $\mathbf{A}'$ as:

$$\mathbf{A}' = \mathbf{A} + \nabla f, \tag{4.3}$$

where f is an arbitrary scalar function. Indeed, from Eq. 4.3:

$$\nabla \times \mathbf{A}' = \nabla \times \mathbf{A} + \nabla \times (\nabla f) = \nabla \times \mathbf{A}. \tag{4.4}$$

Since we have defined the vector potential in terms of its curl such that $\mathbf{B} = \nabla \times \mathbf{A}$, its divergence is arbitrary. The choice of the value of $\nabla \cdot \mathbf{A}$ defines the particular gauge within which to study a particular process. For the investigation of electron–photon interactions, it is common practice to adopt the so-called *radiation or Coulomb gauge*, assuming that $\nabla \cdot \mathbf{A} = 0$. If we calculate the divergence of the vectors in Eq. 4.3, we obtain:

$$\nabla \cdot \mathbf{A}' = \nabla \cdot \mathbf{A} + \nabla \cdot \nabla f = \nabla \cdot \mathbf{A} + \nabla^2 f, \qquad (4.5)$$

it is always possible to choose a particular function f in such a way that $\nabla^2 f + \nabla \cdot \mathbf{A} = 0$. Within this gauge, the solutions for the vector potential are represented by transverse electromagnetic waves. From Maxwell's equations, we have:

$$\nabla \times \mathbf{E} = -\frac{\partial \mathbf{B}}{\partial t} = -\frac{\partial}{\partial t}(\nabla \times \mathbf{A}) = -\nabla \times \left(\frac{\partial \mathbf{A}}{\partial t}\right), \qquad (4.6)$$

from which we obtain:

$$\nabla \times \left(\mathbf{E} + \frac{\partial \mathbf{A}}{\partial t}\right) = 0. \qquad (4.7)$$

Since the vectorial field $\mathbf{E} + \partial \mathbf{A}/\partial t$ is characterized by a curl equal to zero, it is conservative and can be written as the gradient of a scalar potential ϕ:

$$\mathbf{E} + \frac{\partial \mathbf{A}}{\partial t} = -\nabla \phi, \qquad \mathbf{E} = -\frac{\partial \mathbf{A}}{\partial t} - \nabla \phi. \qquad (4.8)$$

Also Eq. 4.8 does not define ϕ in a unique way. If the vector potential is written as in 4.3, the scalar potential can be written as:

$$\phi' = \phi - \frac{\partial f}{\partial t}, \qquad (4.9)$$

without producing any effect on the electric and magnetic fields. In a region where $\mathbf{J} = 0$ and $\rho = 0$, we can assume a constant background potential $\phi = 0$ and the fields are given by:

$$\mathbf{E} = -\frac{\partial \mathbf{A}}{\partial t}, \qquad \mathbf{B} = \nabla \times \mathbf{A}. \qquad (4.10)$$

In a region free from carriers and currents, the vector potential satisfies the homogeneous wave equation:

$$\nabla^2 \mathbf{A} - \frac{1}{c^2}\frac{\partial^2 \mathbf{A}}{\partial t^2} = 0, \qquad (4.11)$$

a plane wave solution of the previous equation is given by:

$$\mathbf{A}(\mathbf{r}, t) = \hat{\boldsymbol{\epsilon}} A_0 \sin(\mathbf{k} \cdot \mathbf{r} - \omega t), \qquad (4.12)$$

where $\hat{\boldsymbol{\epsilon}}$ is the polarization unit vector, $\mathbf{k}$ is the propagation vector of the electromagnetic field, ω is the angular frequency, which is related to $\mathbf{k}$ by the following relation:

$$\omega = kc. \qquad (4.13)$$

The Coulomb gauge $\nabla \cdot \mathbf{A} = 0$ is satisfied if:

$$\mathbf{k} \cdot \hat{\boldsymbol{\epsilon}} = 0, \qquad (4.14)$$

that is, $\mathbf{k}$ is perpendicular to the polarization direction, so that the wave is transverse. The corresponding electric field is:

$$\mathbf{E}(\mathbf{r}, t) = \hat{\boldsymbol{\epsilon}} \omega A_0 \cos(\mathbf{k} \cdot \mathbf{r} - \omega t) = \hat{\boldsymbol{\epsilon}} E_0 \cos(\mathbf{k} \cdot \mathbf{r} - \omega t). \tag{4.15}$$

Since $\mathbf{B} = \nabla \times \mathbf{A}$ and $\omega = kc$, the magnetic field can be written as:

$$\mathbf{B}(\mathbf{r}, t) = \frac{E_0}{c} \left(\frac{\mathbf{k}}{k} \times \hat{\boldsymbol{\epsilon}} \right) \cos(\mathbf{k} \cdot \mathbf{r} - \omega t) = \frac{E_0}{c} (\hat{\mathbf{k}} \times \hat{\boldsymbol{\epsilon}}) \cos(\mathbf{k} \cdot \mathbf{r} - \omega t), \tag{4.16}$$

where $\hat{\mathbf{k}} = \mathbf{k}/k$ is the unit propagation vector. $\mathbf{k}$, $\mathbf{E}$, and $\mathbf{B}$ are mutually perpendicular and $|\mathbf{E}|/|\mathbf{B}| = c$.

The energy density of an electromagnetic field is given by:

$$u(\mathbf{r}, t) = \frac{1}{2} \left\{ \epsilon_0 [\mathbf{E}(\mathbf{r}, t)]^2 + \frac{1}{\mu_0} [\mathbf{B}(\mathbf{r}, t)]^2 \right\}. \tag{4.17}$$

The rate of energy flow through a unit area normal to the propagation direction is given by the Poynting vector:

$$\mathbf{S} = \frac{1}{\mu_0} (\mathbf{E} \times \mathbf{B}). \tag{4.18}$$

Using the expressions for the electric (Eq. 4.15) and magnetic (Eq. 4.16) fields in Eqs. 4.17 and 4.18, we have:

$$u(\mathbf{r}, t) = \epsilon_0 [E_0 \cos(\mathbf{k} \cdot \mathbf{r} - \omega t)]^2 \tag{4.19}$$

and

$$\mathbf{S}(\mathbf{r}, t) = \epsilon_0 c [E_0 \cos(\mathbf{k} \cdot \mathbf{r} - \omega t)]^2 \hat{\mathbf{k}}, \tag{4.20}$$

so that

$$\mathbf{S}(\mathbf{r}, t) = c\, u(\mathbf{r}, t) \hat{\mathbf{k}}. \tag{4.21}$$

If we now average the energy density and the Poynting vector over a period $T = 2\pi/\omega$, we obtain:

$$u = \frac{1}{2} \epsilon_0 E_0^2 = \frac{1}{2} \epsilon_0 \omega^2 A_0^2 \tag{4.22}$$

and

$$\langle \mathbf{S}(t) \rangle_t = \frac{1}{2} \epsilon_0 c E_0^2 \hat{\mathbf{k}} \tag{4.23}$$

and the intensity of the electromagnetic field is:

$$I = \frac{1}{2} \epsilon_0 c E_0^2. \tag{4.24}$$

It is now simple to find a relation between the amplitude of the electric field and the photon flux associated to an electromagnetic wave. If $N(\omega)$ is the number of photons with energy $\hbar \omega$ within a volume V, the energy density can be written as:

$$u = \frac{\hbar \omega N(\omega)}{V}. \tag{4.25}$$

By using Eqs. 4.22 and 4.25, we obtain:

$$E_0^2 = \frac{2\hbar\omega N(\omega)}{\epsilon_0 V}, \qquad A_0^2 = \frac{2\hbar N(\omega)}{\epsilon_0 \omega V}. \qquad (4.26)$$

After this brief summary of the properties of electromagnetic field, we can analyze the interaction of an electron with such a field.

4.3 Electrons in an Electromagnetic Field

Let us consider a particle with mass m and electric charge q in an electromagnetic field. The Lagrangian, $\mathcal{L}$, of the particle can be written as:

$$\mathcal{L} = \frac{1}{2}m\dot{\mathbf{r}}^2 - q\phi(\mathbf{r}, t) + q\dot{\mathbf{r}} \cdot \mathbf{A}(\mathbf{r}, t). \qquad (4.27)$$

The last term is inserted in the expression of $\mathcal{L}$ since the electromagnetic field is not conservative. The Lagrange's equation is:

$$\frac{d}{dt}\left(\frac{\partial \mathcal{L}}{\partial \dot{q}_i}\right) - \frac{\partial \mathcal{L}}{\partial q_i} = 0, \qquad (4.28)$$

where q_i are generalized coordinates and the dot over a variable indicates a total derivative with respect to time. The term in parentheses is the ith component of the *canonical momentum*:

$$p_i = \frac{\partial \mathcal{L}}{\partial \dot{q}_i}. \qquad (4.29)$$

Assuming Cartesian coordinates, we have:

$$\mathcal{L} = \frac{1}{2}m(\dot{x}^2 + \dot{y}^2 + \dot{z}^2) - q\phi + q(\dot{x}A_x + \dot{y}A_y + \dot{z}A_z). \qquad (4.30)$$

Therefore:

$$\frac{\partial \mathcal{L}}{\partial \dot{x}} = m\dot{x} + qA_x[t, x(t), y(t), z(t)] \qquad (4.31)$$

$$\frac{d}{dt}\left(\frac{\partial \mathcal{L}}{\partial \dot{x}}\right) = m\ddot{x} + q\left(\frac{\partial A_x}{\partial t} + \dot{x}\frac{\partial A_x}{\partial x} + \dot{y}\frac{\partial A_x}{\partial y} + \dot{z}\frac{\partial A_x}{\partial z}\right) \qquad (4.32)$$

$$\frac{\partial \mathcal{L}}{\partial x} = -q\frac{\partial \phi}{\partial x} + q\left(\dot{x}\frac{\partial A_x}{\partial x} + \dot{y}\frac{\partial A_y}{\partial x} + \dot{z}\frac{\partial A_z}{\partial x}\right) \qquad (4.33)$$

Using these expression in the Lagrange's equation, we obtain:

$$m\ddot{x} = qE_x + q\left[\dot{y}\left(\frac{\partial A_y}{\partial x} - \frac{\partial A_x}{\partial y}\right) + \dot{z}\left(\frac{\partial A_z}{\partial x} - \frac{\partial A_x}{\partial z}\right)\right], \qquad (4.34)$$

where

$$E_x = -\frac{\partial \phi}{\partial x} - \frac{\partial A_x}{\partial t} \qquad (4.35)$$

is the x-component of the applied electric field. Using the fact that $\mathbf{B} = \nabla \times \mathbf{A}$, Eq. 4.34 can be written as:

$$m\ddot{x} = qE_x + q(\mathbf{v} \times \mathbf{B})_x, \qquad (4.36)$$

which is the equation of motion of a particle of mass m and charge q in an electromagnetic field. Therefore, the expression 4.27 was a good choice for the Lagrangian of the particle. The canonical momentum of the particle is:

$$p_x = \frac{\partial \mathcal{L}}{\partial \dot{x}} = m\dot{x} + qA_x \rightarrow \mathbf{p} = m\dot{\mathbf{r}} + q\mathbf{A} = m\mathbf{v} + q\mathbf{A}. \tag{4.37}$$

The Hamiltonian associated to the Lagrangian 4.27 can be calculated by employing the following Legendre transformation:

$$H(q,p) = \dot{\mathbf{q}} \cdot \mathbf{p} - \mathcal{L} = \sum_i \dot{q}_i p_i - \mathcal{L}(q, \dot{q}, t). \tag{4.38}$$

Therefore, we have:

$$
\begin{aligned}
H(p,q) &= \dot{\mathbf{r}} \cdot (m\dot{\mathbf{r}} + q\mathbf{A}) - \frac{1}{2}m\dot{r}^2 + q\phi - q\dot{\mathbf{r}} \cdot \mathbf{A} = \\
&= \frac{[\mathbf{p} - q\mathbf{A}(\mathbf{r}, t)]^2}{2m} + q\phi(\mathbf{r}, t).
\end{aligned}
\tag{4.39}
$$

If the particle is in a region where also an effective potential, $V(\mathbf{r})$, is present, the Hamiltonian can be written as:

$$H = \frac{[\mathbf{p} - q\mathbf{A}(\mathbf{r}, t)]^2}{2m} + V(\mathbf{r}) + q\phi(\mathbf{r}, t). \tag{4.40}$$

If we also consider the energy of the electromagnetic field, the total Hamiltonian can be written as:

$$H_t = H + H_{field}, \tag{4.41}$$

where H is given by Eq. 4.40 and

$$H_{field} = \frac{1}{2}\epsilon_0 \int (E^2 + c^2 B^2) \, d\tau. \tag{4.42}$$

In several cases of practical interest, the electromagnetic energy is much larger than the interaction energy between field and charged particles. Therefore, we can completely neglect the variation of the field energy during the interaction. In this case, the fields are virtually not influenced by the interaction and can be treated in a classical way (i.e., we do not have to quantize the field). This is at the heart of the semiclassical approximation. In this case, $H_t = H$ and can be written as:

$$H = H_0 + H', \tag{4.43}$$

where:

$$H_0 = \frac{\mathbf{p}^2}{2m} + V(\mathbf{r}) = -\frac{\hbar^2}{2m}\nabla^2 + V(\mathbf{r}) \tag{4.44}$$

is the unperturbed Hamiltonian and

$$
\begin{aligned}
H' &= -\frac{q}{2m}(\mathbf{p} \cdot \mathbf{A} + \mathbf{A} \cdot \mathbf{p}) + \frac{q^2 A^2}{2m} + q\phi = \\
&= i\hbar\frac{q}{2m}(\nabla \cdot \mathbf{A} + \mathbf{A} \cdot \nabla) + \frac{q^2 A^2}{2m} + q\phi
\end{aligned}
\tag{4.45}
$$

is the interaction Hamiltonian. As already pointed out, we can assume $\phi = 0$; moreover, for many applications, we can neglect the term $q^2 A^2/2m$ with respect to the term $(q/2m)\mathbf{p} \cdot \mathbf{A}$. This approximation is valid when the intensity of the electromagnetic field is not too high. As shown in Ex. 4.1, even at an intensity of 1 MW/cm^2, it is still possible to neglect the term proportional to A^2 since $|\mathbf{p}| \gg |e\mathbf{A}|$ (*weak-field limit*).

Exercise 4.1   Consider an electromagnetic field with intensity $I = 10^6$ W/cm^2 and a photon energy of 1 eV. Assuming an electron velocity of 5×10^6 cm/s, calculate the ratio eA/p.

The amplitude of the electric field amplitude can be calculated by using Eq. 4.24:

$$E_0 = \sqrt{2I/\epsilon_0 c} = 2.7 \times 10^6 \text{ V/m}.$$

In the case of a photon energy of 1 eV (angular frequency $\omega = 1.52 \times 10^{15}$ rad/s), the amplitude of the vector potential is:

$$A_0 = \frac{E_0}{\omega} = 1.8 \times 10^{-9} \text{ Vs/m}.$$

If we now calculate the ratio between the terms $q^2 A^2/2m$ and $qpA/2m$ in the case of electrons ($q = -e = -1.6 \times 10^{-19}$ C), assuming an electron velocity of 5×10^6 cm/s, we obtain:

$$eA/p \simeq 6 \times 10^{-3}.$$

∎

From the physical point of view, the term proportional to $\mathbf{A}$ is related to transitions involving the emission or the absorption of a single photon, while the term proportional to A^2 is related to transitions involving the emission or the absorption of two photons, which, in the case of low excitation intensity, are much less probable.

Note that the operators $\mathbf{p} = -i\hbar\nabla$ and $\mathbf{A}$ commute. Indeed, in the Coulomb gauge ($\nabla \cdot \mathbf{A} = 0$), we have:

$$\nabla \cdot (\mathbf{A}|\psi\rangle) = (\nabla \cdot \mathbf{A})|\psi\rangle + \mathbf{A} \cdot \nabla|\psi\rangle = \mathbf{A} \cdot \nabla|\psi\rangle, \tag{4.46}$$

so that

$$\nabla \cdot \mathbf{A} = \mathbf{A} \cdot \nabla \tag{4.47}$$

and the commutator $[-i\hbar\nabla, \mathbf{A}] = 0$. The interaction Hamiltonian given by Eq. 4.45 can be written as:

$$H' = -\frac{q}{m} \mathbf{A} \cdot \mathbf{p}. \tag{4.48}$$

In the case of electrons $q = -e$ and

$$H' = \frac{e}{m} \mathbf{A} \cdot \mathbf{p}. \tag{4.49}$$

4.4 Electric Dipole Approximation

Let us assume that the wavelength λ of the electromagnetic field is large compared to the size of the atomic system under consideration and that the field intensity is not too high. When these two conditions are satisfied, the *dipole approximation* can be used. Within this approximation, the spatial variation of the electromagnetic field across the atom can be neglected. This situation is usually met since the size of typical atomic systems is of the order of 0.1–10 nm, much smaller than the wavelength of electromagnetic radiation of interest in photonic and electronic applications (typically in the ultraviolet-visible-infrared spectral regions). Note that the electric dipole approximation works well even in the case of electromagnetic fields extending into the far ultraviolet. Within this approximation, if we consider an atom whose nucleus is located at the position $\mathbf{R}$, the vector potential can be assumed as spatially homogeneous so that:

$$\mathbf{A}(\mathbf{r}, t) \approx \mathbf{A}(\mathbf{R}, t). \tag{4.50}$$

This means that we can neglect the spatial variation of $\mathbf{A}$ over atomic dimensions.

In this case, the Lagrangian given by Eq. 4.27 can be written as:

$$\mathcal{L} = \frac{1}{2}m\dot{\mathbf{r}}^2 - V(\mathbf{r}) - q\phi + q\dot{\mathbf{r}} \cdot \mathbf{A}(\mathbf{R}, t). \tag{4.51}$$

It is a general result of calculus that if a total time derivative is added to a Lagrangian, the equations of motion derived from it do not change (in other words, two Lagrangian which differ by a total time derivative, $\mathcal{L}' = \mathcal{L} + \frac{d}{dt}F(q, t)$, are equivalent). Therefore, we may add to the Lagrangian 4.51 the following total time derivative:

$$-q\frac{d}{dt}[\mathbf{r} \cdot \mathbf{A}(\mathbf{R}, t)], \tag{4.52}$$

in this way an equivalent Lagrangian is obtained:

$$\mathcal{L} = \frac{1}{2}m\dot{\mathbf{r}}^2 - V(\mathbf{r}) - q\phi - q\mathbf{r} \cdot \dot{\mathbf{A}}(\mathbf{R}, t). \tag{4.53}$$

The canonical momentum $p_i = \partial\mathcal{L}/\partial\dot{q}_i$, is given by:

$$\mathbf{p} = m\dot{\mathbf{r}} = m\frac{d\mathbf{r}}{dt} \tag{4.54}$$

and the corresponding Hamiltonian, which can be calculated by using Eq. 4.38, can be written as follows:

$$H = \frac{\mathbf{p}^2}{2m} + V(\mathbf{r}) + q\phi + q\mathbf{r} \cdot \frac{d\mathbf{A}(\mathbf{R}, t)}{dt}. \tag{4.55}$$

In a region free from charged particles, we can assume $\phi = 0$ and Eq. 4.55 can be rewritten as:

$$H = \frac{\mathbf{p}^2}{2m} + V(\mathbf{r}) - \boldsymbol{\mu} \cdot \mathbf{E}(\mathbf{R}, t), \tag{4.56}$$

where:

$$\boldsymbol{\mu} = q\mathbf{r} \tag{4.57}$$

is the *electric dipole moment* and we have used the fact that, over atomic dimensions:

$$\frac{d\mathbf{A}(\mathbf{R}, t)}{dt} = -\mathbf{E}(\mathbf{R}, t). \tag{4.58}$$

Therefore, in the framework of the electric dipole approximation, the interaction Hamiltonian is given by:

$$H' = -\boldsymbol{\mu} \cdot \mathbf{E}. \tag{4.59}$$

The matrix elements of the interaction Hamiltonian H' given by Eq. 4.59 are:

$$H'_{mn} = \langle u_m | H' | u_n \rangle = -\langle u_m | \boldsymbol{\mu} \cdot \mathbf{E} | u_n \rangle, \tag{4.60}$$

where $|u_i\rangle$ are the eigenvectors of the unperturbed Hamiltonian H_0. Since in the electric dipole approximation, the electric field is a function only of time, in Eq. 4.60 $\mathbf{E}$ can be taken outside the scalar product. Therefore:

$$H'_{mn} = -\langle u_m | \boldsymbol{\mu} | u_n \rangle \cdot \mathbf{E} = -\boldsymbol{\mu}_{mn} \cdot \mathbf{E}. \tag{4.61}$$

If $H_0(\mathbf{r}) = H_0(-\mathbf{r})$, that is, if the Hamiltonian is invariant under the transformation $\mathbf{r} \to -\mathbf{r}$:

$$|u_k(-\mathbf{r})\rangle = \pm |u_k(\mathbf{r})\rangle, \tag{4.62}$$

that is, the eigenstates have definite parity. When the eigenfunction is an even function of $\mathbf{r}$ it is called eigenfunction of even parity and the associated state is called state with even parity. If $|u_k(-\mathbf{r})\rangle = -|u_k(\mathbf{r})\rangle$, the wavefunction has odd parity and the associated state is called state with odd parity. For example, $H_0(\mathbf{r}) = H_0(-\mathbf{r})$ when the system is characterized by a center of symmetry, or when the potential energy $V(\mathbf{r})$ is a function only of the distance between particles. Therefore, if two eigenstates $|u_m\rangle$ and $|u_n\rangle$ with different eigenvalues, $\mathcal{E}_m$ and $\mathcal{E}_n$, have the same parity, the matrix elements μ_{mn} are zero since in the integral $\int u_m^*(\mathbf{r}) \mathbf{r} \, u_n(\mathbf{r}) \, d\tau = \langle u_m | \mathbf{r} | u_n \rangle$ the integrand is an odd function. The consequence is that the transition between states with the same parity, which is directly related to the matrix element H'_{mn}, is forbidden in the electric dipole approximation. The transition can be possible but in higher-order processes, for example as the electric quadrupole transition, which however is much less probable. Therefore, we can conclude that electric dipole transitions can occur only between states with opposite parity. If the eigenstates $|u_m\rangle$ and $|u_n\rangle$ have definite parity and are nondegenerate, the diagonal matrix elements of $\boldsymbol{\mu}$ are zero.

In the following section, we will apply the density matrix formalism to calculate the linear optical susceptibility in the case of a two-level system excited by a plane monochromatic electromagnetic radiation, in the framework of the dipole approximation. Then, in Chapter 5, we will generalize the results to the case of a semiconductor.

4.5 Linear Optical Susceptibility

Let us consider a two-level system characterized by the unperturbed Hamiltonian H_0. We will consider two eigenstates $|u_1\rangle$ and $|u_2\rangle$ with energies $\mathcal{E}_1$ and $\mathcal{E}_2$, respectively: $H_0|u_i\rangle =$

$\mathcal{E}_i|u_i\rangle$ ($i = 1, 2$). The temporal evolution of the matrix elements of the density operator can be obtained by using Eq. 3.133 and Eq. 3.141:

$$\frac{\partial \rho_{21}}{\partial t} = -i\omega_0 \rho_{21} - \frac{i}{\hbar}[H', \rho]_{21} - \frac{\rho_{21}}{T_2} \tag{4.63}$$

$$\frac{\partial \rho_{11}}{\partial t} = -\frac{i}{\hbar}[H', \rho]_{11} - \frac{\rho_{11} - \rho_{11}^e}{T_1} \tag{4.64}$$

$$\frac{\partial \rho_{22}}{\partial t} = -\frac{i}{\hbar}[H', \rho]_{22} - \frac{\rho_{22} - \rho_{22}^e}{T_1}, \tag{4.65}$$

where $\omega_0 = \omega_{21}$ and $\rho_{12} = \rho_{21}^*$. In the electric dipole approximation, the interaction Hamiltonian is given by Eq. 4.59:

$$H' = -\boldsymbol{\mu} \cdot \mathbf{E} = -\mu E, \tag{4.66}$$

where μ is the component of the electric dipole moment on the polarization direction of the electric field $\mathbf{E}$. Assuming that the states $|u_i\rangle$ have definite parity (we assume that the system is symmetric with respect to the center), the matrix elements of the dipole moment operator are given by:

$$\mu = \begin{pmatrix} 0 & \mu_{12} \\ \mu_{21} & 0 \end{pmatrix}, \tag{4.67}$$

where $\mu_{21} = \mu_{12}^*$. We can assume that μ_{12} and μ_{21} are real and therefore equal. The average value of the dipole moment can be calculated by using the density matrix formalism:

$$\langle \mu \rangle = \text{Tr}(\rho \mu) = \mu(\rho_{12} + \rho_{21}) = 2\mu \text{Re}\{\rho_{21}\}. \tag{4.68}$$

The commutator $[H', \rho]$ is given by:

$$
\begin{aligned}
[H', \rho] &= \begin{pmatrix} 0 & -\mu E \\ -\mu E & 0 \end{pmatrix}\begin{pmatrix} \rho_{11} & \rho_{12} \\ \rho_{21} & \rho_{22} \end{pmatrix} + \\
&\quad - \begin{pmatrix} \rho_{11} & \rho_{12} \\ \rho_{21} & \rho_{22} \end{pmatrix}\begin{pmatrix} 0 & -\mu E \\ -\mu E & 0 \end{pmatrix} = \\
&= \begin{pmatrix} \mu E(\rho_{12} - \rho_{21}) & \mu E(\rho_{11} - \rho_{22}) \\ \mu E(\rho_{22} - \rho_{11}) & \mu E(\rho_{21} - \rho_{12}) \end{pmatrix}.
\end{aligned}
\tag{4.69}
$$

Therefore:

$$\frac{\partial \rho_{21}}{\partial t} = -i\omega_0 \rho_{21} + \frac{i}{\hbar}\mu E(t)(\rho_{11} - \rho_{22}) - \frac{\rho_{21}}{T_2} \tag{4.70}$$

$$\frac{\partial \rho_{11}}{\partial t} = \frac{i}{\hbar}\mu E(t)(\rho_{21} - \rho_{21}^*) - \frac{\rho_{11} - \rho_{11}^e}{T_1} \tag{4.71}$$

$$\frac{\partial \rho_{22}}{\partial t} = -\frac{i}{\hbar}\mu E(t)(\rho_{21} - \rho_{21}^*) - \frac{\rho_{22} - \rho_{22}^e}{T_1}. \tag{4.72}$$

From the last two equations, we obtain:

$$\frac{\partial}{\partial t}(\rho_{11} - \rho_{22}) = 2i\frac{\mu}{\hbar}E(t)(\rho_{21} - \rho_{21}^*) - \frac{(\rho_{11} - \rho_{22}) - (\rho_{11}^e - \rho_{22}^e)}{T_1}. \tag{4.73}$$

4.5.1 Harmonic Perturbation

Let us consider a monochromatic electromagnetic field with electric field:

$$E(t) = E_0 \cos \omega t = \text{Re}\{E_0 e^{-i\omega t}\} = \frac{E_0}{2}(e^{-i\omega t} + e^{i\omega t}). \tag{4.74}$$

We will assume that $\omega \approx \omega_0$. Since in the field-free case, the element ρ_{21} oscillates with frequency ω_0, as shown by Eq. 4.70, it is useful to transform into the rotating frame ($\omega \approx \omega_0$):

$$\rho_{21}(t) = \sigma_{21}(t)\, e^{-i\omega t} \tag{4.75}$$

$$\rho_{12}(t) = \rho_{21}^*(t) = \sigma_{21}^*\, e^{i\omega t}, \tag{4.76}$$

where $\sigma_{21}(t)$ is a slowly varying function. Equation 4.70 can be rewritten as:

$$\frac{d\sigma_{21}}{dt} e^{-i\omega t} - i\omega\sigma_{21}e^{-i\omega t} =$$
$$= -i\omega_0\sigma_{21}e^{-i\omega t} + \frac{\mu}{\hbar}(\rho_{11} - \rho_{22})\frac{E_0}{2}(e^{i\omega t} + e^{-i\omega t}) - \frac{\sigma_{21}}{T_2}e^{-i\omega t}. \tag{4.77}$$

If we consider only the terms with $e^{-i\omega t}$, we have:

$$\frac{d\sigma_{21}}{dt} = i(\omega - \omega_0)\sigma_{21} + i\frac{\mu E_0}{2\hbar}(\rho_{11} - \rho_{22}) - \frac{\sigma_{21}}{T_2}. \tag{4.78}$$

Equation 4.73 can be written as:

$$\frac{d}{dt}(\rho_{11} - \rho_{22}) = \frac{i\mu E_0}{\hbar}(\sigma_{21} - \sigma_{21}^*) - \frac{(\rho_{11} - \rho_{22}) - (\rho_{11} - \rho_{22})_{eq}}{T_1}, \tag{4.79}$$

where we have kept only the terms with no exponential time dependence. We have neglected the terms with $e^{2i\omega t}$ and $e^{-2i\omega t}$. From the physical point of view, this is reasonable since the contribution of these terms averages to zero on a temporal interval shorter than that of interest.

The average value of the dipole moment (Eq. 4.68) is given by:

$$\langle \mu \rangle = 2\mu \text{Re}\{\rho_{21}\} = 2\mu \text{Re}\{\sigma_{21}e^{-i\omega t}\} =$$
$$= 2\mu[\text{Re}\sigma_{21}(t)\cos \omega t + \text{Im}\sigma_{21}(t)\sin \omega t]. \tag{4.80}$$

We can now calculate the steady-state solutions of Eqs. 4.78 and 4.79.

4.5.1.1 Steady-State Solutions

To find the steady-state solutions of Eqs. 4.78 and 4.79, we put:

$$\frac{d\sigma_{21}}{dt} = \frac{d}{dt}(\rho_{11} - \rho_{22}) = 0. \tag{4.81}$$

Therefore:

$$i(\omega - \omega_0)\sigma_{21} + \frac{i\mu E_0}{2\hbar}(\rho_{11} - \rho_{22}) - \frac{\sigma_{21}}{T_2} = 0 \tag{4.82}$$

$$\frac{i\mu E_0}{\hbar}(\sigma_{21} - \sigma_{21}^*) - \frac{(\rho_{11} - \rho_{22}) - (\rho_{11} - \rho_{22})_{eq}}{T_1} = 0 \tag{4.83}$$

If we take the complex conjugate of Eq. 4.82:

$$-i(\omega - \omega_0)\sigma_{21}^* - \frac{i\mu E_0}{2\hbar}(\rho_{11} - \rho_{22}) - \frac{\sigma_{21}^*}{T_2} = 0, \tag{4.84}$$

and we sum Eqs. 4.82 and 4.84, we obtain:

$$i(\omega - \omega_0)(\sigma_{21} - \sigma_{21}^*) - \frac{\sigma_{21} + \sigma_{21}^*}{T_2} = 0. \tag{4.85}$$

If we subtract Eqs. 4.82 and 4.84, we obtain:

$$i(\omega - \omega_0)(\sigma_{21} + \sigma_{21}^*) + \frac{i\mu E_0}{\hbar}(\rho_{11} - \rho_{22}) - \frac{\sigma_{21} - \sigma_{21}^*}{T_2} = 0. \tag{4.86}$$

These last two equations can be written as:

$$i(\omega - \omega_0)(2i \,\mathrm{Im}\sigma_{21}) - \frac{2\,\mathrm{Re}\sigma_{21}}{T_2} = 0 \tag{4.87}$$

$$i(\omega - \omega_0)2\,\mathrm{Re}\sigma_{21} + \frac{i\mu E_0}{\hbar}(\rho_{11} - \rho_{22}) - \frac{2i\,\mathrm{Im}\sigma_{21}}{T_2} = 0. \tag{4.88}$$

With suitable calculations:

$$\mathrm{Re}\sigma_{21} = -(\omega - \omega_0)T_2\,\mathrm{Im}\sigma_{21} \tag{4.89}$$

$$i(\omega - \omega_0)[-2(\omega - \omega_0)T_2\,\mathrm{Im}\sigma_{21}] + \frac{i\mu E_0}{\hbar}(\rho_{11} - \rho_{22}) - \frac{2i}{T_2}\,\mathrm{Im}\sigma_{21} = 0, \tag{4.90}$$

from which it is possible to obtain:

$$\mathrm{Im}\sigma_{21} = \frac{\mu E_0(\rho_{11} - \rho_{22})/\hbar}{2[(\omega - \omega_0)^2 T_2^2 + 1]/T_2} = \frac{\Omega T_2(\rho_{11} - \rho_{22})}{1 + (\omega - \omega_0)^2 T_2^2}, \tag{4.91}$$

where $\Omega = \mu E_0/2\hbar$. We can now rewrite Eq. 4.83 as:

$$\frac{i\mu E_0}{\hbar}2i\,\mathrm{Im}\sigma_{21} - \frac{(\rho_{11} - \rho_{22}) - (\rho_{11} - \rho_{22})_{eq}}{T_1} = 0, \tag{4.92}$$

from which we obtain:

$$\rho_{11} - \rho_{22} = -2\frac{\mu E_0}{\hbar}\,\mathrm{Im}\sigma_{21}\,T_1 + (\rho_{11} - \rho_{22})_{eq}, \tag{4.93}$$

which can be used in Eq. 4.91, so that :

$$\mathrm{Im}\sigma_{21} = \frac{\Omega\,T_2(\rho_{11} - \rho_{22})_{eq}}{1 + (\omega - \omega_0)^2 T_2^2 + 4\Omega^2 T_2 T_1} \tag{4.94}$$

$$\mathrm{Re}\sigma_{21} = \frac{(\omega_0 - \omega)T_2^2\Omega(\rho_{11} - \rho_{22})_{eq}}{1 + (\omega - \omega_0)^2 T_2^2 + 4\Omega^2 T_2 T_1} \tag{4.95}$$

$$(\rho_{11} - \rho_{22}) = (\rho_{11} - \rho_{22})_{eq}\frac{1 + (\omega - \omega_0)^2 T_2^2}{1 + (\omega - \omega_0)^2 T_2^2 + 4\Omega^2 T_2 T_1}. \tag{4.96}$$

The macroscopic polarization can be written as follows:

$$P = N\langle\mu\rangle = \frac{\mu^2 \Delta N_0 T_2}{\hbar}E_0\frac{\sin\omega t + (\omega_0 - \omega)T_2\,\cos\omega t}{1 + (\omega - \omega_0)^2 T_2^2 + 4\Omega^2 T_2 T_1}, \tag{4.97}$$

and the population difference per unit volume is:

$$\Delta N = \Delta N_0\frac{1 + (\omega - \omega_0)^2 T_2^2}{1 + (\omega - \omega_0)^2 T_2^2 + 4\Omega^2 T_2 T_1}, \tag{4.98}$$

where $\Delta N_0 = N(\rho_{11} - \rho_{22})_{eq}$ is the unperturbed population difference. The polarization P can be written as:

$$P = \frac{\pi \mu^2}{\hbar} \Delta N \, E_0 \, g(\omega - \omega_0)[\sin \omega t + (\omega_0 - \omega)T_2 \cos \omega t], \qquad (4.99)$$

where:

$$g(\omega - \omega_0) = \frac{T_2}{\pi[1 + (\omega - \omega_0)^2 T_2^2]} \qquad (4.100)$$

is the normalized Lorentzian lineshape function. If we now write the atomic susceptibility χ as:

$$\chi = \chi' + i\chi'', \qquad (4.101)$$

then:

$$P(t) = \mathrm{Re}\{\epsilon_0 \chi E_0 e^{-i\omega t}\} = E_0(\epsilon_0 \chi' \cos \omega t + \epsilon_0 \chi'' \sin \omega t). \qquad (4.102)$$

From Eqs. 4.99 and 4.102, we obtain:

$$\chi'(\omega) = \frac{\pi \mu^2}{\epsilon_0 \hbar}(\omega_0 - \omega)T_2 \Delta N g(\omega - \omega_0) \qquad (4.103)$$

$$\chi''(\omega) = \frac{\pi \mu^2}{\epsilon_0 \hbar} \Delta N g(\omega - \omega_0). \qquad (4.104)$$

4.5.1.2 Lorentzian Function

A few remarks about the lineshape function $g(\omega - \omega_0)$. If we want to write the lineshape function as a function of frequency ν, we have to recall that $g(\omega)d\omega = g(\nu)d\nu$, so that:

$$g(\nu - \nu_0) = \frac{2T_2}{1 + 4\pi^2(\nu - \nu_0)^2 T_2^2}. \qquad (4.105)$$

The full-width at half-maximum of the functions $g(\omega)$ and $g(\nu)$ are:

$$\Delta \omega = \frac{2}{T_2} \qquad \Delta \nu = \frac{1}{\pi T_2}, \qquad (4.106)$$

thus showing that the transverse relaxation time is related to the linewidth of the transition. Moreover, the lineshape function is normalized:

$$\int_{-\infty}^{\infty} g(\nu) \, d\nu = 1. \qquad (4.107)$$

Figure 4.1 shows the real and imaginary components of the atomic susceptibility as a function of the frequency detuning $(\omega - \omega_0)$ in the limit $4\Omega^2 T_2 T_1 \ll 1$.

4.6 From Optical Susceptibility to Absorption Coefficient

It is well known that the absorption coefficient of a medium can be obtained from the atomic susceptibility. The macroscopic polarization $\mathbf{P} = N\langle \boldsymbol{\mu} \rangle$ can be decomposed in two components:

$$\mathbf{P}_T = \mathbf{P}_{nr} + \mathbf{P}_r, \qquad (4.108)$$

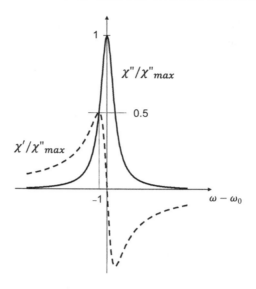

Figure 4.1 Real (dashed line) and imaginary (solid line) components of the atomic susceptibility as a function of the frequency detuning $\omega - \omega_0$. Both functions are normalized to the maximum value of the imaginary component.

where $\mathbf{P}_r$ is the polarization vector of the system near resonance and $\mathbf{P}_{nr}$ is the nonresonant contribution, which corresponds to all the off-resonant contributions:

$$\mathbf{P}_{nr} = \epsilon_0 \chi_{nr} \mathbf{E}, \tag{4.109}$$

where χ_{nr} is the nonresonant component of the atomic susceptibility.

$$\mathbf{P}_r = \epsilon_0 \chi_r(\omega)\mathbf{E} = \epsilon_0 \chi(\omega)\mathbf{E}, \tag{4.110}$$

where $\chi(\omega)$ is the resonant component of the atomic susceptibility, which we have calculated in the previous section. While χ_{nr} is weakly dependent on frequency, $\chi(\omega)$ strongly depends on ω, as shown in Fig. 4.1. Therefore:

$$\mathbf{P}_T = \epsilon_0[\chi_{nr} + \chi(\omega)]\mathbf{E}. \tag{4.111}$$

The displacement vector $\mathbf{D}$ is given by:

$$\mathbf{D} = \epsilon_0\mathbf{E} + \mathbf{P}_T = \epsilon_0[1 + \chi_{nr} + \chi(\omega)]\mathbf{E}. \tag{4.112}$$

We can define the total dielectric constant, ϵ_{tot}, as:

$$\epsilon_{tot} = \epsilon_0[1 + \chi_{nr} + \chi(\omega)] = \epsilon_0\epsilon_{r,tot}. \tag{4.113}$$

The total refractive index is given by:

$$n_{tot}^2 = \epsilon_{r,tot} = (1 + \chi_{nr}) + \chi(\omega) = n^2 + \chi(\omega) = n^2\left[1 + \frac{\chi(\omega)}{n^2}\right], \tag{4.114}$$

where n is the refractive index due to all the off-resonant transitions (i.e., it is the refractive index far from the resonance). Therefore:

$$k_{tot}^2 = \left(\frac{\omega n_{tot}}{c}\right)^2 = \left(\frac{\omega n}{c}\right)^2\left[1 + \frac{\chi(\omega)}{n^2}\right]. \tag{4.115}$$

If $\chi(\omega)/n^2 \ll 1$:

$$k_{tot} \simeq \frac{\omega n}{c}\left[1 + \frac{\chi(\omega)}{2n^2}\right], \qquad n_{tot} \simeq n\left[1 + \frac{\chi(\omega)}{2n^2}\right] \qquad (4.116)$$

$$k_{tot} = k' + ik'' \simeq k\left[1 + \frac{\chi'(\omega)}{2n^2}\right] + i\frac{k\chi''(\omega)}{2n^2}, \qquad (4.117)$$

where $k = \omega n/c$. The electric field of the electromagnetic field, which is propagating in the medium is:

$$
\begin{aligned}
E(z,t) &= \mathrm{Re}\{E_0 e^{i(k_{tot}z - \omega t)}\} && (4.118)\\
&= \mathrm{Re}\left\{E_0 \exp i\left[k\left(1 + \frac{\chi'(\omega)}{2n^2}\right)z - \omega t\right]\exp\left(-\frac{k\chi''(\omega)}{2n^2}z\right)\right\}.
\end{aligned}
$$

Therefore, the polarization of the medium gives rise to a variation of the phase delay per unit length from k to $k + \Delta k$, where:

$$\Delta k = \frac{k\chi'(\omega)}{2n^2}, \qquad (4.119)$$

and an exponential variation of the amplitude, $\exp(-\alpha z/2)$, where:

$$\alpha(\omega) = \frac{k\chi''(\omega)}{n^2}. \qquad (4.120)$$

By using the expression (Eq. 4.104) of $\chi''(\omega)$ obtained by using the density matrix formalism in Eq. 4.120, the absorption coefficient can be written as:

$$\alpha(\omega) = \frac{k\mu_{12}^2\pi}{\epsilon_0 n^2 \hbar}\Delta N g(\omega - \omega_0), \qquad (4.121)$$

we recall that $\Delta N = N_1 - N_2$. If $N_2 > N_1$ the system is characterized by a negative absorption, that is, it presents an optical gain. In Chapter 5, we will apply the main results of this chapter to semiconductors.

4.7 Momentum of an Electron in a Periodic Crystal

In this last paragraph, we will come back to the concept of momentum of an electron in a periodic crystal. In particular, the expectation value of the momentum operator in a periodic crystal will be directly calculated.

We have already pointed out that the expectation value of the momentum of an electron in free space is simply given by $\hbar\mathbf{k}$. Indeed, since the electron wavefunction is given by

$$\psi_{\mathbf{k}}(\mathbf{r}) = \frac{1}{\sqrt{V}}e^{i\mathbf{k}\cdot\mathbf{r}}, \qquad (4.122)$$

the expectation value of the electron momentum can be easily calculated as:

$$\langle\mathbf{p}\rangle = \langle\psi_{\mathbf{k}}| - i\hbar\nabla|\psi_{\mathbf{k}}\rangle, \qquad (4.123)$$

where:

$$\nabla \equiv \frac{\partial}{\partial x}\mathbf{u}_x + \frac{\partial}{\partial y}\mathbf{u}_y + \frac{\partial}{\partial z}\mathbf{u}_z. \qquad (4.124)$$

Therefore:

$$\langle \mathbf{p} \rangle = \frac{1}{V} \langle e^{-i\mathbf{k}\cdot\mathbf{r}}| - i\hbar\nabla|e^{i\mathbf{k}\cdot\mathbf{r}}\rangle = \hbar\mathbf{k}\frac{1}{V}\langle e^{-i\mathbf{k}\cdot\mathbf{r}}|e^{i\mathbf{k}\cdot\mathbf{r}}\rangle = \hbar\mathbf{k}. \tag{4.125}$$

We will now calculate the momentum of an electron in a given eigenstate inside a periodic crystal. The wave equation is:

$$H|\psi_{n,\mathbf{k}}\rangle = \mathcal{E}_{n,\mathbf{k}}|\psi_{n,\mathbf{k}}\rangle. \tag{4.126}$$

If we apply the gradient in the $\mathbf{k}$-space, $\nabla_{\mathbf{k}}$, to Eq. 4.126, since the Hamiltonian H is not a function of $\mathbf{k}$, we obtain:

$$\nabla_{\mathbf{k}}H|\psi_{n,\mathbf{k}}\rangle = \nabla_{\mathbf{k}}\left(\mathcal{E}_{n,\mathbf{k}}|\psi_{n,\mathbf{k}}\rangle\right)$$
$$H\nabla_{\mathbf{k}}|\psi_{n,\mathbf{k}}\rangle = \mathcal{E}_{n,\mathbf{k}}\nabla_{\mathbf{k}}|\psi_{n,\mathbf{k}}\rangle + |\psi_{n,\mathbf{k}}\rangle\nabla_{\mathbf{k}}\mathcal{E}_{n,\mathbf{k}}. \tag{4.127}$$

If we now calculate the scalar product of the previous equation with $|\psi_{n,\mathbf{k}}\rangle$, we obtain:

$$\langle \psi_{n,\mathbf{k}}|(H - \mathcal{E}_{n,\mathbf{k}})\nabla_{\mathbf{k}}|\psi_{n,\mathbf{k}}\rangle = \nabla_{\mathbf{k}}\mathcal{E}_{n,\mathbf{k}}\langle \psi_{n,\mathbf{k}}|\psi_{n,\mathbf{k}}\rangle, \tag{4.128}$$

where $\langle \psi_{n,\mathbf{k}}|\psi_{n,\mathbf{k}}\rangle = 1$ in the right-hand side of the previous equation. We observe that:

$$\begin{aligned}\nabla_{\mathbf{k}}|\psi_{n,\mathbf{k}}\rangle &= \nabla_{\mathbf{k}}\left(\frac{1}{\sqrt{V}}e^{i\mathbf{k}\cdot\mathbf{r}}|u_{n,\mathbf{k}}\rangle\right) = \frac{1}{\sqrt{V}}i\mathbf{r}e^{i\mathbf{k}\cdot\mathbf{r}}|u_{n,\mathbf{k}}\rangle + \frac{1}{\sqrt{V}}e^{i\mathbf{k}\cdot\mathbf{r}}\nabla_{\mathbf{k}}|u_{n,\mathbf{k}}\rangle \\ &= i\mathbf{r}|\psi_{n,\mathbf{k}}(\mathbf{r})\rangle + \frac{1}{\sqrt{V}}\nabla_{\mathbf{k}}|u_{n,\mathbf{k}}\rangle. \end{aligned} \tag{4.129}$$

By using the previous result, we obtain:

$$\begin{aligned}\langle \psi_{n,\mathbf{k}}|(H - \mathcal{E}_{n,\mathbf{k}})\nabla_{\mathbf{k}}|\psi_{n,\mathbf{k}}\rangle &= \langle \psi_{n,\mathbf{k}}|(H - \mathcal{E}_{n,\mathbf{k}})i\mathbf{r}|\psi_{n,\mathbf{k}}\rangle + \\ &+ \frac{1}{\sqrt{V}}\langle \psi_{n,\mathbf{k}}|(H - \mathcal{E}_{n,\mathbf{k}})e^{i\mathbf{k}\cdot\mathbf{r}}\nabla_{\mathbf{k}}|u_{n,\mathbf{k}}\rangle. \end{aligned} \tag{4.130}$$

Let us first consider the second term on the right-hand side of the previous equation:

$$\langle \psi_{n,\mathbf{k}}|(H - \mathcal{E}_{n,\mathbf{k}})e^{i\mathbf{k}\cdot\mathbf{r}}\nabla_{\mathbf{k}}|u_{n,\mathbf{k}}\rangle = \int \psi_{n,\mathbf{k}}^*(H - \mathcal{E}_{n,\mathbf{k}})e^{i\mathbf{k}\cdot\mathbf{r}}\nabla_{\mathbf{k}}u_{n,\mathbf{k}}\,d\tau. \tag{4.131}$$

Since $(H - \mathcal{E}_{n,\mathbf{k}})$ is Hermitian and the functions in the integrand of the previous integral obey the periodic boundary conditions, the integral in Eq. 4.131 can be written as follows:

$$\int \psi_{n,\mathbf{k}}^*(H - \mathcal{E}_{n,\mathbf{k}})e^{i\mathbf{k}\cdot\mathbf{r}}\nabla_{\mathbf{k}}u_{n,\mathbf{k}}\,d\tau = \int \left[(H - \mathcal{E}_{n,\mathbf{k}})\psi_{n,\mathbf{k}}\right]^* e^{i\mathbf{k}\cdot\mathbf{r}}\nabla_{\mathbf{k}}u_{n,\mathbf{k}}\,d\tau = 0, \tag{4.132}$$

since from Eq. 4.126, we also have:

$$(H - \mathcal{E}_{n,\mathbf{k}})|\psi_{n,\mathbf{k}}\rangle = 0, \tag{4.133}$$

the integral of Eq. 4.131 is zero. Note that the same mathematical procedure cannot be applied to the first term in Eq. 4.129 since the function $i\mathbf{r}\psi_{n,\mathbf{k}}(\mathbf{r})$ is not periodic. This term can be written in a different way. We can first write:

$$H\,i\mathbf{r}|\psi_{n,\mathbf{k}}\rangle = \left[-\frac{\hbar^2}{2m}\nabla^2 + V(\mathbf{r})\right]i\mathbf{r}|\psi_{n,\mathbf{k}}\rangle, \tag{4.134}$$

where the term $\nabla^2(\mathbf{r}\psi_{n,\mathbf{k}})$ can be easily calculated by using the following vector identity:

$$\nabla^2(f\mathbf{v}) = \mathbf{v}\nabla^2 f + 2(\nabla f \cdot \nabla)\mathbf{v} + f\nabla^2\mathbf{v}. \tag{4.135}$$

In our case, $f = \psi_{n,\mathbf{k}}(\mathbf{r})$ and $\mathbf{v} = \mathbf{r}$. Note that:

$$\nabla^2 \mathbf{r} = 0 \tag{4.136}$$

and

$$
\begin{aligned}
\left(\nabla\psi_{n,\mathbf{k}} \cdot \nabla\right)\mathbf{r} &= \left(\frac{\partial\psi_{n,\mathbf{k}}}{\partial x}\frac{\partial}{\partial x} + \frac{\partial\psi_{n,\mathbf{k}}}{\partial y}\frac{\partial}{\partial y} + \frac{\partial\psi_{n,\mathbf{k}}}{\partial z}\frac{\partial}{\partial z}\right)(x\mathbf{u}_x + y\mathbf{u}_y + z\mathbf{u}_z) = \\
&= \frac{\partial\psi_{n,\mathbf{k}}}{\partial x}\mathbf{u}_x + \frac{\partial\psi_{n,\mathbf{k}}}{\partial y}\mathbf{u}_y + \frac{\partial\psi_{n,\mathbf{k}}}{\partial z}\mathbf{u}_z = \nabla\psi_{n,\mathbf{k}}. \tag{4.137}
\end{aligned}
$$

Therefore:

$$\nabla^2\left(\mathbf{r}|\psi_{n,\mathbf{k}}\rangle\right) = \mathbf{r}\nabla^2\psi_{n,\mathbf{k}} + 2\nabla\psi_{n,\mathbf{k}}, \tag{4.138}$$

so that Eq. 4.134 becomes:

$$H\, i\mathbf{r}\psi_{n,\mathbf{k}} = -\frac{\hbar^2}{2m}i\left(\mathbf{r}\nabla^2\psi_{n,\mathbf{k}} + 2\nabla\psi_{n,\mathbf{k}}\right) + i\mathbf{r}V\psi_{n,\mathbf{k}} \tag{4.139}$$

$$H\, i\mathbf{r}\psi_{n,\mathbf{k}} = i\mathbf{r}\left(-\frac{\hbar^2}{2m}\nabla^2 + V\right)\psi_{n,\mathbf{k}} + \frac{\hbar}{m}(-i\hbar\nabla)\psi_{n,\mathbf{k}}, \tag{4.140}$$

which can be written as:

$$H\, i\mathbf{r}|\psi_{n,\mathbf{k}}\rangle = i\mathbf{r}H|\psi_{n,\mathbf{k}}\rangle + \frac{\hbar}{m}(-i\hbar\nabla)|\psi_{n,\mathbf{k}}\rangle. \tag{4.141}$$

If we subtract the term $\mathcal{E}_{n,\mathbf{k}}i\mathbf{r}|\psi_{n,\mathbf{k}}\rangle$ on the left- and right-hand side of the previous equation, we obtain:

$$(H - \mathcal{E}_{n,\mathbf{k}})\, i\mathbf{r}|\psi_{n,\mathbf{k}}\rangle = i\mathbf{r}(H - \mathcal{E}_{n,\mathbf{k}})|\psi_{n,\mathbf{k}}\rangle + \frac{\hbar}{m}(-i\hbar\nabla)|\psi_{n,\mathbf{k}}\rangle, \tag{4.142}$$

where the first term on the right-hand side is zero (see Eq. 4.133). If we now perform the scalar product with $|\psi_{n,\mathbf{k}}\rangle$, we have:

$$\langle\psi_{n,\mathbf{k}}|(H - \mathcal{E}_{n,\mathbf{k}})\, i\mathbf{r}|\psi_{n,\mathbf{k}}\rangle = \langle\psi_{n,\mathbf{k}}|\frac{\hbar}{m}(-i\hbar\nabla)|\psi_{n,\mathbf{k}}\rangle, \tag{4.143}$$

where the term at the right-hand side can be written in a different way:

$$\frac{\hbar}{m}\langle\psi_{n,\mathbf{k}}| -i\hbar\nabla|\psi_{n,\mathbf{k}}\rangle = \frac{\hbar}{m}\langle\psi_{n,\mathbf{k}}|\mathbf{p}|\psi_{n,\mathbf{k}}\rangle, \tag{4.144}$$

$\mathbf{p} = -i\hbar\nabla$ is the momentum operator. Substituting Eq. 4.144 into Eq. 4.130 gives:

$$\langle\psi_{n,\mathbf{k}}|(H - \mathcal{E}_{n,\mathbf{k}})\nabla_{\mathbf{k}}|\psi_{n,\mathbf{k}}\rangle = \frac{\hbar}{m}\langle\psi_{n,\mathbf{k}}|\mathbf{p}|\psi_{n,\mathbf{k}}\rangle = \frac{\hbar}{m}\langle\mathbf{p}\rangle. \tag{4.145}$$

From the previous equation and from Eq. 4.128, we have:

$$\langle\mathbf{p}\rangle = \frac{m}{\hbar}\nabla_{\mathbf{k}}\mathcal{E}_{n,\mathbf{k}}, \tag{4.146}$$

which is our last and important result. Previous equation gives the expectation value of the momentum of an electron in a given energy eigenstate in terms of the $\mathbf{k}$-space derivative of the corresponding eigenvalue, $\mathcal{E}_{n,\mathbf{k}}$. For the dipole approximation, the canonical momentum equals the particle momentum (see Eq. 4.54):

$$\mathbf{p} = m\dot{\mathbf{r}}, \tag{4.147}$$

therefore:

$$\langle \mathbf{v} \rangle = \frac{1}{m} \langle \mathbf{p} \rangle = \frac{1}{\hbar} \nabla_{\mathbf{k}} \mathcal{E}_{n,\mathbf{k}}, \tag{4.148}$$

as we have already written (see Eq. 1.53).

4.8 Exercises

Exercise 4.1 The intensity of a monochromatic ($\lambda = 800$ nm) electromagnetic radiation is $I = 1/\mu$W/m^2. Determine:
a) the amplitude of the electric field;
b) the amplitude of the vector potential;
c) the photon density.

Exercise 4.2 Demonstrate that two Lagrangian which differ by a total time derivative, $\mathcal{L}$ and $\mathcal{L}' = \mathcal{L} + \frac{d}{dt} F(q, t)$, where q is the generalized coordinate, are equivalent.

Exercise 4.3 The Hamiltonian of a particle with mass m and charge q in an electromagnetic field can be written as:

$$H = \frac{1}{2m} [\mathbf{p} - q\mathbf{A}(\mathbf{r}, t)]^2 + q\phi(\mathbf{r}, t).$$

The average value of the velocity operator $\mathbf{v}$ can be written as:

$$\langle \mathbf{v} \rangle = \frac{d}{dt} \langle \mathbf{r} \rangle.$$

By using the Ehrenfest's theorem:

$$\frac{d}{dt} \langle x \rangle = \frac{i}{\hbar} \langle [H, x] \rangle + \left\langle \frac{\partial x}{\partial t} \right\rangle$$

demonstrate that:

$$\mathbf{v} = \frac{\mathbf{p} - q\mathbf{A}(\mathbf{r}, t)}{m}.$$

Exercise 4.4 Consider the Hamiltonian of Exercise 4.3 and the gauge transformation:

$$\mathbf{A}' = \mathbf{A} + \nabla f \qquad \phi' = \phi - \frac{\partial f}{\partial t}.$$

Demonstrate that if the wavefunction $\psi(\mathbf{r}, t)$ obeys the time-dependent Schrödinger equation

$$i\hbar \frac{\partial \psi}{\partial t} = H(\mathbf{A}, \phi)\psi,$$

the wavefunction $\psi'(\mathbf{r}, t)$ defined as

$$\psi'(\mathbf{r}, t) = \exp\left(i\frac{q}{\hbar} t\right) \psi(\mathbf{r}, t)$$

obeys the equation

$$i\hbar\frac{\partial\psi'}{\partial t} = H(\mathbf{A}',\phi')\psi'.$$

Exercise 4.5   For a two-level system, the population difference per unit volume is:

$$\Delta N = \Delta N_0 \frac{1+(\omega-\omega_0)^2 T_2^2}{1+(\omega-\omega_0)^2 T_2^2 + 4\Omega^2 T_1 T_2}.$$

a) Show that ΔN can be written as:

$$\Delta N = \Delta N_0 \left[1 + \frac{I}{I_{sat}}\frac{g(\omega-\omega_0)}{T_2/\pi}\right]^{-1},$$

where: $g(\omega-\omega_0)$ is the Lorentzian line-shape

$$g(\omega-\omega_0) = \frac{T_2}{\pi[1+(\omega-\omega_0)^2 T_2^2]}$$

and I_{sat} is the so-called saturation intensity, given by:

$$I_{sat} = \frac{\epsilon_0 c n \hbar^2}{2\mu^2 T_1 T_2}.$$

b) Assuming that the electromagnetic wave is at resonance, plot $\Delta N/\Delta N_0$ as a function of intensity, I. Which is the physical meaning of I_{sat}?

Exercise 4.6   a) Using the results of Exercise 4.5, write the expression for the absorption coefficient when saturation effects become important.
b) Show that the peak absorption can be written as:

$$\alpha_p = \frac{\alpha_{unsat,p}}{1+I/I_{sat}},$$

where $\alpha_{unsat,p}$ is the peak of the unsaturated absorption (i.e., the absorption coefficient with $I=0$).
c) Plot the absorption coefficient for $I=0$, $I=I_{sat}$, and $I=2I_{sat}$.

Exercise 4.7   Calculate the full-width at half maximum of the absorption coefficient under saturating condition, showing that:

$$\Delta\omega_{FWHM} = \Delta\omega_{unsat,FWHM}\sqrt{1+\frac{I}{I_{sat}}}.$$

Optical Properties of Semiconductors

5.1 Stimulated Transitions: Selection Rules

In Chapter 3, we have obtained the Fermi's Golden rule, which, in the case of a two-level system ($m = 1$, $k = 2$), can be written as:

$$W_{12} = \frac{2\pi}{\hbar^2} |H'_{21}|^2 \delta(\omega - \omega_0), \tag{5.1}$$

where $\hbar\omega_0 = \mathcal{E}_2 - \mathcal{E}_1$. Within the electric dipole approximation, the interaction Hamiltonian can be written as:

$$H' = -\boldsymbol{\mu} \cdot \mathbf{E}, \tag{5.2}$$

and the matrix element H'_{21} can be calculated as follows:

$$H'_{21} = \langle \psi_2 | - \boldsymbol{\mu} \cdot \mathbf{E} | \psi_1 \rangle, \tag{5.3}$$

where $|\psi_1\rangle$ and $|\psi_2\rangle$ are the Bloch wavefunctions of the states with energy $\mathcal{E}_1$ and $\mathcal{E}_2$, respectively:

$$|\psi_{1,2}(\mathbf{r})\rangle = |u_{1,2}(\mathbf{r})\rangle e^{i\mathbf{k}_{1,2} \cdot \mathbf{r}}. \tag{5.4}$$

$\mathbf{E}(\mathbf{r}, t)$ is the electric field, which will be assumed polarized along the direction indicated by the unit vector $\hat{e}$:

$$\mathbf{E}(\mathbf{r}, t) = \frac{E(\mathbf{r}_0)}{2} \hat{e} \, e^{i(\mathbf{k}_{opt} \cdot \mathbf{r} - \omega t)}, \tag{5.5}$$

where $\mathbf{k}_{opt}$ is the field wavevector. The matrix element H'_{21} is given by:

$$H'_{21} = \langle u_2 e^{-i\mathbf{k}_2 \cdot \mathbf{r}} | - \boldsymbol{\mu} \cdot \frac{E(\mathbf{r}_0)}{2} \hat{e} \, e^{i(\mathbf{k}_{opt} \cdot \mathbf{r} - \omega t)} | u_1 e^{i\mathbf{k}_1 \cdot \mathbf{r}} \rangle. \tag{5.6}$$

Since the term $\exp{[i(\mathbf{k}_1 + \mathbf{k}_{opt} - \mathbf{k}_2) \cdot \mathbf{r}]}$ oscillates rapidly with $\mathbf{r}$, H'_{21} is zero unless

$$\mathbf{k}_2 = \mathbf{k}_{opt} + \mathbf{k}_1. \tag{5.7}$$

Since $\hbar\mathbf{k}_{1,2}$ are the electron momenta in the valence and conduction band and $\hbar\mathbf{k}_{opt}$ is the photon momentum, Eq. 5.7 shows that total momentum is conserved in the transition. If we assume that the refractive index of the semiconductor is $n = 3.5$ and $\lambda = 1$ μm, we have $k_{opt} = 2\pi n/\lambda \simeq 2 \times 10^5$ cm^{-1}. On the other hand, typical values for $k_{1,2} = k_{v,c}$ are of the order of $10^7 - 10^8$ cm^{-1} (see Exercise 5.1). Therefore, $k_{opt} \ll k_{c,v}$ so that the momentum conservation 5.7 can be written as:

$$\mathbf{k}_1 = \mathbf{k}_2. \tag{5.8}$$

This equation is often referred to as the **k**-selection or **k**-conservation rule, and it indicates that stimulated transitions must occur vertically in the **k**-space. Note that the **k**-selection rule can be derived also considering the interaction Hamiltonian $H' = (e/m)\mathbf{A} \cdot \mathbf{p}$.

Since the transition rate depends on $\delta(\omega - \omega_0)$, we can conclude that $\omega = \omega_0$. This means

$$\hbar\omega = \mathcal{E}_2 - \mathcal{E}_1, \tag{5.9}$$

which represents the energy conservation rule. Finally, we note that the electromagnetic wave does not interact with the electron spin, so that the spin is not involved in the interaction Hamiltonian. We can conclude that the spin cannot change in the transition and the spin-selection rule can be written as:

$$\Delta S = 0. \tag{5.10}$$

Exercise 5.1   Calculate typical values of **k** for a thermal electron in bulk GaAs ($m_c = 0.067\ m_0$, $m_v = m_{hh} = 0.45\ m_0$).

For an electron in the conduction band having thermal velocity v_{th}, we have (see Section 2.8):

$$\frac{1}{2}m_c v_{th}^2 = \frac{3}{2}k_B T,$$

where T is the electron temperature. We also have

$$p = \hbar k_c = m_c v_{th}.$$

Combining the two previous expressions, we obtain

$$k_c = (3m_c k_B T)^{1/2}/\hbar.$$

In the case of bulk GaAs at $T = 300$ K ($k_B T = 26$ meV), we obtain $k_c = 2.6 \times 10^6$ cm^{-1} and $k_v = (m_v/m_c)^{1/2}k_c = 6.7 \times 10^6$ cm^{-1}.   ∎

5.2 Joint Density of States

Let us consider two energy levels, $\mathcal{E}_1$ (in the valence band) and $\mathcal{E}_2$ (in the conduction band), with energy difference $\mathcal{E}_2 - \mathcal{E}_1 = \mathcal{E}_0 = \hbar\omega_0$, where ω_0 is the angular frequency of the transition. By using the parabolic approximation for the dispersion curve, we can write:

$$\mathcal{E}_1 = -\frac{\hbar^2 k^2}{2m_v} \tag{5.11}$$

$$\mathcal{E}_2 = \mathcal{E}_g + \frac{\hbar^2 k^2}{2m_c}, \tag{5.12}$$

where $\mathcal{E}_g$ is the energy gap. In Eqs. 5.11 and 5.12, we have assumed $k_c = k_v = k$. The energy difference is:

$$\mathcal{E}_0 = \mathcal{E}_2 - \mathcal{E}_1 = \mathcal{E}_g + \frac{\hbar^2 k^2}{2}\left(\frac{1}{m_c} + \frac{1}{m_v}\right) = \mathcal{E}_g + \frac{\hbar^2 k^2}{2m_r}, \tag{5.13}$$

where m_r is the *reduced mass*:

$$\frac{1}{m_r} = \frac{1}{m_c} + \frac{1}{m_v}.$$

(5.14)

The energies $\mathcal{E}_1$ and $\mathcal{E}_2$ are related to the transition energy $\mathcal{E}_0$ by the following expressions:

$$\mathcal{E}_1 = -\frac{m_r}{m_v}(\mathcal{E}_0 - \mathcal{E}_g) = -\frac{m_r}{m_v}(\hbar\omega_0 - \mathcal{E}_g)$$

(5.15)

$$\mathcal{E}_2 = \mathcal{E}_g + \frac{m_r}{m_c}(\mathcal{E}_0 - \mathcal{E}_g) = \mathcal{E}_g + \frac{m_r}{m_c}(\hbar\omega_0 - \mathcal{E}_g).$$

(5.16)

The *joint density of states* $\rho_j(\mathcal{E}_0)$ is defined as follows: $\rho_j(\mathcal{E}_0)d\mathcal{E}_0$ is the number of transitions per unit volume with transition energy in the range between $\mathcal{E}_0$ and $\mathcal{E}_0 + d\mathcal{E}_0$. As a consequence of the selection rules, a given state in the valence band is associated to a single state in the conduction band, with the same k value and the same spin. Therefore, the number of transitions is equal to the number of the corresponding states in the valence or in the conduction bands:

$$\rho_j(\mathcal{E}_0)d\mathcal{E}_0 = \rho(k)dk = \frac{k^2}{\pi^2}\, dk,$$

(5.17)

where we have used Eq. 2.22. From Eq. 5.13, we have:

$$d\mathcal{E}_0 = \frac{\hbar^2 k}{m_r}dk,$$

(5.18)

so that:

$$\rho_j(\mathcal{E}_0) = \frac{m_r k}{\pi^2 \hbar^2}.$$

(5.19)

By using the expression of k from Eq. 5.13 in Eq. 5.19, we have:

$$\rho_j(\mathcal{E}_0) = \frac{1}{2\pi^2}\left(\frac{2m_r}{\hbar^2}\right)^{3/2}(\mathcal{E}_0 - \mathcal{E}_g)^{1/2}.$$

(5.20)

The joint density of states as a function of the transition frequency can be calculated by writing: $\rho_j(\omega_0)d\omega_0 = \rho_j(\mathcal{E}_0)d\mathcal{E}_0$, where $d\mathcal{E}_0 = \hbar d\omega_0$, so that:

$$\rho_j(\omega_0) = \frac{1}{2\pi^2\hbar^2}(2m_r)^{3/2}(\hbar\omega_0 - \mathcal{E}_g)^{1/2}.$$

(5.21)

In order to have an absorption transition, the state with energy $\mathcal{E}_1$ must be occupied by an electron, while the state with energy $\mathcal{E}_2$ must be empty. Therefore, the number of absorption transitions per unit volume is given by:

$$dN_{12} = \rho_j(\omega_0)f_v(\mathcal{E}_1)[1 - f_c(\mathcal{E}_2)]d\omega_0.$$

(5.22)

We assume that thermal equilibrium has been achieved (separately) in the conduction and valence bands so that quasi-Fermi levels are used in Eq. 5.22. Stimulated emission events have to be taken into account from the state $\mathcal{E}_2$ to the state $\mathcal{E}_1$. The number of stimulated emission transitions per unit volume is given by:

$$dN_{21} = \rho_j(\omega_0)f_c(\mathcal{E}_2)[1 - f_v(\mathcal{E}_1)]d\omega_0.$$

(5.23)

Therefore, the net number of absorption transitions per unit volume is:

$$dN = dN_{12} - dN_{21} = \rho_j(\omega_0)[f_v(\mathcal{E}_1) - f_c(\mathcal{E}_2)]d\omega_0.$$

(5.24)

Since $\mathcal{E}_1$ and $\mathcal{E}_2$ are determined by $\mathcal{E}_0$ (see Eqs. 5.15 and 5.16), the previous equation can be written as:

$$dN = \rho_j(\omega_0)[f_v(\hbar\omega_0) - f_c(\hbar\omega_0)]d\omega_0. \tag{5.25}$$

5.3 Susceptibility and Absorption Coefficient in a Semiconductor

In Chapter 4, we have seen that the absorption coefficient of a medium is related to the imaginary part of the optical susceptibility by the following expression:

$$\alpha(\omega) = \frac{k\chi''(\omega)}{n^2} = \frac{\omega\chi''(\omega)}{nc}, \tag{5.26}$$

where $\chi''(\omega)$ is given by (see Eq. 4.104)

$$\chi''(\omega) = \frac{\pi\mu_{21}^2}{\epsilon_0\hbar}\Delta N g(\omega - \omega_0). \tag{5.27}$$

In the case of a semiconductor, ΔN must be replaced by the net number of absorption transitions per unit volume, dN, given by Eq. 5.25, so that the absorption coefficient due to the transitions with transition frequency in the range between ω_0 and $\omega_0 + d\omega_0$ is:

$$d\alpha = \frac{\omega\pi}{nc\epsilon_0\hbar}\mu_{21}^2\rho_j(\omega_0)g(\omega - \omega_0)[f_v(\hbar\omega_0) - f_c(\hbar\omega_0)]d\omega_0. \tag{5.28}$$

The absorption coefficient of the semiconductor can be obtained from Eq. 5.28 by integrating over all possible transition frequencies ω_0:

$$\alpha(\omega) = \frac{\omega\pi\mu_{21}^2}{nc\epsilon_0\hbar}\int_{\mathcal{E}_g/\hbar}^{\infty}\rho_j(\omega_0)\,[f_v(\hbar\omega_0) - f_c(\hbar\omega_0)]g(\omega - \omega_0)\,d\omega_0. \tag{5.29}$$

In Eq. 5.29, the matrix element μ_{21} has been assumed independent of ω_0. This integral can be easily calculated if we assume that the Lorentzian function $g(\omega - \omega_0)$ behaves as a delta-function $\delta(\omega - \omega_0)$ in comparison to the slowly varying functions $\rho_j(\omega_0)$ and $[f_v(\hbar\omega_0) - f_c(\hbar\omega_0)]$, so that:

$$\alpha(\omega) = \frac{\omega\pi}{cn\epsilon_0\hbar}\mu_{21}^2\rho_j(\omega)\,[f_v(\hbar\omega) - f_c(\hbar\omega)] = \alpha_0(\omega)[f_v(\hbar\omega) - f_c(\hbar\omega)]. \tag{5.30}$$

In a semiconductor at thermal equilibrium at $T = 0$ K, quasi-Fermi levels coincide with the Fermi level, $f_v(\mathcal{E}_1) = 1$ and $f_c(\mathcal{E}_2) = 0$, so that $\alpha(\omega) = \alpha_0(\omega)$, which represents the maximum absorption coefficient at a given frequency. Note that in the case of an intrinsic semiconductor, assuming $\mathcal{E}_g \gg k_B T$, $f_v(\mathcal{E}_1) \simeq 1$, and $f_c(\mathcal{E}_2) \simeq 0$, so that $\alpha(\omega) \simeq \alpha_0(\omega)$ even at room temperature.

The matrix element μ_{21} can be calculated as follows. We have first to write the relation between the matrix elements $\mathbf{r}_{cv}$ and $\mathbf{p}_{cv}$, given by the following expressions:

$$\mathbf{r}_{cv} = \langle u_{c,\mathbf{k}}|\mathbf{r}|u_{v,\mathbf{k}}\rangle \tag{5.31}$$

$$\mathbf{p}_{cv} = \langle u_{c,\mathbf{k}}|\mathbf{p}|u_{v,\mathbf{k}}\rangle. \tag{5.32}$$

Table 5.1 Kane energy ($\mathcal{E}_P$) and matrix element r_{cv} for various III–V semiconductors at $T = 0\,\text{K}$ [5].

	GaAs	InP	InAs
$\mathcal{E}_g$ (eV)	1.52	1.42	0.43
$\mathcal{E}_P$ (eV)	22.71	17	21.11
r_{cv} (Å)	6.12	5.67	20.9

These two matrix elements are related to the interaction Hamiltonians $-\boldsymbol{\mu}\cdot\mathbf{E}$ and $-(q/m)\mathbf{A}\cdot\mathbf{p}$, respectively. It is possible to demonstrate (see the following gray box) that:

$$\mathbf{r}_{cv} = -\frac{i\hbar}{m_0(\mathcal{E}_c - \mathcal{E}_v)}\mathbf{p}_{cv}, \tag{5.33}$$

where m_0 is the free electron mass. We have already mentioned in Chapter 1 that the Kane energy $\mathcal{E}_P$ is given by (see Eq. 1.126):

$$\mathcal{E}_P = \frac{2}{m_0}|\mathbf{p}_{cv}|^2, \tag{5.34}$$

which is largely constant in the case of III–V semiconductors (values for the Kane energy for various semiconductors are given in Table 5.1).

To first order in k, the matrix elements $\mathbf{p}_{cv}$ are constant. By using Eqs. 5.33 and 5.34, we can write:

$$|\mathbf{r}_{cv}| = \frac{\hbar}{\mathcal{E}_g}\sqrt{\frac{\mathcal{E}_P}{2m_0}}. \tag{5.35}$$

In order to take into account the fact that in semiconductors, only two valence bands out of three effectively contribute to optical transitions, namely the heavy- and light-hole bands (since the split-off energy Δ_{SO} is generally much larger than $k_B T$) it is common practice to use the matrix element x_{cv}, defined in the following way:

$$x_{cv}^2 = \frac{2}{3}|\mathbf{r}_{cv}|^2 = \frac{\hbar^2\mathcal{E}_P}{3m_0\mathcal{E}_g^2}. \tag{5.36}$$

The matrix element of the electric dipole moment can be written as:

$$\mu_{21}^2 = e^2 x_{cv}^2. \tag{5.37}$$

The absorption coefficient given by Eq. 5.30 can be written as follows:

$$\alpha(\omega) = \frac{\omega\pi e^2}{cn\epsilon_0\hbar}x_{cv}^2\rho_j(\omega)[f_v(\hbar\omega) - f_c(\hbar\omega)]. \tag{5.38}$$

The absorption coefficient can be written as a function of the photon energy, $\hbar\omega$:

$$\alpha(\hbar\omega) = \frac{\omega\pi e^2}{cn\epsilon_0}x_{cv}^2\rho_j(\hbar\omega)[f_v(\hbar\omega) - f_c(\hbar\omega)]. \tag{5.39}$$

5.3.1   Relation between $\mathbf{r}_{cv}$ and $\mathbf{p}_{cv}$.

We will assume that the electric field $\mathbf{E}$ and the vector potential $\mathbf{A}$ are polarized along the $z-$axis.

$$\mathbf{A}(t) = \frac{A(\mathbf{r}_0)}{2}\mathbf{u}_z\, e^{i(\mathbf{k}_{opt}\cdot\mathbf{r}-\omega t)}. \tag{5.40}$$

The matrix element, H'_{21}, of the interaction Hamiltonian $H' = (e/m)\mathbf{A}\cdot\mathbf{p}$ is given by:

$$H'_{21} = \langle\psi_2|\frac{e}{2m_0}A(\mathbf{r}_0)e^{i\mathbf{k}_{opt}\cdot\mathbf{r}}\mathbf{u}_z\cdot\mathbf{p}|\psi_1\rangle. \tag{5.41}$$

By taking into account the $k-$selection rule ($\mathbf{k}_2 = \mathbf{k}_1 + \mathbf{k}_{opt}$), Eq. 5.41 can be rewritten as:

$$H'_{21} = \frac{e}{2m_0}A(\mathbf{r}_0)\langle u_2|p_z|u_1\rangle, \tag{5.42}$$

where $p_z = \mathbf{p}\cdot\mathbf{u}_z$ is the component of $\mathbf{p}$ on the polarization direction.

In the same way, the matrix element of the interaction Hamiltonian $H' = -\boldsymbol{\mu}\cdot\mathbf{E}$ is given by:

$$H'_{21} = \frac{e}{2}E(\mathbf{r}_0)\langle u_2|z|u_1\rangle, \tag{5.43}$$

so that

$$|H'_{21}|^2 = \frac{E^2(\mathbf{r}_0)}{4}|\mu_{21}|^2. \tag{5.44}$$

The relation between $\langle u_2|z|u_1\rangle$ and $\langle u_2|p_z|u_1\rangle$ can be calculated by using the following procedure. We first calculate the commutator $[z, H_0]$:

$$\begin{aligned}[z, H_0] &= \left[z, \frac{p_z^2}{2m_0}\right] = \frac{1}{2m_0}(zp_z^2 - p_z^2 z) = \frac{1}{2m_0}(zp_z^2 - p_z z p_z + p_z z p_z - p_z^2 z) = \\ &= \frac{1}{2m_0}([z, p_z]p_z + p_z[z, p_z]), \end{aligned} \tag{5.45}$$

where the commutator $[z, p_z]$ is simply given by:

$$[z, p_z] = i\hbar, \tag{5.46}$$

so that, from Eq. 5.45:

$$[z, H_0] = i\hbar\frac{p_z}{m_0}, \qquad p_z = \frac{m_0}{i\hbar}[z, H_0]. \tag{5.47}$$

We can now calculate the matrix element $\langle u_2|p_z|u_1\rangle$:

$$\begin{aligned}\langle u_2|p_z|u_1\rangle &= \frac{m_0}{i\hbar}\langle u_2|[z, H_0]|u_1\rangle = \frac{m_0}{i\hbar}\left[\langle u_2|zH_0|u_1\rangle - \langle u_2|H_0 z|u_1\rangle\right] = \\ &= \frac{m_0}{i\hbar}(\mathcal{E}_1 - \mathcal{E}_2)\langle u_2|z|u_1\rangle \end{aligned} \tag{5.48}$$

from this last equation we obtain Eq. 5.33:

$$\langle u_2|z|u_1\rangle = -\frac{i\hbar}{m_0(\mathcal{E}_2 - \mathcal{E}_1)}\langle u_2|p_z|u_1\rangle. \tag{5.49}$$

From this result, we can also determine the relation between the two expressions of the interaction Hamiltonian. Indeed:

$$
\frac{e}{2m_0}A(\mathbf{r}_0)\langle u_2|p_z|u_1\rangle = i\frac{e}{2m_0}\frac{E(\mathbf{r}_0)}{\omega}\frac{m_0(\mathcal{E}_2-\mathcal{E}_1)}{\hbar}\langle u_2|z|u_1\rangle =
$$
$$
= i\frac{\omega_0}{\omega}\left[\frac{e}{2}E(\mathbf{r}_0)\langle u_2|z|u_1\rangle\right], \tag{5.50}
$$

where $\hbar\omega_0 = \mathcal{E}_2 - \mathcal{E}_1$. Therefore, the two expressions of the interaction Hamiltonian are equivalent only at resonance. If $\omega = \omega_0$, we obtain:

$$
|p_{21}|^2 = \frac{m_0^2\omega^2|\mu_{21}|^2}{e^2}. \tag{5.51}
$$

By using Eqs. 5.51 and 5.37, the absorption coefficient, $\alpha(\hbar\omega)$, can be written as:

$$
\alpha(\hbar\omega) = \frac{\pi e^2}{cn\epsilon_0 m_0^2\omega}|\hat{e}\cdot\mathbf{p}_{cv}|^2\rho_j(\hbar\omega)[f_v(\hbar\omega)-f_c(\hbar\omega)], \tag{5.52}
$$

where $\hat{e}$ gives the polarization direction of the electric field. It is sometimes also useful to use the bulk momentum matrix element M_b given by:

$$
M_b^2 = \frac{1}{3}|\hat{e}\cdot\mathbf{p}_{cv}|^2 = \frac{m_0}{6}\mathcal{E}_p, \tag{5.53}
$$

where the Kane's parameter $\mathcal{E}_p$ is taken from experimental data.

Exercise 5.2   Plot the absorption coefficient of bulk GaAs at room temperature assuming the following data: $\mathcal{E}_g = 1.424$ eV, $m_c = 0.067\,m_0$, $m_v = 0.45\,m_0$, $n = 3.64$. In order to fit the experimental data, it is worth to assume $x_{cv} = 3.2$ Å.

From Eq. 5.39, we have:

$$
\alpha_0(\hbar\omega) = \frac{\omega e^{5/2}x_{cv}^2}{2\pi cn\epsilon_0}\left(\frac{2m_r}{\hbar^2}\right)^{3/2}\sqrt{\frac{\hbar\omega-\mathcal{E}_g}{e}} = \mathcal{A}\sqrt{\frac{\hbar\omega-\mathcal{E}_g}{e}},
$$

where

$$
\mathcal{A} = \frac{\omega e^{5/2}x_{cv}^2}{2\pi cn\epsilon_0}\left(\frac{2m_r}{\hbar^2}\right)^{3/2} = \frac{e^{5/2}x_{cv}^2}{\lambda_0 n\epsilon_0\hbar^3}(2m_r)^{3/2}.
$$

The reduced mass is $m_r = 0.059\,m_0$. For the calculation of $\mathcal{A}$, we can assume $\omega \simeq \mathcal{E}_g/\hbar = 2.16 \times 10^{15}$ rad/s, so that

$$
\mathcal{A} = 1.12 \times 10^4 \text{ cm}^{-1}\text{eV}^{-1/2}.
$$

Figure 5.1 shows the calculated absorption coefficient as a function of energy: When the photon energy exceeds the energy gap of just 50 meV, the absorption coefficient is already of the order of 2,500 cm^{-1}. Figure 5.2 shows, for comparison, the measured absorption spectra of a few semiconductors.

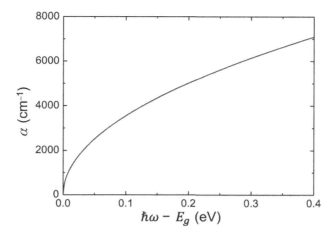

Figure 5.1 Plot of the absorption coefficient as a function of the difference between the photon energy, $\hbar\omega$, and the energy gap, $\mathcal{E}_g$, for an intrinsic GaAs bulk semiconductor calculated by using Eq. 5.39.

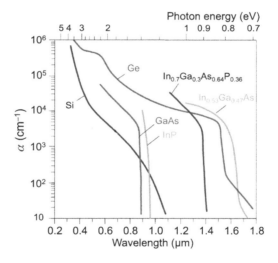

Figure 5.2 Measured absorption spectra of a few semiconductors.

5.3.2 Direct Calculation of Absorption Coefficient

It is now instructive to derive the expression of the absorption coefficient by using a different approach. Since the net number of absorption transitions is given by Eq. 5.25, the absorbed electromagnetic power is given by:

$$dP_{abs} = W_{12}\hbar\omega_0 dN = W_{12}\hbar\omega_0\rho_j(\omega_0)[f_v(\hbar\omega_0) - f_c(\hbar\omega_0)]d\omega_0, \qquad (5.54)$$

where W_{12} is the transition rate corresponding to the transition with frequency ω_0. In other words, dP_{abs} is the electromagnetic power absorbed per unit volume when the transition frequency is in the range between ω_0 and $\omega_0 + d\omega_0$. Since the absorption coefficient is given by the ratio between the absorbed power per unit volume and the input intensity $I = \epsilon_0 n c E_0^2 / 2$, we have:

$$d\alpha = \frac{2 d P_{abs}}{\epsilon_0 n c E_0^2} = \frac{2 \hbar \omega_0}{\epsilon_0 n c E_0^2} W_{12} \rho_j(\omega_0)[f_v(\mathcal{E}_1) - f_c(\mathcal{E}_2)] d\omega_0, \tag{5.55}$$

where the transition rate W_{12} is given by:

$$W_{12} = \frac{2\pi}{\hbar^2} |H'_{21}|^2 g(\omega - \omega_0).$$

The absorption coefficient can be calculated from Eq. 5.55 by integrating over all possible transition frequencies ω_0:

$$\alpha(\omega) = \int_{\mathcal{E}_g / \hbar}^{\infty} \frac{\omega_0 \pi \mu_{21}^2}{n c \epsilon_0 \hbar} \rho_j(\omega_0) g(\omega - \omega_0)[f_v(\hbar\omega_0) - f_c(\hbar\omega_0)] d\omega_0, \tag{5.56}$$

where we have used Eq. 5.44. If $g(\omega - \omega_0)$ is replaced by the delta-function $\delta(\omega - \omega_0)$, we find the same result obtained by using the approach based on the optical susceptibility (see Eq. 5.30).

5.4 Gain Coefficient and Bernard–Duraffourg Condition

The gain coefficient is given by $g(\omega) = -\alpha(\omega)$, therefore:

$$g(\omega) = \alpha_0(\omega)[f_c(\hbar\omega) - f_v(\hbar\omega)]. \tag{5.57}$$

Therefore, the condition for optical gain is given by:

$$f_c(\hbar\omega) - f_v(\hbar\omega) = f_c(\mathcal{E}_2) - f_v(\mathcal{E}_1) > 0, \tag{5.58}$$

that is:

$$1 + \exp\left(\frac{\mathcal{E}_2 - \mathcal{E}_{Fc}}{k_B T}\right) < 1 + \exp\left(\frac{\mathcal{E}_1 - \mathcal{E}_{Fv}}{k_B T}\right), \tag{5.59}$$

which implies:

$$\mathcal{E}_2 - \mathcal{E}_1 < \mathcal{E}_{Fc} - \mathcal{E}_{Fv}, \tag{5.60}$$

which represents the *Bernard–Duraffourg condition*: Only those photons with energy $\hbar\omega = \mathcal{E}_2 - \mathcal{E}_1$ smaller than the energy separation between the two quasi-Fermi levels present a positive gain and are amplified. Since $\hbar\omega = \mathcal{E}_2 - \mathcal{E}_1 > \mathcal{E}_g$, in order to have net gain we must have:

$$\mathcal{E}_g < \hbar\omega < \mathcal{E}_{Fc} - \mathcal{E}_{Fv}. \tag{5.61}$$

The limiting case, $\mathcal{E}_{Fc} - \mathcal{E}_{Fv} = \mathcal{E}_g$, defines the *transparency condition*. In this case, $g = 0$ at $\omega = \mathcal{E}_g / \hbar$. To reach the transparency condition, electrons must be injected into the conduction band: The corresponding density is called transparency density, N_{tr}. In the case

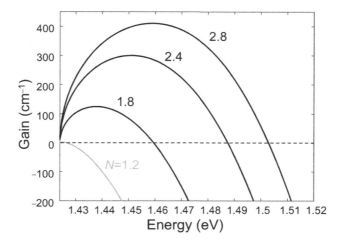

Gain coefficient in GaAs as a function of photon energy calculated for different values of the injected carrier density, N (in units of 10^{18} cm^{-3}).

of GaAs $N_{tr} \simeq 10^{18}$ cm^{-3}. When the density of the electrons injected into the conduction band exceeds the transparency density, $\mathcal{E}_{Fc} - \mathcal{E}_{Fv} > \mathcal{E}_g$ and net gain is present in the photon energy range between $\mathcal{E}_g$ and $\mathcal{E}_{Fc} - \mathcal{E}_{Fv}$. Figure 5.3 shows the gain coefficient $g(\omega)$ in the case of GaAs, for different values of the injected electron density. As shown in Fig. 5.3, even at high injection levels, the gain bandwidth is only a small fraction of the energy gap. The peak of the gain curve increases upon increasing the density of the injected electrons. For typical gain values ($20 \leq g \leq 80$ cm^{-1}), the gain peak increases almost linearly with N:

$$g = \sigma (N - N_{tr}), \tag{5.62}$$

where σ is called *differential gain coefficient* ($\sigma \simeq 1.5 \times 10^{-16}$ cm^2 in the case of GaAs).

5.5 Spontaneous Emission

In this section, we will consider the process of spontaneous emission, without entering into too many details. The semiclassical approach adopted so far, in which the radiation is treated classically (i.e., by using the Maxwell's equations), while matter is quantized and therefore treated using quantum mechanics, cannot be used to analyze the process of spontaneous emission, which has to be treated by using quantum electrodynamics, in which both matter and radiation are quantized. A very powerful approach, which does not require field quantization, is based on Einstein thermodynamic treatment. By using the Einstein's coefficient A_{cv}, the spontaneous emission rate, $r_{sp}(\mathcal{E})$, for a given transition from the level $\mathcal{E}_2$ in the conduction band to the level $\mathcal{E}_1$ in the valence band, with the same wavevector $\mathbf{k}$, is independent of the photon density in the semiconductor and can be written as follows:

$$r_{sp}(\mathcal{E}) = A_{cv}f_c(\mathcal{E}_2)[1 - f_v(\mathcal{E}_1)], \tag{5.63}$$

where A_{cv} is the spontaneous transition rate in the volume V. $r_{sp}(\mathcal{E})$ is proportional to the probability that the state with energy $\mathcal{E}_2$ in the conduction band is filled by an electron and to the probability that the corresponding state $\mathcal{E}_1$ (i.e., the state with the same $\mathbf{k}$) in the valence band is empty. The Einstein's coefficient A_{cv} can be written as:

$$A_{cv} = 1/\tau_r, \tag{5.64}$$

where τ_r is the radiative lifetime (or spontaneous emission lifetime) given by:

$$\tau_r = \frac{\pi \epsilon_0 \hbar c^3}{e^2 x_{vc}^2 n \omega_{vc}^3}, \tag{5.65}$$

where $\hbar\omega_{vc} = \mathcal{E}_c(\mathbf{k}) - \mathcal{E}_v(\mathbf{k})$. Spontaneous emission is often negligible in the middle- and far-infrared, where nonradiative decay usually dominates. When one considers the x-ray region ($\lambda < 5$ nm), τ_r becomes exceedingly short ($\sim 10 - 100$ fs). In this case, spontaneous emission becomes the predominant decay mechanism and natural broadening becomes the predominant broadening mechanism ($\Delta \nu_{nat} = 1/2\pi\tau_r$).

Due to energy conservation, we must have:

$$\hbar\omega_{vc} = \mathcal{E}_c(\mathbf{k}) - \mathcal{E}_v(\mathbf{k}) = \mathcal{E}_g + \frac{\hbar^2 k^2}{2m_r}. \tag{5.66}$$

If we now consider the joint density of states, $\rho_j(\mathcal{E})$, the total spontaneous emission rate $R_{sp}(\hbar\omega)$ can be calculated as:

$$R_{sp}(\hbar\omega) = \int r_{sp}(\mathcal{E})\rho_j(\mathcal{E})g(\mathcal{E} - \hbar\omega)d\mathcal{E}. \tag{5.67}$$

$r_{sp}(\mathcal{E})\rho_j(\mathcal{E})$ is slowly varying compared to the lineshape function; therefore, in Eq. 5.67, we can replace $g(\mathcal{E} - \hbar\omega)$ with $\delta(\mathcal{E} - \hbar\omega)$, so that Eq. 5.67 simplifies to:

$$R_{sp}(\hbar\omega) = r_{sp}(\hbar\omega)\rho_j(\hbar\omega). \tag{5.68}$$

From Eqs. 5.68, 5.63 and 5.64, we can write:

$$R_{sp}(\hbar\omega) = \frac{1}{\tau_r}\rho_j(\hbar\omega)f_c(\hbar\omega)[1 - f_v(\hbar\omega)]. \tag{5.69}$$

We note that at thermal equilibrium with background thermal radiation, the ratio between the spontaneous emission rate, $R_{sp}(\hbar\omega)$, and the absorption $\alpha(\hbar\omega)$ (see Eq. 5.39):

$$\alpha(\hbar\omega) = \frac{\omega\pi e^2 x_{cv}^2}{cn\epsilon_0}\rho_j(\hbar\omega)[f_v(\hbar\omega) - f_c(\hbar\omega)], \tag{5.70}$$

is given by the following relation, called *van Roosbroeck–Shockley equation*

$$\frac{R_{sp}(\hbar\omega)}{\alpha(\hbar\omega)} = \frac{(\hbar\omega)^2 n^2}{\pi^2 c^2 \hbar^3}\frac{1}{\exp(\hbar\omega/k_B T) - 1}. \tag{5.71}$$

The "non-equilibrium" van Roosbroeck–Shockley equation is given by:

$$\frac{R_{sp}(\hbar\omega)}{\alpha(\hbar\omega)} = \frac{(\hbar\omega)^2 n^2}{\pi^2 c^2 \hbar^3}\frac{f_c(\hbar\omega)[1 - f_v(\hbar\omega)]}{f_v(\hbar\omega) - f_c(\hbar\omega)}, \tag{5.72}$$

which reduces to Eq. 5.71 at thermal equilibrium (when $\mathcal{E}_{Fc} = \mathcal{E}_{Fv} = \mathcal{E}_F$).

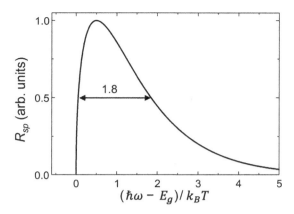

Figure 5.4 Spectral distribution of the spontaneous emission rate as a function of $(\hbar\omega - \mathcal{E}_g)/k_BT$, calculated by using Eq. 5.79.

If the Fermi–Dirac distribution functions can be approximated by the Boltzmann distribution function (i.e., if the quasi-Fermi levels are far from the band edges), we can write:

$$f_c(\mathcal{E}_2) \approx \exp\left(-\frac{\mathcal{E}_2 - \mathcal{E}_{Fc}}{k_BT}\right), \tag{5.73}$$

$$1 - f_v(\mathcal{E}_1) \approx \exp\left(-\frac{\mathcal{E}_{Fv} - \mathcal{E}_1}{k_BT}\right), \tag{5.74}$$

so that:

$$f_c(\hbar\omega)[1 - f_v(\hbar\omega)] \approx \exp\left(-\frac{\hbar\omega}{k_BT} + \frac{\Delta\mathcal{E}_F}{k_BT}\right), \tag{5.75}$$

where $\Delta\mathcal{E}_F = \mathcal{E}_{Fc} - \mathcal{E}_{Fv}$. With these approximations, $R_{sp}(\hbar\omega)$ can be written as follows:

$$R_{sp}(\hbar\omega) = \frac{1}{\tau_r}\rho_j(\hbar\omega)\exp\left(-\frac{\hbar\omega}{k_BT}\right)\exp\left(\frac{\Delta\mathcal{E}_F}{k_BT}\right), \tag{5.76}$$

where $\rho_j(\hbar\omega) = \rho_j(\mathcal{E})$ is given by Eq. 5.20 so that:

$$R_{sp}(\hbar\omega) = \frac{1}{\tau_r}\frac{1}{2\pi^2}\left(\frac{2m_r}{\hbar^2}\right)^{3/2}(\hbar\omega - \mathcal{E}_g)^{1/2}\exp\left(-\frac{\hbar\omega - \mathcal{E}_g}{k_BT}\right)\exp\left(\frac{\Delta\mathcal{E}_F - \mathcal{E}_g}{k_BT}\right). \tag{5.77}$$

If we now introduce the parameter $\mathfrak{R}_{sp}$ defined as:

$$\mathfrak{R}_{sp} = \frac{1}{2\pi^2\tau_r}\left(\frac{2m_r}{\hbar^2}\right)^{3/2}\exp\left(\frac{\Delta\mathcal{E}_F - \mathcal{E}_g}{k_BT}\right). \tag{5.78}$$

Equation 5.77 can be rewritten as follows:

$$R_{sp}(\hbar\omega) = \mathfrak{R}_{sp}\,(\hbar\omega - \mathcal{E}_g)^{1/2}\exp\left(-\frac{\hbar\omega - \mathcal{E}_g}{k_BT}\right). \tag{5.79}$$

Figure 5.4 shows the spontaneous emission rate $R_{sp}(\hbar\omega)$ as a function of $(\hbar\omega - \mathcal{E}_g)/k_BT$.

Exercise 5.3 Determine the photon energy that corresponds to the peak value of the spontaneous emission rate.

We have to calculate the position of the peak of $R_{sp}(\hbar\omega)$, by solving the equation $dR_{sp}/d\omega = 0$. If we use the variable $z = (\hbar\omega - \mathcal{E}_g)/k_B T$, Eq. 5.79 can be written as:

$$R_{sp} = \Re_{sp} \sqrt{k_B T} \sqrt{z} e^{-z}. \tag{5.80}$$

If we now equate to zero dR_{sp}/dz, we obtain:

$$\Re_{sp} \sqrt{k_B T} e^{-z} \left(\frac{1}{2\sqrt{z}} - \sqrt{z} \right) = 0, \tag{5.81}$$

which gives $z = 1/2$. Therefore, the photon energy corresponding to the spontaneous emission peak is given by:

$$\hbar\omega = \mathcal{E}_g + \frac{k_B T}{2}. \tag{5.82}$$

∎

Exercise 5.4 Determine the full-width at half-maximum (FWHM) of the spontaneous emission spectrum. Assuming an emission wavelength $\lambda = 580$ nm, calculate the FWHM of the spontaneous emission spectrum (write the result in nanometers.)

The peak of the function $R_{sp}(\hbar\omega)$ can be calculated from Eq. 5.80 by using $z = 1/2$:

$$R_{sp,max} = \Re_{sp} \sqrt{\frac{k_B T}{2e}}, \tag{5.83}$$

where e is the Neper number. The FWHM can be calculated by solving the following equation:

$$2\sqrt{2e}\sqrt{z} = e^z. \tag{5.84}$$

By using a simple graphical method (see Fig. 5.5), we obtain the FWHM:

$$\Delta\mathcal{E} \approx 1.8 k_B T. \tag{5.85}$$

Since $\Delta\mathcal{E} = h\Delta\nu$ and $\Delta\lambda/\lambda = \Delta\nu/\nu$, the FWHM of the spontaneous emission spectrum can be written in terms of the wavelength as follows:

$$\Delta\lambda \approx \frac{1.8\lambda^2 k_B T}{hc}. \tag{5.86}$$

For example, assuming that the wavelength corresponding to the maximum emission is $\lambda = 580$ nm at $T = 300$ K, the FWHM of the spectrum is given by $\Delta\lambda \simeq 12.6$ nm. ∎

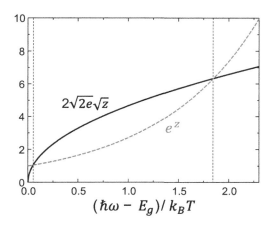

Figure 5.5 Graphical solution of Eq. 5.84.

Exercise 5.5 Calculate the spontaneous lifetime of GaAs by using the numerical data reported in Exercise 5.2.

From Eq. 5.65, we have:

$$\tau_r = \frac{\pi \epsilon_0 \hbar c^3}{e^2 x_{vc}^2 n \omega_{vc}^3} = \frac{\pi \epsilon_0 \hbar^4 c^3}{e^5 x_{vc}^2 n (\hbar \omega_{vc}/e)^3},$$

where $\hbar \omega_{vc} \simeq \mathcal{E}_g$ so that $\hbar \omega_{vc}/e = 1.424$ eV. By substituting the numerical values, we obtain:

$$\tau_r \simeq 0.8 \text{ ns}.$$

■

5.5.1 Bimolecular Recombination

We can now calculate the total radiative recombination rate for photons of all energies by considering the following integral:

$$R_{sp,T} = \frac{1}{\tau_r} \int_{\mathcal{E}_g}^{\infty} \rho_j(\hbar \omega) f_c(\hbar \omega) [1 - f_v(\hbar \omega)] d(\hbar \omega). \tag{5.87}$$

By using Eq. 5.76, we can write:

$$R_{sp,T} = \frac{e^{\Delta \mathcal{E}_F/k_B T}}{\tau_r} \int_{\mathcal{E}_g}^{\infty} \rho_j(\hbar \omega) e^{-\hbar \omega/k_B T} d(\hbar \omega). \tag{5.88}$$

If we now use the expression of $\rho_j(\hbar \omega) = \rho_j(\mathcal{E})$ given by Eq. 5.20, we obtain:

$$\begin{aligned}
R_{sp,T} &= \frac{e^{\Delta \mathcal{E}_F/k_B T}}{\tau_r} \frac{1}{2\pi^2} \left(\frac{2m_r}{\hbar^2} \right)^{3/2} \int_{\mathcal{E}_g}^{\infty} (\hbar \omega - \mathcal{E}_g)^{1/2} e^{-\hbar \omega/k_B T} d(\hbar \omega) = \\
&= \frac{e^{(\Delta \mathcal{E}_F - \mathcal{E}_g)/k_B T}}{\tau_r} \frac{1}{2\pi^2} \left(\frac{2m_r k_B T}{\hbar^2} \right)^{3/2} \int_0^{\infty} \frac{\sqrt{x}}{e^x} dx, \tag{5.89}
\end{aligned}$$

Table 5.2 Bimolecular recombination coefficients for various semiconductors at 300 K. M. Grundmann, The Physics of Semiconductors, Springer, 2006.

	B (cm^{3}s^{-1})
GaAs	1.0×10^{-10}
AlAs	7.5×10^{-11}
InP	6.0×10^{-11}
InAs	2.1×10^{-11}

where $x = (\hbar\omega - \mathcal{E}_g)/k_B T$. Since:

$$\int_0^\infty \frac{\sqrt{x}}{e^x}\, dx = \frac{\sqrt{\pi}}{2}, \tag{5.90}$$

from Eq. 5.89, one obtains:

$$R_{sp,T} = \frac{1}{\tau_r} N_r \exp\left(\frac{\Delta\mathcal{E}_F - \mathcal{E}_g}{k_B T}\right), \tag{5.91}$$

where, in analogy with the effective density of states N_c and N_v introduced in Chapter 2, we have used the effective density of states N_r defined as:

$$N_r = \frac{1}{4}\left(\frac{2m_r k_B T}{\pi \hbar^2}\right)^{3/2}. \tag{5.92}$$

Since $n = N_c e^{-\frac{\mathcal{E}_c - \mathcal{E}_{Fc}}{k_B T}}$ and $p = N_v e^{-\frac{\mathcal{E}_{Fv} - \mathcal{E}_v}{k_B T}}$, Eq. 5.91 can be rewritten as follows:

$$R_{sp,T} = \frac{1}{\tau_r}\frac{N_r}{N_c N_v} np = Bnp, \tag{5.93}$$

where B is called *bimolecular recombination coefficient*

$$B = \frac{1}{\tau_r}\frac{N_r}{N_c N_v} = \frac{1}{\tau_r N_c}\left(\frac{m_r}{m_v}\right)^{3/2}, \tag{5.94}$$

with typical values in the range $10^{-11} - 10^{-9}$ cm^3/s for III–V semiconductors (Table 5.2).

At thermal equilibrium, the recombination rate is equal to the thermal generation rate, G_0, so that we can write:

$$R_0 = G_0 = Bn_0 p_0 = Bn_i^2, \tag{5.95}$$

where n_0 and p_0 are the equilibrium carrier densities and we have used the mass-action law $n_0 p_0 = n_i^2$. In the presence of excess electron–hole pairs $\Delta n = \Delta p$ (to maintain charge neutrality), the nonequilibrium carrier densities are given by:

$$n = n_0 + \Delta n, \qquad p = p_0 + \Delta p. \tag{5.96}$$

These excess carriers can be generated by absorption of light or by an injection current. In this situation, the net radiative recombination rate is given by:

$$R_r = Bnp - R_0 = B(np - n_0 p_0). \tag{5.97}$$

When carriers are injected into the semiconductor, the product np is larger than in the case of thermodynamic equilibrium, so that $np > n_0 p_0$ and the recombination rate is positive. The recombination dynamics can be described by the following equation:

$$\frac{dn}{dt} = \frac{dp}{dt} = R_0 - Bnp = -B(np - n_0 p_0) = -B(np - n_i^2), \tag{5.98}$$

where $dn/dt = d\Delta n/dt$ and $dp/dt = d\Delta p/dt$ (moreover we assume $\Delta n = \Delta p$). Equation 5.98 can be rewritten as:

$$\frac{d\Delta n}{dt} = -B(n_0 \Delta p + p_0 \Delta n + \Delta n \Delta p). \tag{5.99}$$

The general solution of Eq. 5.99 is given by:

$$\Delta n = \frac{(n_0 + p_0)\Delta n(0)}{[n_0 + p_0 + \Delta n(0)]\exp[B(n_0 + p_0)t] - \Delta n(0)}, \tag{5.100}$$

where $\Delta n(0)$ is the excess electron density at $t = 0$. In the case of low injection $\Delta n = \Delta p \ll n_0, p_0$, so that Eq. 5.99 can be written as:

$$\frac{d\Delta n}{dt} = -B(n_0 + p_0)\Delta n, \tag{5.101}$$

whose solution is:

$$\Delta n = \Delta n(0)\exp[-B(n_0 + p_0)t] = \Delta n(0)\exp\left(-\frac{t}{\tau_{rad}}\right), \tag{5.102}$$

where

$$\tau_{rad} = \frac{1}{B(n_0 + p_0)} \tag{5.103}$$

is the radiative recombination time. In the case of low injection, the net radiative recombination rate of Eq. 5.97 can be written as:

$$R_r = B(np - n_0 p_0) \simeq B(n_0 + p_0)\Delta n, \tag{5.104}$$

which can be rewritten by using the definition of τ_{rad}:

$$R_r = \frac{\Delta n}{\tau_{rad}}. \tag{5.105}$$

In a n-type semiconductor $n_0 \gg p_0$ so that the radiative recombination time can be written as:

$$\tau_{rad} = \frac{1}{Bn_0} = \frac{1}{BN_D}, \tag{5.106}$$

where N_D is the donor density. In the same way, in a p-type semiconductor $p_0 \gg n_0$ and the radiative recombination time is given by:

$$\tau_{rad} = \frac{1}{Bp_0} = \frac{1}{BN_A}, \tag{5.107}$$

where N_A is the acceptor density.

In the case of strong injection, $n \approx p \gg n_0, p_0$ and Eq. 5.99 can be written as follows:

$$\frac{d\Delta n}{dt} = -B(\Delta n)^2, \tag{5.108}$$

and the solution is:

$$\Delta n(t) = \frac{\Delta n(0)}{1 + B t \, \Delta n(0)}. \tag{5.109}$$

Exercise 5.6   Consider a GaAs bulk semiconductor at $T = 300$ K n-doped with Silicon ($N_D = 10^{17}$ cm^{-3}); the spontaneous emission lifetime is $\tau_r = 0.8$ ns. Calculate the bimolecular recombination coefficient and the radiative recombination time, τ_{rad}, assuming low injection levels.

In the case of low-injection in an n-type semiconductor, the radiative lifetime is given by Eq. 5.106, where the bimolecular recombination coefficient is given by Eq. 5.94. We have first to calculate the effective density of states N_c:

$$N_c = \frac{1}{4} \left(\frac{2m_c k_B T}{\pi \hbar^2} \right)^{3/2} = 4.4 \times 10^{17} \text{ cm}^{-3}.$$

Since the reduced mass is $m_r = 0.059 \, m_0$, the bimolecular recombination coefficient is given by:

$$B = \frac{1}{\tau_r N_c} \left(\frac{m_r}{m_v} \right)^{3/2} = 1.3 \times 10^{-10} \text{ cm}^3\text{s}^{-1}$$

and

$$\tau_{rad} = \frac{1}{B N_D} = 77 \text{ ns.}$$

■

5.6 Nonradiative Recombination

There are processes which lead to electron–hole recombination without photon emission. In these cases, the electron energy is converted to vibrational energy of the lattice, that is, the recombination generates phonons instead of photons. Nonradiative recombination can be determined by the presence of defects in the semiconductor, which generally lead to the formation of electronic energy levels within the band gap of the semiconductor. Other origins of nonradiative recombination are the Auger process and the recombination at surface.

5.6.1 Shockley–Read–Hall Model

Energy levels within the gap of semiconductors generated by the presence of impurities are efficient recombination centers, which are particularly effective when they are close to the middle of the band gap. In the framework of the *Shockley–Read–Hall (SRH) model*, four processes involving trap states are possible: (i) an electron from the conduction band can be trapped in an energy trap; (ii) a filled trap can emit an electron in the conduction band;

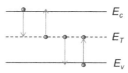

Figure 5.6   Nonradiative processes involving trap states.

(iii) an empty trap can emit a hole in the valence band, or, in other words, an electron from the valence band can be trapped in an empty trap; and (iv) a filled trap can emit an electron in the valence band. These processes are schematically shown in Fig. 5.6. In the following, the essential ingredients of the SRH model will be outlined.

Let us assume that trap states are present in the gap at energy $\mathcal{E}_T$, with a density N_T (number of traps per unit volume). The density of filled traps is given by:

$$N_T f(\mathcal{E}_T), \tag{5.110}$$

where $f(\mathcal{E}_T)$ is the distribution function giving the probability of occupation of the trap state. The density of empty traps is:

$$N_T [1 - f(\mathcal{E}_T)]. \tag{5.111}$$

It is evident that the capture rate of electrons in the trap states is proportional to the product of the number of empty traps and of the electron density in the conduction band. It is generally assumed that the proportionality constant can be written as $C_n = v_{th}\sigma_n$, where v_{th} is the thermal velocity (see Exercise 5.1) $v_{th} = (3k_BT/m^*)^{1/2}$ ($v_{th} \simeq 10^7$ cm/s at room temperature) and σ_n is called electron capture cross-section, which turns out to be of the order of the square of the lattice constant and is typically $\sigma_n \simeq 10^{-15}$ cm^2. Therefore, the capture rate for electrons can be written as follows:

$$R_{cn} = C_n n N_T [1 - f(\mathcal{E}_T)] = v_{th}\sigma_n n N_T [1 - f(\mathcal{E}_T)]. \tag{5.112}$$

The emission rate from filled traps is proportional to the number of filled traps and can be written as:

$$R_{en} = e_n N_T f(\mathcal{E}_T). \tag{5.113}$$

If the semiconductor is in thermodynamic equilibrium, the capture rate for electrons must equal the emission rate for electrons so that:

$$R_{cn}^0 = R_{en}^0 \tag{5.114}$$

thus giving:

$$e_n = C_n n_0 \frac{1 - f_0(\mathcal{E}_T)}{f_0(\mathcal{E}_T)}, \tag{5.115}$$

where n_0 is the equilibrium electron density and $f_0(\mathcal{E}_T)$ is the equilibrium distribution function:

$$f_0(\mathcal{E}_T) = \frac{1}{1 + e^{\frac{\mathcal{E}_T - \mathcal{E}_F}{k_B T}}}. \tag{5.116}$$

Therefore, Eq. 5.115 can be rewritten as:

$$e_n = C_n n_0 \exp\left(\frac{\mathcal{E}_T - \mathcal{E}_F}{k_B T}\right). \tag{5.117}$$

From Eqs. 2.47 and 2.55, the equilibrium electron density n_0 can be written as:

$$n_0 = n_i \exp\left(\frac{\mathcal{E}_F - \mathcal{E}_i}{k_B T}\right) = N_c \exp\left(\frac{\mathcal{E}_F - \mathcal{E}_c}{k_B T}\right), \tag{5.118}$$

where $\mathcal{E}_i$ is the intrinsic Fermi level, equal to the Fermi level in the intrinsic semiconductor. By using Eq. 5.118, Eq. 5.117 can be written as:

$$e_n = C_n n_i \exp\left(\frac{\mathcal{E}_T - \mathcal{E}_i}{k_B T}\right) = C_n N_c \exp\left(\frac{\mathcal{E}_T - \mathcal{E}_c}{k_B T}\right) = C_n n_1, \tag{5.119}$$

where:

$$n_1 = n_i \exp\left(\frac{\mathcal{E}_T - \mathcal{E}_i}{k_B T}\right) = N_c \exp\left(\frac{\mathcal{E}_T - \mathcal{E}_c}{k_B T}\right). \tag{5.120}$$

If the semiconductor is not in thermodynamic equilibrium, capture rates and emission rates are not equal. In this case, the net trap capture rate is given by the difference between the capture and emission rates:

$$R_n = R_{cn} - R_{en}. \tag{5.121}$$

By using Eqs. 5.112–5.113 and Eq. 5.119, R_n can be written as:

$$R_n = C_n N_T \{n[1 - f(\mathcal{E}_T)] - n_1 f(\mathcal{E}_T)\}, \tag{5.122}$$

where $f(\mathcal{E}_T)$ is a nonequilibrium distribution function, whose expression will be calculated below.

We can repeat the same discussion for the derivation of the net hole recombination rate R_p, given by the difference between the hole capture and emission rates. The hole capture rate must be proportional to the hole density and to the number of filled traps:

$$R_{cp} = C_p p N_T f(\mathcal{E}_T) = v_{th} \sigma_p p N_T f(\mathcal{E}_T), \tag{5.123}$$

where we have used the cross-section for hole capture, σ_p. The hole emission rate is related to the number of empty traps: An electron in the valence band is captured by an empty trap, thus leaving a hole in the valence band. Since there are plenty of electrons in the valence band, the hole emission is not limited by the electron concentration in the valence band. Therefore:

$$R_{ep} = e_p N_T [1 - f(\mathcal{E}_T)]. \tag{5.124}$$

By following the same procedure used in the case of electrons, the parameter e_p can be written as follows:

$$e_p = C_p p_1, \tag{5.125}$$

where p_1 is given by:

$$p_1 = n_i \exp\left(-\frac{\mathcal{E}_T - \mathcal{E}_i}{k_B T}\right) = N_v \exp\left(-\frac{\mathcal{E}_T - \mathcal{E}_v}{k_B T}\right). \tag{5.126}$$

Therefore, the net hole recombination rate R_p is:

$$R_p = R_{cp} - R_{ep} = C_p N_T \{ p f(\mathcal{E}_T) - p_1 [1 - f(\mathcal{E}_T)] \}. \tag{5.127}$$

In steady-state, the net electron capture rate must be equal to the net hole recombination rate, $R_n = R_p$. By equating Eqs. 5.122 and 5.127, the nonequilibrium distribution function $f(\mathcal{E}_T)$ can be obtained:

$$f(\mathcal{E}_T) = \frac{C_n n + C_p p_1}{C_n (n + n_1) + C_p (p + p_1)}. \tag{5.128}$$

We note that from Eqs. 5.120 and 5.126, the product $n_1 p_1$ is simply given by:

$$n_1 p_1 = n_i^2, \tag{5.129}$$

therefore, the recombination rates $R_n = R_p$ can be written as follows:

$$R_n = R_p = \frac{C_n C_p N_T (np - n_i^2)}{C_n (n + n_1) + C_p (p + p_1)}. \tag{5.130}$$

From the previous equation, it is evident that when the semiconductor is in thermodynamic equilibrium $np = n_i^2$ so that $R_n = R_p = 0$, as we have already pointed out. If there is an excess carrier density in the semiconductor, np turns out to be larger than n_i^2 and the system reacts trying to establish an equilibrium condition by exploiting recombination processes: Indeed, Eq. 5.130 gives a positive recombination rate. In the opposite case, when there is a net carrier depletion, $np < n_i^2$ so that $R_n = R_p < 0$: In this case, we have a net generation rate.

It is common practice to introduce the electron and hole lifetimes, τ_n and τ_p, respectively, defined by the following expressions:

$$\tau_n = \frac{1}{C_n N_T}, \qquad \tau_p = \frac{1}{C_p N_T}. \tag{5.131}$$

In terms of τ_n and τ_p, the recombination rates can be written as:

$$R_n = R_p = \frac{np - n_i^2}{\tau_p (n + n_1) + \tau_n (p + p_1)}. \tag{5.132}$$

Now, the nonequilibrium electron and hole concentrations are given by:

$$n = n_0 + \Delta n, \qquad p = p_0 + \Delta p, \tag{5.133}$$

so that the recombination rate can be rewritten as:

$$R_n = R_p = \frac{(n_0 + \Delta n)(p_o + \Delta p) - n_i^2}{\tau_p (n_0 + \Delta n + n_1) + \tau_n (p_0 + \Delta p + p_1)}. \tag{5.134}$$

Let us now consider a n-type semiconductor, so that $n_0 \gg p_0$; moreover, we will assume that the excess carrier concentration is much smaller than the equilibrium electron concentration $\Delta n \ll n_0$. Moreover, if we further assume that the trap levels are close to the intrinsic Fermi level, $\mathcal{E}_T \approx \mathcal{E}_i$, we have:

$$n_1 \simeq n_i, \qquad p_1 \simeq n_i. \tag{5.135}$$

With these assumptions, Eq. 5.134 becomes:

$$R = \frac{n_0 \Delta p}{\tau_p(n_0 + \Delta n + n_i)}, \tag{5.136}$$

indeed, in the denominator of the ratio in Eq. 5.134, the second term is much smaller than the first one (all terms are much smaller than n_0) and it can be neglected. In Eq. 5.136, we can also neglect Δn and n_i with respect to n_0 so that:

$$R = \frac{\Delta p}{\tau_p}, \tag{5.137}$$

which gives the recombination rate in terms of the excess minority carrier concentration Δp and lifetime τ_p.

By using Eq. 5.130, it is simple to deduce the position of the trap levels within the band gap which maximizes the recombination rate R. The energy $\mathcal{E}_T$ can be calculated by equating to zero the derivative of R with respect to $\mathcal{E}_T$:

$$\frac{\partial R}{\partial \mathcal{E}_T} = 0, \tag{5.138}$$

so that

$$C_n \frac{dn_1}{d\mathcal{E}_T} + C_p \frac{dp_1}{d\mathcal{E}_T} = 0. \tag{5.139}$$

From Eqs. 5.120 and 5.126, we obtain:

$$\mathcal{E}_T = \frac{\mathcal{E}_c + \mathcal{E}_v}{2} + \frac{k_B T}{2} \log \frac{C_p N_v}{C_n N_c}, \tag{5.140}$$

so that we can conclude that a recombination center is most effective when it is located close to the middle of the band gap.

5.6.2 Auger Recombination

Another important nonradiative recombination mechanism is related to the Auger process, in which the energy released by an electron–hole recombination event is not released in the form of photons, but it is transferred to another electron or hole. The energy is eventually transferred in a nonradiative way from this third particle to the semiconductor lattice by phonon emission. In the direct band-to-band Auger process, three particles are involved; thus, it is a relevant recombination mechanism at high carrier concentration. This is generally the case of small band gap semiconductors or in systems which are far from thermodynamic equilibrium, for example in semiconductor lasers. Without entering into too many details, there are various possibilities, schematically shown in Fig. 5.7.

(i) Electron capture. An electron from the conduction band recombines with a hole in the valence band, and the excess energy is transferred to an electron in the conduction band.

(ii) Hole capture. Also in this second process, an electron from the conduction band recombines with a hole in the valence band, but the excess energy is now transferred to an additional hole in the valence band.

(iii) Electron emission. A high-energy electron in the conduction band transfers its energy to an electron in the valence band, which is thus excited in the conduction band leading to the generation of an electron–hole pair.

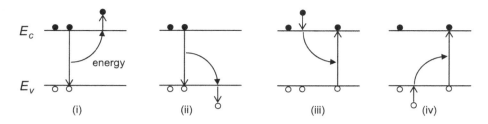

Figure 5.7   Various band-to-band Auger processes in a semiconductor (see text).

(iv) Hole emission. A high-energy hole in the valence band transfers its energy to an electron in the valence band, which is excited into the conduction band thus generating an electron–hole pair.

A simple theory describing the Auger process can be introduced by using arguments similar to those introduced in the previous section in the case of the SRH model. We can first associate to the four processes described above a suitable rate. In particular, in the case of electron capture, two electrons in the conduction band and one hole in the valence band are required, so that an electron capture rate R_{ec} can be written as:

$$R_{ec} = C_{ec}n^2p, \tag{5.141}$$

where C_{ec} is a suitable proportionality constant. In a similar way, in the case of hole capture, two holes in the valence band and one electron in the conduction band are required, so that a hole capture rate R_{hc} can be written:

$$R_{hc} = C_{hc}np^2. \tag{5.142}$$

In the case of electron emission, only one electron in the conduction band is required so that:

$$R_{ee} = C_{ee}n, \tag{5.143}$$

in the same way, for hole emission, we can write:

$$R_{he} = C_{he}p. \tag{5.144}$$

In overall thermodynamic equilibrium, the electron capture rate must be equal to the electron emission rate and the same for hole capture and emission rates, so that:

$$R_{ec}^0 = R_{ee}^0, \qquad R_{hc}^0 = R_{he}^0. \tag{5.145}$$

Therefore, in thermodynamic equilibrium, we have:

$$C_{ec}n_i^2 = C_{ee}, \qquad C_{hc}n_i^2 = C_{he}. \tag{5.146}$$

If the semiconductor is not in thermodynamic equilibrium, the net Auger recombination rate is given by:

$$R_{Aug} = (R_{ec} - R_{ee}) + (R_{hc} - R_{he}) = (C_{ec}n + C_{hc}p)(np - n_i^2), \tag{5.147}$$

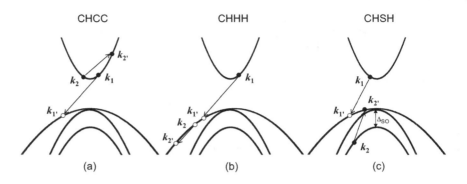

Figure 5.8 Three possible band-to-band direct Auger recombination processes: (a) CHCC, (b) CHHH, and (c) CHSH.

where Eqs. 5.146 were used. As a particular example, in the case of GaAs, $C_{ec} = 5 \times 10^{-30}$ cm^6/s and $C_{hc} = 3 \times 10^{-30}$ cm^6/s.

Figure 5.8 schematically shows three possible Auger recombination processes and illustrates the energy and momentum conservation rules. In Fig. 5.8(a), an electron with wavevector $\mathbf{k}_1$ recombines with a hole with wavevector $\mathbf{k}_{1'}$ and the released energy is transferred to a second electron with initial wavevector $\mathbf{k}_2$, which is excited to a state with wavevector $\mathbf{k}_{2'}$. This process is called Conduction-Heavy hole-Conduction-Conduction (CHCC). Figure 5.8(b) shows a process where an electron with wavevector $\mathbf{k}_1$ recombines with a hole with wavevector $\mathbf{k}_{1'}$ and the released energy is transferred to a second hole in the valence band with an initial wavevector $\mathbf{k}_2$ an final wavevector $\mathbf{k}_{2'}$. This process is called CHHH. Other Auger recombination processes are possible, involving also the light-hole valence band (CHLH processes) and the split-off band (CHSH processes). In Fig. 5.8(c), an electron with wavevector $\mathbf{k}_1$ recombines with a hole with wavevector $\mathbf{k}_{1'}$ and the released energy is transferred to an electron in the split-off band with wavevector $\mathbf{k}_2$, which is excited to a state in the heavy-hole valence band with wavevector $\mathbf{k}_{2'}$. In this case, the energy gap of the semiconductor is larger than the split-off energy: $\mathcal{E}_g > \Delta_{SO}$.

Figure 5.8 refers to the simplest Auger process: the phononless band-to-band Auger process in a direct semiconductor. In this case, the energy and momentum conservation must be satisfied by the electrons and holes involved in the process. For this reason, it is not possible to have a recombination between an electron at the bottom of the conduction band and a hole at the top of the valence band ($\Delta k = 0$, $\Delta \mathcal{E} = \mathcal{E}_g$) since the excited third particle cannot make a vertical transition. Therefore, there must be an activation energy, which is related to the size of the energy gap. In the case of direct band gap semiconductors with large gap, this activation energy is large, thus leading to a decrease of the process probability. The activation energy is equal to the energy required to place the participating carriers in the appropriate threshold states for an Auger transition to occur.

The dependence of the band-to-band Auger process on the band structure, temperature, and carrier density can be understood by using a relatively simple model. Let us assume that the dispersion relations $\mathcal{E}(\mathbf{k})$ can be well approximated by parabolic expressions both in the conduction and in the valence bands and that the semiconductor is nondegenerate. The Fermi–Dirac distribution function can be approximated by the Boltzmann distribution function, so that the probability of occupation of a state in the conduction band can be written as:

$$f_c(\mathbf{k}) = \exp\left(-\frac{\mathcal{E}_c(\mathbf{k}) - \mathcal{E}_{Fc}}{k_B T}\right), \tag{5.148}$$

while the probability that a valence state is not occupied can be written as:

$$\bar{f}_v(\mathbf{k}) = \exp\left(-\frac{\mathcal{E}_{Fv} - \mathcal{E}_v(\mathbf{k})}{k_B T}\right), \tag{5.149}$$

where $\mathcal{E}_{Fc}$ and $\mathcal{E}_{Fv}$ are the quasi-Fermi energies in the conduction and valence bands, respectively. $\mathcal{E}_c(\mathbf{k})$ and $\mathcal{E}_v(\mathbf{k})$ are given by:

$$\mathcal{E}_c(k) = \mathcal{E}_c + \frac{\hbar^2 k^2}{2m_c}, \qquad \mathcal{E}_v(k) = \mathcal{E}_v - \frac{\hbar^2 k^2}{2m_v}, \tag{5.150}$$

where $\mathcal{E}_c$ corresponds to the bottom of the conduction band and $\mathcal{E}_v$ to the top of the valence band. It is evident that the probability of a CHCC process is related to the probability that the states in the conduction band corresponding to the wavevectors $\mathbf{k}_1$ and $\mathbf{k}_2$ are each occupied by an electron and to the probability that the state with wavevector $\mathbf{k}_{1'}$ in the valence band is empty; the state in the conduction band with wavevector $\mathbf{k}_{2'}$ can be assumed empty. Therefore, we can write a probability P for the CHCC process as:

$$P(\mathbf{k}_1, \mathbf{k}_2, \mathbf{k}_{1'}) = f_c(\mathbf{k}_1) f_c(\mathbf{k}_2) \bar{f}_v(\mathbf{k}_{1'}) = \tag{5.151}$$

$$= \exp\left[-\frac{\mathcal{E}_c(k_1) - \mathcal{E}_{Fc} + \mathcal{E}_c(k_2) - \mathcal{E}_{Fc} + \mathcal{E}_{Fv} - \mathcal{E}_v(k_{1'})}{k_B T}\right].$$

By using Eq. 5.150, we can write Eq. 5.151 as follows:

$$P(\mathbf{k}_1, \mathbf{k}_2, \mathbf{k}_{1'}) = \exp\left(-2\frac{\mathcal{E}_c - \mathcal{E}_{Fc}}{k_B T}\right) \exp\left(-\frac{\mathcal{E}_{Fv} - \mathcal{E}_v}{k_B T}\right)$$

$$\cdot \exp\left[-\frac{\hbar^2 [k_1^2 + k_2^2 + (m_c/m_v)k_{1'}^2]}{2m_c k_B T}\right]. \tag{5.152}$$

In nondegenerate semiconductors, the electrons and hole concentrations can be calculated as:

$$n = N_c \exp\left(-\frac{\mathcal{E}_c - \mathcal{E}_{Fc}}{k_B T}\right)$$

$$p = N_v \exp\left(-\frac{\mathcal{E}_{Fv} - \mathcal{E}_v}{k_B T}\right), \tag{5.153}$$

so that the probability $P(\mathbf{k}_1, \mathbf{k}_2, \mathbf{k}_{1'})$ can be written as:

$$P(\mathbf{k}_1, \mathbf{k}_2, \mathbf{k}_{1'}) = \frac{n^2 p}{N_c^2 N_v} \exp\left[-\frac{\hbar^2 [k_1^2 + k_2^2 + (m_c/m_v)k_{1'}^2]}{2m_c k_B T}\right]. \tag{5.154}$$

The momentum conservation requires that

$$\mathbf{k}_1 + \mathbf{k}_2 = \mathbf{k}_{1'} + \mathbf{k}_{2'}, \tag{5.155}$$

and the energy conservation gives:

$$\mathcal{E}_c(k_1) + \mathcal{E}_c(k_2) = \mathcal{E}_v(k_{1'}) + \mathcal{E}_c(k_{2'}), \tag{5.156}$$

which can be also written as:

$$\frac{\hbar^2}{2m_c}\left[k_1^2 + k_2^2 + \left(\frac{m_c}{m_v}\right)k_{1'}^2 - k_{2'}^2\right] + \mathcal{E}_g = 0, \tag{5.157}$$

where $\mathcal{E}_g = \mathcal{E}_c - \mathcal{E}_v$ is the energy gap.

The most probable process is therefore obtained by maximizing $P(\mathbf{k}_1, \mathbf{k}_2, \mathbf{k}_{1'})$ with the requirement that momentum and energy are conserved. This can be obtained by using the method of Lagrange multipliers (see the following gray box). The values of the wavevectors $\mathbf{k}_1, \mathbf{k}_2$, and $\mathbf{k}_{1'}$ which correspond to the most probable process are given by:

$$k_2 = k_1, \qquad k_{1'} = -\frac{m_v}{m_c}k_1, \tag{5.158}$$

so that $k_{2'}$ is given by:

$$k_{2'} = k_1 + k_2 - k_{1'} = \left(2 + \frac{m_v}{m_c}\right)k_1. \tag{5.159}$$

On the other hand, the value k_1 can be obtained from the energy conservation (see Eq. 5.157):

$$\frac{\hbar^2 k_1^2}{2m_c} = \frac{m_c^2}{(m_c + m_v)(2m_c + m_v)}\mathcal{E}_g, \tag{5.160}$$

so that the maximum probability is given by the following formula:

$$P_{max} = \frac{n^2 p}{N_c^2 N_v}\exp\left(-\frac{m_c}{m_c + m_v}\frac{\mathcal{E}_g}{k_B T}\right). \tag{5.161}$$

It is instructive to rewrite Eq. 5.161 by using Eq. 5.153:

$$\begin{aligned}
P_{max} &= \frac{n}{N_c}\frac{np}{N_c N_v}\exp\left(-\frac{m_c}{m_c + m_v}\frac{\mathcal{E}_g}{k_B T}\right) = \\
&= \frac{n}{N_c}\exp\left(\frac{\mathcal{E}_{Fc} - \mathcal{E}_{Fv}}{k_B T}\right)\exp\left(-\frac{\mathcal{E}_g}{k_B T}\right)\exp\left(-\frac{m_c}{m_c + m_v}\frac{\mathcal{E}_g}{k_B T}\right) = \\
&= \frac{n}{N_c}\exp\left(\frac{\mathcal{E}_{Fc} - \mathcal{E}_{Fv}}{k_B T}\right)\exp\left(-\frac{2m_c + m_v}{m_c + m_v}\frac{\mathcal{E}_g}{k_B T}\right).
\end{aligned} \tag{5.162}$$

If the semiconductor is in thermodynamic equilibrium ($\mathcal{E}_{Fc} - \mathcal{E}_{Fv} = 0$) and $n = n_0$, we have:

$$P_{max} = \frac{n_0}{N_c}\exp\left(-\frac{2m_c + m_v}{m_c + m_v}\frac{\mathcal{E}_g}{k_B T}\right). \tag{5.163}$$

It is evident that a CHHH process can be analyzed using the same procedure. In this case, the maximum probability can be written by using Eq. 5.163 replacing n_0 with p_0, m_c with m_v and N_c with N_v, so that:

$$P_{max} = \frac{p_0}{N_v}\exp\left(-\frac{m_c + 2m_v}{m_c + m_v}\frac{\mathcal{E}_g}{k_B T}\right). \tag{5.164}$$

As an additional example, we can consider the CHSH process, which leads to the generation of a hot hole in the split-off band, schematically shown in Fig. 5.8(c). In this case, the
exponential factor in the expression of the process probability is given by:

$$P_{CHSH} \propto \exp\left(-\frac{m_{SO}}{m_c + 2m_v - m_{SO}} \frac{\mathcal{E}_g - \Delta_{SO}}{k_B T}\right), \tag{5.165}$$

where m_{SO} is the split-off effective mass and Δ_{SO} is the split-off energy.

It is therefore evident that the probability of Auger processes increases upon decreasing
the energy gap. Moreover, due to the presence of the exponential factor in the Auger
probability, the Auger process turns out to be very sensitive to the band structure (which
directly determine the values of the effective masses) and to the temperature.

5.6.3   Calculation of the Values of $k_1, k_2,$ and $k_{1'}$ Which Maximize the Probability of the CHCC Auger Process

We have to find the maximum value of the function $P(k_1, k_2, k_{1'})$, given by Eq. 5.154,
subject to the constraints imposed by the momentum and energy conservation expressed
by Eqs. 5.155 and 5.157. We can use the method of Lagrange undetermined multipliers.
To maximize $P(k_1, k_2, k_{1'})$, we require:

$$dP = \frac{\partial P}{\partial k_1} dk_1 + \frac{\partial P}{\partial k_2} dk_2 + \frac{\partial P}{\partial k_{1'}} dk_{1'} = 0. \tag{5.166}$$

If dk_1, dk_2, and $dk_{1'}$ were independent, we could conclude that:

$$\frac{\partial P}{\partial k_1} = \frac{\partial P}{\partial k_2} = \frac{\partial P}{\partial k_{1'}} = 0,$$

however, they are not independent but constrained because of the momentum and
energy conservation, so that by differentiating Eqs. 5.155 and 5.157, we obtain:

$$dk_1 + dk_2 - dk_{1'} = 0 \tag{5.167}$$

$$\frac{\hbar^2}{m_c} k_1 dk_1 + \frac{\hbar^2}{m_c} k_2 dk_2 + \frac{\hbar^2}{m_c} \frac{m_c}{m_v} k_{1'} dk_{1'} = 0. \tag{5.168}$$

On the other hand, equating dP to zero one obtains:

$$\frac{\hbar^2 k_1}{m_c k_B T} P(k_1, k_2, k_{1'}) dk_1 + \frac{\hbar^2 k_2}{m_c k_B T} P(k_1, k_2, k_{1'}) dk_2 +$$

$$+ \frac{m_c}{m_v} \frac{k_{1'}}{m_c k_B T} P(k_1, k_2, k_{1'}) dk_{1'} = 0. \tag{5.169}$$

Multiplying Eq. 5.167 by an as yet unknown number λ and Eq. 5.168 by μ (λ and μ
are the so called Lagrange undetermined multipliers) and adding them to Eq. 5.169 we
obtain an equation where dk_1, dk_2, and $dk_{1'}$ are independent and arbitrary, so that we
must choose λ and μ such that:

$$-\frac{\hbar^2 k_1}{m_c k_B T} P + \lambda + \mu \frac{\hbar^2}{m_c} k_1 = 0 \tag{5.170}$$

$$-\frac{\hbar^2 k_2}{m_c k_B T}P + \lambda + \mu \frac{\hbar^2}{m_c}k_2 = 0 \tag{5.171}$$

$$-\frac{m_c}{m_v}\frac{\hbar^2 k_{1'}}{m_c k_B T}P - \lambda + \mu \frac{\hbar^2}{m_c}\frac{m_c}{m_v}k_{1'} = 0, \tag{5.172}$$

which can be rewritten as follows:

$$-\frac{\hbar^2 k_1}{m_c k_B T}(P - \mu k_B T) + \lambda = 0 \tag{5.173}$$

$$-\frac{\hbar^2 k_2}{m_c k_B T}(P - \mu k_B T) + \lambda = 0 \tag{5.174}$$

$$-\frac{m_c}{m_v}\frac{\hbar^2 k_{1'}}{m_c k_B T}(P - \mu k_B T) - \lambda = 0. \tag{5.175}$$

These equations, together with the constraints given by momentum and energy conservation, are sufficient to determine the five unknown, λ, μ and the values of k_1, k_2, and $k_{1'}$ at the stationary point. Indeed, by subtracting Eq. 5.174 from Eq. 5.173, we have:

$$-\frac{\hbar^2}{m_c k_B T}(P - \mu k_B T)(k_1 - k_2) = 0, \tag{5.176}$$

from which we obtain:

$$k_1 = k_2. \tag{5.177}$$

If we now sum Eq. 5.173 and Eq. 5.175, we obtain:

$$-\frac{\hbar^2}{m_c k_B T}(P - \mu k_B T)\left(k_1 + \frac{m_c}{m_v}k_{1'}\right) = 0, \tag{5.178}$$

which gives

$$k_{1'} = -\frac{m_v}{m_c}k_1. \tag{5.179}$$

5.6.4 Surface Recombination

Recombination at surfaces and interfaces in semiconductor structures can be quite important and can limit the performances of the semiconductor devices. As a consequence of the sudden termination of the semiconductor lattice structure at the surface, dangling bonds are generated at the surface and defects are very likely to form at surfaces. For example, oxygen is always found on a semiconductor surface and can produce surface states. The surface states that are neutral when occupied by electrons and positively charged when they are not occupied are called donor states, while states that are negatively charged when occupied and neutral when empty are called acceptor states. On the other hand, the interface between semiconductor and external environment cannot be abrupt on the atomic scale: It consists of a layer of intermediate materials and impurities with a thickness of several atomic layers. In this region, there are surface states which are very close to the bulk semiconductor and which remain in thermal equilibrium with the bulk also when the potential

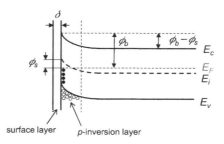

surface layer   p-inversion layer

Figure 5.9   Schematic band diagram for a semiconductor showing a thin surface layer containing acceptor-type surface states.

is changed. Such states are called *fast surface states* since the electrons which occupy these states reach the equilibrium configuration rather quickly (on the order of microseconds or less). In addition, there are surface states which are in the intermediate layer, relatively far from the bulk semiconductor, so that they can reach thermal equilibrium with bulk states in relatively longer times (which can be of the order of seconds or more). For this reason, they are called *slow surface states*.

The band structure of the semiconductor surface is strongly modified by the presence of surface states, as schematically shown in Fig. 5.9, where, to account for surface effects, a surface layer with a thickness δ has been considered between the bulk semiconductor and the external environment (e.g., vacuum). The figure refers to the case of an n-type semiconductor with a thin surface layer containing acceptor-type surface states: The semiconductor is depleted of electrons near the surface. It is also possible that the hole density at the surface, p_s, may become much larger than the electron density at the surface, n_s ($p_s \gg n_s$): In this case, the surface becomes of p-type (this situation is shown in Fig. 5.9).

Surface states may become effective recombination centers. Surface recombination can be analyzed by using the same procedure employed in the case of bulk nonradiative recombination. In particular, within the SRH model, the surface recombination rate can be written by using Eq. 5.130, where n is repaced by the surface electron density n_s, p by the surface hole density p_s, and N_T by the surface state density per unit area, N_{Ts}:

$$R_s = \frac{C_n C_p N_{Ts}(n_s p_s - n_i^2)}{C_n(n_s + n_1) + C_p(p_s + p_1)}, \tag{5.180}$$

where n_1 and p_1 are given by Eqs. 5.120 and 5.126, respectively ($\mathcal{E}_T$ in this case is the energy level of the fast surface states), and the electron and hole density at surface are related to the electron and hole density by:

$$n_s = n \exp\left[-\frac{q(\phi_b - \phi_s)}{k_B T}\right] \tag{5.181}$$

$$p_s = p \exp\left[\frac{q(\phi_b - \phi_s)}{k_B T}\right], \tag{5.182}$$

where ϕ_b and ϕ_s are reported in Fig. 5.9. Moreover:

$$n_s p_s = np = (n_0 + \Delta n)(p_0 + \Delta p) \tag{5.183}$$

where Δn and Δp are the excess electron and hole densities, respectively, so that:

$$R_s = \frac{C_n C_p N_{Ts}[(n_0 + \Delta n)(p_0 + \Delta p) - n_i^2]}{C_n(n_s + n_1) + C_p(p_s + p_1)}.$$
(5.184)

In the case of small excess carrier density ($\Delta n = \Delta p \ll n_0$), the term in the square brackets at the numerator of Eq. 5.184 can be approximated as follows ($n_0 p_0 = n_i^2$):

$$(n_0 + \Delta n)(p_0 + \Delta p) - n_i^2 = \Delta n(n_0 + p_0).$$
(5.185)

In the denominator of Eq. 5.184, the surface carrier densities n_s and p_s can be approximated by the surface carrier densities at equilibrium, n_{s0} and p_{s0}, given by:

$$n_{s0} = n_i \exp\left(\frac{\mathcal{E}_F - \mathcal{E}_{is}}{k_B T}\right) = n_i \exp\left(\frac{q\phi_s}{k_B T}\right)$$
(5.186)

$$p_{s0} = n_i \exp\left(-\frac{\mathcal{E}_F - \mathcal{E}_{is}}{k_B T}\right) = n_i \exp\left(-\frac{q\phi_s}{k_B T}\right),$$
(5.187)

where $\mathcal{E}_{is}$ is the intrinsic Fermi level at the semiconductor surface and ϕ_s is the surface potential (see Fig. 5.9). With these approximations, the surface recombination velocity, defined as:

$$v_s = \frac{R_s}{\Delta n} = \frac{R_s}{\Delta p}$$
(5.188)

can be written as follows:

$$v_s = \frac{N_{Ts}\sqrt{C_n C_p}(n_0 + p_0)/2n_i}{\cosh\left[(\mathcal{E}_T - \mathcal{E}_i - q\phi_0)/k_B T\right] + \cosh\left[q(\phi_s - \phi_0)/k_B T\right]},$$
(5.189)

where:

$$q\phi_0 = \frac{k_B T}{2} \log \frac{C_p}{C_n}.$$
(5.190)

Therefore, the surface recombination velocity depends on the surface potential ϕ_s, which is very sensitive to the environment conditions at the surface: This is why in semiconductor devices, it is important to maintain stable surface conditions. Moreover, v_s is directly related to the surface state density.

5.7 Competition between Radiative and Nonradiative Recombination

In a real semiconductor, it is impossible to avoid completely nonradiative bulk and surface recombination and Auger recombination, which therefore will compete with the radiative recombination processes. The total recombination probability is given by the sum of the radiative and nonradiative recombination probabilities:

$$\frac{1}{\tau} = \frac{1}{\tau_{rad}} + \frac{1}{\tau_{nr}},$$
(5.191)

so that the relative probability of radiative recombination, which is also called *radiative efficiency*, is given by:

$$\eta_r = \frac{\tau}{\tau_{rad}} = \frac{\tau_{nr}}{\tau_{rad} + \tau_{nr}},$$ (5.192)

the radiative efficiency is the ratio between the number of photons emitted inside a semiconductor device and the total number of charge quanta, which undergo recombination.

5.8 Exercises

Exercise 5.1   InP is a direct band gap semiconductor with a band gap of 1.35 eV at room temperature. A thin wafer of InP with a thickness of 1 μm is antireflection coated on both surfaces. Assuming that the absorption coefficient is 3.5×10^6 m^{-1} at 775 nm, estimate the transmission of the plate at 620 nm.

Exercise 5.2   GaAs is characterized by the following band structure parameters: $\mathcal{E}_g = 1.519$ eV at $T = 0$ K, $\mathcal{E}_g = 1.424$ eV at $T = 300$ K, $m_{hh}^* = 0.5\ m_0$, $m_{lh}^* = 0.08\ m_0$, $m_{so}^* = 0.15\ m_0$, $m_c^* = 0.067\ m_0$.
a) Calculate the modulus of the **k** vector of an electron excited at $T = 300$ K from the heavy-hole band to the conduction band by photons with energy 1.6 eV. Calculate the same quantity in the case of a transition from the light-hole band.
b) Calculate the ratio of the joint density of states for the heavy- and light-hole transitions.
c) Determine the wavelength at which transitions from the split-off band become possible.

Exercise 5.3   Repeat the same calculations of Exercise 5.2 in the case of InP and InAs, whose band structure parameters are listed in Table 5.3.

Exercise 5.4   Determine the density of states of a free electron in a bulk material. Calculate the density of states for free electrons with energy 0.1 eV.

Exercise 5.5   Estimate the thermal velocity of electrons and holes in silicon at room temperature.

Exercise 5.6   In a semiconductor out of equilibrium, quasi-Fermi levels can be used. Since electrons have typically smaller effective masses than holes, $\mathcal{E}_{Fc} - \mathcal{E}_c$ is usually larger than $\mathcal{E}_v - \mathcal{E}_{Fv}$, as shown in Fig. 5.10. Demonstrate that in conditions of charge neutrality, when an electron at energy $\mathcal{E}_{Fc}$ recombines with a hole and emits a photon, the hole energy is $\mathcal{E}_{Fv}$.

Exercise 5.7   The absorption coefficient of bulk GaAs at room temperature is given by:

$$\alpha(\hbar\omega) = \alpha_0(\hbar\omega)[f_v(\hbar\omega) - f_c(\hbar\omega)],$$

where:

$$\alpha_0(\hbar\omega) = 1.12 \times 10^4 \sqrt{\frac{\hbar\omega - \mathcal{E}_g}{e}}\ [\text{cm}^{-1}].$$

Table 5.3   Band structure parameters for InP and InAs. The effective masses are normalized to the free electron mass m_0.

	$\mathcal{E}_g$ (eV) (T = 0 K)	$\mathcal{E}_g$ (eV) (T = 300 K)	Δ_{so} (eV)	m_c^*	m_{hh}^*	m_{lh}^*	m_{so}^*
InP	1,42	1.34	0.11	0.077	0.6	0.12	0.12
InAs	0.42	0.35	0.38	0.022	0.4	0.026	0.14

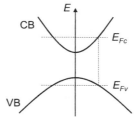

Figure 5.10   Quasi-Fermi levels in a semiconductor

Plot the gain coefficient at $T = 10$ K and 300 K for injected carrier density of $N = 0.2\, N_{tr}$, $N = N_{tr}$, and $N = 3\, N_{tr}$, where N_{tr} is the transparency carrier density.

Exercise 5.8   In bulk GaAs, calculate the gain at a photon energy exceeding the band gap energy by $0.45 k_B T$ for a carrier injection $N = 1.6 \times 10^{18}$ cm^{-3}.

Exercise 5.9   Calculate and plot the quasi-Fermi levels of the conduction band, $\mathcal{E}_{Fc}/k_B T$, as a function of the concentration of injected carriers, N, normalized to the effective electron density, N_c.

Exercise 5.10   Calculate and plot the quasi-Fermi level difference, $(\mathcal{E}_{Fc} - \mathcal{E}_{Fv}) - \mathcal{E}_g$, as a function of the injected carrier density in bulk GaAs.

Exercise 5.11   A monochromatic laser beam at $\lambda = 700$ nm impinges onto a GaAs target, with an intensity $I = 2$ kW cm^{-2}. Assuming that the electron–hole recombination time is 100 ns and considering the absorption coefficient shown in Figure 5.2, calculate:
a) the excess electron and hole densities produced by light;
b) the position of the quasi-Fermi levels for electrons and holes in the presence of light;
c) the conductivity of GaAs with and without light, assuming that the electron and hole mobilities are $\mu_n = 8500$ cm^{2}V^{-1}s^{-1} and $\mu_p = 400$ cm^{2}V^{-1}s^{-1}.

Exercise 5.12   Consider p-type silicon at room temperature, doped with $N_A = 10^{17}$ cm^{-3}. The electron mobility is $\mu_n = 300$ cm^{2}V^{-1}s^{-1} and $\tau_n = 1$ μs. By using the minority carrier diffusion equation for electrons:

$$\frac{\partial \Delta n}{\partial t} = D_n \frac{d^2 \Delta n}{dx^2} - \frac{\Delta n}{\tau_n} + G_L,$$

determine the steady-state excess minority carrier concentration, and the position of the quasi-Fermi levels in the following cases:

a) Uniformly illuminated sample, with an optical generation rate $G_L = 10^{20}$ cm^{-3}s^{-1}.

b) After illumination with $G_L = 10^{20}$ cm^{-3}s^{-1}, at time $t = 0$ the light is switched off.

c) Uniformly illuminated sample with $G_L = 10^{20}$ cm^{-3}s^{-1}. The minority carriers lifetime is $\tau_n = 1$ μs, except for a 10-nm-thick layer near $x = 0$, where $\tau_n = 0.1$ μs.

Hint: consider the initial thin layer as a boundary condition.

Exercise 5.13   Consider InP at $T = 300$ K: $\mathcal{E}_g = 1.344$ eV, $m_c = 0.077$ m_0, $m_{hh} = 0.6$ m_0, refractive index $n = 3.5$, Kane energy $\mathcal{E}_P = 20.7$ eV. Assuming that only the heavy-hole band is populated in the valence band, estimate the transparency carrier density.

Exercise 5.14   Consider InP at $T = 300$ K, as in Exercise 5.13. If $\mathcal{E}_{Fc} - \mathcal{E}_{Fv} = 1.5$ eV, plot the function $f_v(\hbar\omega) - f_c(\hbar\omega)$ at $T = 0$ K and $T = 300$ K for photon energies between 1 and 1.7 eV.

Exercise 5.15   As in Exercise 5.14, plot the gain coefficient $g(\hbar\omega)$ as a function of energy.

Exercise 5.16   Show that the general solution of the equation:

$$\frac{d\Delta n}{dt} = -B(n_0 \Delta p + p_0 \Delta n + \Delta n \Delta p)$$

is

$$\Delta n = \frac{(n_0 + p_0)\Delta n(0)}{[n_0 + p_0 + \Delta n(0)] \exp[B(n_0 + p_0)t - \Delta n(0)]}.$$

Exercise 5.17   Consider GaAs bulk and assume a carrier density $N = 4 \times 10^{18}$ cm^{-3}. Calculate and plot the gain (in cm^{-1}) at four different temperatures, $T = 250$ K, 300 K, 350 K, and 400 K, including the heavy-hole to conduction band transitions and the light-hole to conduction band transitions, showing that the peak gain decreases upon increasing the temperature. Motivate physically this behavior.

Exercise 5.18   By using the Joyce-Dixon approximation, the Fermi energy for the conduction band can be calculated as:

$$\mathcal{E}_F = \mathcal{E}_c + k_B T \left[\ln \frac{n}{N_c} + \frac{1}{\sqrt{8}} \frac{n}{N_c} \right] = \mathcal{E}_v - k_B T \left[\ln \frac{p}{N_v} + \frac{1}{\sqrt{8}} \frac{p}{N_v} \right],$$

where N_c and N_v are the effective density of states in the conduction and valence band, respectively. Calculate the transparency density in GaAs at 77 K and at 300 K.

Exercise 5.19   Show that if a semiconductor is heavily doped and $n \gg p$, the total radiative recombination rate can be well approximated by the following expression:

$$R_{sp,T} \approx \frac{1}{\tau_r} \left(\frac{m_r}{m_v} \right)^{3/2} p.$$

Exercise 5.20   Show that in the regime of strong injection, that is, when a high density of both electrons and holes is injected, the total radiative recombination rate can be well approximated by the following expression:

$$R_{sp,T} \approx \frac{n}{\tau_r} = \frac{p}{\tau_r}.$$

Quantum Wells

6.1 Introduction

So far we have always considered bulk semiconductors, that is semiconductors with spatial dimensions much larger than the de Broglie wavelength of the electron involved in the relevant physical processes. In a quantum well (QW), a very thin layer of a semiconductor, with energy gap $\mathcal{E}_{g1}$, is grown between two semiconductors with energy gap $\mathcal{E}_{g2} > \mathcal{E}_{g1}$. If the thickness, d, of the layer with smaller band gap is of the order of a few nanometers (for example, $d < 20$ nm), the confined electrons and holes show relevant quantum effects. The band structure of the material is greatly modified, thus leading to significant changes of all physical properties with respect to the bulk case, with very important advantages for various applications.

Another important difference with respect to the bulk case is the following. In the case of bulk semiconductors, it is possible to have an epitaxial growth of one semiconductor onto another one only if the lattice spacings are almost the same for the two semiconductors, with maximum relative differences of the order of 0.1% of the lattice constant (e.g., the relative lattice period variation, $\Delta a/a$, is $< 0.12\%$ in the case of GaAs/AlGaAs). In this way, it is possible to fabricate semiconductor heterostructures (i.e., structures composed by semiconductors with different energy gaps). A GaAs/AlGaAs heterostructure can be easily obtained since the lattice period of GaAs (564 pm) is almost equal to the lattice period of AlAs (566 pm), as shown in Fig. 6.1, which displays the low-temperature energy band gap of a few semiconductors as a function of their lattice constants. The gray vertical regions mark the groups of semiconductors with similar lattice constants. Semiconductor contained in the same gray region can, in principal, be combined to form heterostructures with a given energy gap offset. Upon using ternary or quaternary alloys, it is possible to increase the possible choice of band gap offsets, thus offering the possibility to tailor the operating wavelength of the device (e.g., the emission wavelength of the laser based on the use of a given heterostructure). The solid lines in Fig. 6.1, which join together some of the semiconductors, indicate that these semiconductors form stable alloys over the entire alloy range (e.g., InGaAs, Al GaAs, and InGaP).

In the case of the quaternary alloy $In_{1-x}Ga_xAs_yP_{1-y}$, lattice matching with InP can be achieved only for a specific ratio y/x. Nearly perfect lattice matching is required to minimize defects at the interface between the two semiconductors. A lattice mismatch leads to mechanical stress of the semiconductor, which may result in the generation of dislocation defects at the interface, with detrimental consequences on the optical and electronic properties of the material. The situation is different in the case of QWs. Due to the very small

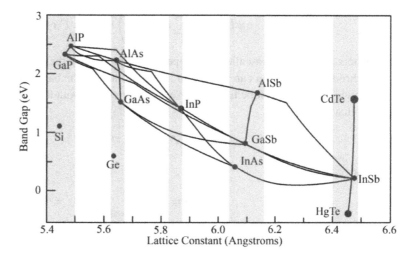

Figure 6.1 Plot of the low temperature energy band gaps of a number of semiconductors versus their lattice constants. The shaded gray regions show several groups of semiconductors with similar lattice constants. Semiconductors joined by solid lines form stable alloys. Adapted from [6].

thickness of the QW layer, the lattice constants of the two semiconductors can be significantly different. As a result, strain develops within the QW layer. As it will be discussed in this chapter, the strain changes the properties of the QW semiconductor, in particular it changes the effective masses and the band edges, with potential remarkable advantages for the applications. We will see that the quantum confinement of electrons and holes in QW semiconductors and, for strained QWs, the variation of effective masses lead to remarkable differences of the optical properties of QWs with respect to bulk semiconductors.

6.2 Electronic States

We will consider a QW consisting of a direct band gap material, so that the bottom of the conduction band is of s-type, while the valence band states are of p-type. We will first consider the conduction band. Let us assume a potential well with thickness L_z (z is the direction of the semiconductor growth). The electronic wavefunction must be spatially localized within the well so that it can be written as a linear combination of Bloch wavefunctions:

$$\psi(\mathbf{r}) = \sum_{\mathbf{k}} F(\mathbf{k}) e^{i\mathbf{k}\cdot\mathbf{r}} u_{\mathbf{k}}(\mathbf{r}). \tag{6.1}$$

In the framework of the *envelope function approximation*, we can assume that the periodic function $u_{\mathbf{k}}(\mathbf{r})$ does not strongly depend on $\mathbf{k}$ around the band edge, so that it can be approximated by its value near the band edge ($\mathbf{k} = 0$), i.e., $u_{\mathbf{k}}(\mathbf{r}) \approx u_0(\mathbf{r})$. With this approximation, the wavefunction $\psi(\mathbf{r})$ can be written as:

$$\psi(\mathbf{r}) = \left(\sum_{\mathbf{k}} F(\mathbf{k}) e^{i\mathbf{k}\cdot\mathbf{r}} \right) u_0(\mathbf{r}) \equiv F(\mathbf{r}) u_0(\mathbf{r}), \tag{6.2}$$

where $F(\mathbf{r})$ is the so-called envelope function, so that the approximate solution of the Schrödinger equation can be written as the product of the band edge Bloch function and a slowly varying envelope function. We can assume that both these functions are normalized so that:

$$\langle F|F \rangle = \int_V F^*F \, d\tau = 1 \tag{6.3}$$

and

$$\langle u_0|u_0 \rangle = \frac{1}{V_u} \int_{\text{unit cell}} u_0^* u_0 \, d\tau = 1, \tag{6.4}$$

where V is the volume of the crystal and V_u is the volume of a unit cell of the crystal. Moreover, the functions $u_{0,c}(\mathbf{r})$ and $u_{0,v}(\mathbf{r})$, for the conduction and valence bands, respectively, are orthonormal:

$$\langle u_{0,v}|u_{0,c} \rangle = 0. \tag{6.5}$$

In the case of a QW extending in the z-direction, the wavefunction $\psi(\mathbf{r})$ can be written as:

$$\psi(\mathbf{r}) = F_c(z) \exp\left[i(k_x x + k_y y)\right] u_0(x, y) = F_c(z) e^{i\mathbf{K}\cdot\mathbf{r}} u_0(x, y), \tag{6.6}$$

where $\mathbf{K} = k_x \mathbf{u}_x + k_y \mathbf{u}_y$ and the envelope function $F_c(z)$ approximately satisfy the following one-dimensional Schrödinger equation:

$$\left[-\frac{\hbar^2}{2m_{c,j}} \frac{\partial^2}{\partial z^2} + V_{c,j}(z) + V_{Kj}(z) \right] F_c(z) = \mathcal{E}_z F_c(z), \tag{6.7}$$

where the subscript j indicates if we are considering the well region ($j = w$) or the barrier region ($j = b$); $V_{c,j}(z)$ is the confining potential and $V_{Kj}(z)$ is given by:

$$V_{Kj}(z) = \frac{\hbar^2}{2m_{c,j}} (k_x^2 + k_y^2) = \frac{\hbar^2 K^2}{2m_{c,j}}. \tag{6.8}$$

In the well $V_{c,w}(z) = 0$, outside the well $V_{c,b}(z) = V_0$, as shown in Fig. 6.2. We will first consider the case of infinite barriers (i.e., $V_0 \to \infty$). As a first step, we will assume that the

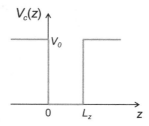

Figure 6.2   Schematic confining potential in a QW with thickness L_z.

electron has no momentum in the plane $x - y$, so that $V_{Kj} = 0$. In this case, the boundary conditions are the following:

$$F_c(0) = F_c(L_z) = 0, \tag{6.9}$$

and Eq. 6.7 reduces to:

$$\frac{d^2 F_c}{dz^2} = -\frac{2m_{c,w}}{\hbar^2} \mathcal{E}_z F_c = -k_z^2 F_c, \tag{6.10}$$

where:

$$k_z^2 = \frac{2m_{c,w}}{\hbar^2} \mathcal{E}_z. \tag{6.11}$$

The general solution is given by:

$$F_c(z) = A \sin{(k_z z)} + B \cos{(k_z z)}. \tag{6.12}$$

Upon taking into account the boundary conditions 6.9, we have:

$$B = 0, \qquad A \sin{(k_z L_z)} = 0, \tag{6.13}$$

so that:

$$k_z = n \frac{\pi}{L_z}, \tag{6.14}$$

where n is an integer. From Eq. 6.11, we obtain:

$$\mathcal{E}_z = n^2 \frac{\hbar^2 \pi^2}{2m_{c,w} L_z^2} = \mathcal{E}_{nc}. \tag{6.15}$$

The coefficient A can be calculated by applying the normalization condition:

$$\int_0^{L_z} |F_c(z)|^2 dz = 1, \tag{6.16}$$

which gives $A = \sqrt{2/L_z}$, so that:

$$F_c(z) = \sqrt{\frac{2}{L_z}} \sin{\left(n \frac{\pi}{L_z} z \right)}. \tag{6.17}$$

The first three functions, corresponding to $n = 1, 2$, and 3, are shown in Fig. 6.3(a). If the barriers of the potential well are not infinite, the wavefunction decays exponentially into the barrier regions. The first three solutions are shown in Fig. 6.3(b).

In Eq. 6.17, each n value corresponds to a two-dimensional subband in the conduction band, since the electron can move in the plane $x - y$ of the well. Indeed, the energy values given by Eq. 6.15 have been obtained upon assuming $V_{Kj} = 0$. In the general case, the total potential is given by $V_{cj} + V_{Kj}$. The simplest case is when the effective masses in the well and in the barrier are equal: $m_w = m_b$. In this case, V_{Kj} does not depend on z and it just produces a change of the reference potential. If $m_w \neq m_b$ V_{Kj} will depend on z and new solutions of Eq. 6.7 have to be calculated. In the limiting case, when $V_{K,w} = V_0 + V_{K,b}$, the total potential profile $V_{cj}(z) + V_{Kj}(z)$ is completely flat, thus removing any confinement effect. In general, we can assume that $|V_{K,w} - V_{K,b}| \ll V_0$, so that V_{Kj} can be treated as a small perturbation of the potential well reported in Fig. 6.2. Therefore, it is reasonable to

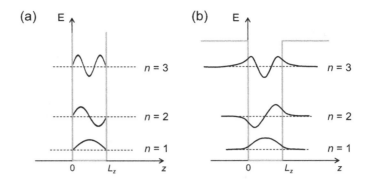

Figure 6.3   Eigenenergies and wavefunctions of the first three bound states of a quantum well in the case of infinite well depth (a) and of a well with finite potential barriers (b).

assume that the solutions of Eq. 6.7 are close to the solutions found in the case $V_{K,j} = 0$, so that we can write $F_c(z) = F_{c,0}(z) + dF_c(z)$ and $\mathcal{E}_{nc} = \mathcal{E}_{nc,0} + d\mathcal{E}_{nc}$, where $F_{c,0}$ and $\mathcal{E}_{nc,0}$ are the unperturbed solutions of Eq. 6.7. These new solutions are then inserted in Eq. 6.7, which is then multiplied by $F_{c,0}^*$ and integrated from $-\infty$ to $+\infty$. After some calculations, the perturbation approach allows one to obtain the following approximated perturbed solutions:

$$\mathcal{E}_{nc} = \mathcal{E}_{nc,0} + \frac{\hbar^2}{2m_{c,eff}}(k_x^2 + k_y^2) = \mathcal{E}_{nc,0} + \frac{\hbar^2 K^2}{2m_{c,eff}}, \tag{6.18}$$

where the effective mass is given by:

$$\frac{1}{m_{c,eff}} = \frac{\Gamma}{m_{cw}} + \frac{1-\Gamma}{m_{cb}}, \tag{6.19}$$

and Γ is the confinement factor given by the following formula:

$$\Gamma = \int_0^{L_z} |F_{c,0}|^2 dz \Big/ \int_{-\infty}^{\infty} |F_{c,0}|^2 dz \tag{6.20}$$

The dispersion relation given by Eq. 6.18 is parabolic with an in-plane effective mass $m_{c,eff}$, which is a weighted average of the bulk effective masses in the well and in the barrier. Each level n gives rise to a two-dimensional subband in the $k_x - k_y$ plane, as shown in Fig. 6.4. The electron wavefunction can be written as:

$$\psi_c(x, y, z) = \sqrt{\frac{2}{L_z}} \sin\left(n\frac{\pi}{L_z}z\right) \exp\left(i\mathbf{K} \cdot \mathbf{r}\right) u_{c0}(\mathbf{r}), \tag{6.21}$$

where $\mathbf{K} \cdot \mathbf{r} = k_x x + k_y y$.

6.2.1 Electronic States in the Valence Band

In the case of the valence band, the determination of the hole states when quantum confinement is present is more complicated. In the case of the electron states in the conduction band, we have restricted our analysis to the case of direct band gap semiconductors, where

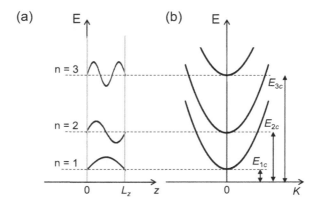

Figure 6.4 (a) Plots of the first three energy levels and of the corresponding eigenfunctions in the conduction band for infinite well depths (as in Fig. 6.3(a)); (b) corresponding energy versus K relations.

the electronic states, at least near the bottom of the band, are described by a simple s-type band. In this case, we can assume that the interaction with other energy bands is weak. As we have discussed in Chapter 1, the valence band states are generated from p-type states, thus leading to the formation of the heavy hole (HH) and light-hole (LH) bands, which are degenerate near the band edge and whose reciprocal interaction cannot be generally treated as a small perturbation. By using the simple model employed in the case of the conduction band and assuming that the valence band may be represented by parabolic bands, the energy states in the well can be written as:

$$\mathcal{E}_{nv,hh} = \mathcal{E}_v - \mathcal{E}_{n,hh} - \frac{\hbar^2(k_x^2 + k_y^2)}{2m_{hh}}, \tag{6.22}$$

$$\mathcal{E}_{nv,lh} = \mathcal{E}_v - \mathcal{E}_{n,lh} - \frac{\hbar^2(k_x^2 + k_y^2)}{2m_{lh}}, \tag{6.23}$$

where $\mathcal{E}_v$ is the top of the valence band for the well semiconductor and the confinement energies $\mathcal{E}_{n,hh}$ and $\mathcal{E}_{n,lh}$ are given by:

$$\mathcal{E}_{n,hh} = n^2 \frac{\hbar^2\pi^2}{2m_{hh}L_z^2}, \tag{6.24}$$

$$\mathcal{E}_{n,lh} = n^2 \frac{\hbar^2\pi^2}{2m_{lh}L_z^2}. \tag{6.25}$$

Equations 6.22 and 6.23 show that in a QW the degeneracy between HH and LH states at the zone center is removed, since their effective mass enters in the confinement energy. The actual situation is much more complex as a result of the strong interaction between the valence bands. Here, we just mention that a good approximation can be obtained by solving the Kohn–Luttinger form of the Schrödinger equation. Within this approach, the effective hole masses along the z direction (which have to be used in Eq. 6.7 with the subscript c replaced by v) can be written in terms of the so-called Luttinger parameters γ_1 and γ_2 as:

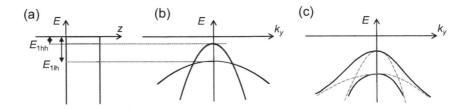

Figure 6.5 (a) The degeneracy between heavy and light hole states at $K = 0$ is removed in a QW. (b) Corresponding dispersion curves with band crossing; (c) high-order terms in the perturbative calculations remove the band crossing in a real QW.

Table 6.1 Luttinger parameters γ_1 and γ_2 for a few semiconductors [1].		
	γ_1	γ_2
GaAs	6.98	2.06
AlAs	3.76	0.82
InAs	20.0	8.5
InP	5.08	1.60
GaN	2.67	0.75
AlN	1.92	0.47

$$\frac{1}{m_{hh}^z} = \frac{\gamma_1 - 2\gamma_2}{m_0} \tag{6.26}$$

$$\frac{1}{m_{lh}^z} = \frac{\gamma_1 + 2\gamma_2}{m_0}. \tag{6.27}$$

The Luttinger parameters γ_1 and γ_2 of a few semiconductors are reported in Table 6.1. Since $m_{lh}^z < m_{hh}^z$, the light-hole states have a larger confinement energy (see Fig. 6.5(a)).

In the $x - y$ plane, the transverse effective masses are given by:

$$\frac{1}{m_{hh}^{xy}} = \frac{\gamma_1 + \gamma_2}{m_0} \tag{6.28}$$

$$\frac{1}{m_{lh}^{xy}} = \frac{\gamma_1 - \gamma_2}{m_0}. \tag{6.29}$$

Note that in the $x - y$ plane, the mass of the HH is smaller than the mass of the LHs, thus leading to the crossing points between the HH and LH bands shown in Fig. 6.5(b). The behavior in real QWs is more complex: Indeed, higher order terms in the perturbation calculations remove such band crossing. Therefore, the valence bands display an anticrossing behavior and can be strongly deformed, as schematically shown in Fig. 6.5(c).

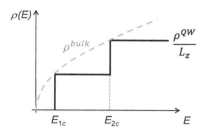

Figure 6.6 Plot of the quantum well density of states in the conduction band normalized to the well thickness, L_z, as a function of energy $\mathcal{E}_c$ (solid line) and plot of the density of states for the corresponding bulk semiconductor, ρ_{bulk} (dashed gray line).

6.3 Density of States

In Section 2.3.2 the density of states has been calculated in the case of a two-dimensional structure:

$$\rho(\mathcal{E})d\mathcal{E} = \frac{K}{\pi}dK, \tag{6.30}$$

where $\mathbf{K} = k_x\mathbf{u}_x + k_y\mathbf{u}_y$. For the conduction subband $n = 1$, assuming a parabolic dispersion curve:

$$\mathcal{E} = \mathcal{E}_{1c} + \frac{\hbar^2 K^2}{2m_{c,w}} \qquad d\mathcal{E} = \frac{\hbar^2 K}{m_{c,w}}dK, \tag{6.31}$$

from Eq. 6.30, it is possible to obtain the density of states:

$$\rho_c(\mathcal{E}) = \frac{m_{c,w}}{\pi\hbar^2}. \tag{6.32}$$

The density of states can be written as a function of frequency, ω:

$$\rho_c(\omega) = \frac{m_{c,w}}{\pi\hbar}. \tag{6.33}$$

The total density of states is given by the sum of the density of states from each subband:

$$\rho(\mathcal{E}) = \sum_n \frac{m_{c,w,n}}{\pi\hbar^2}H(\mathcal{E} - \mathcal{E}_{nc}), \tag{6.34}$$

where $H(\mathcal{E})$ is the Heaviside step function, as shown in Fig. 6.6.

It is interesting to note, as shown in Fig. 6.6, that the increase of the step-like function $\rho^{QW}(\mathcal{E})/L_z$ follows the function $\rho^{bulk}(\mathcal{E})$, that is, the density of states of the bulk semiconductor, given by:

$$\rho^{bulk}(\mathcal{E}) = \frac{1}{2\pi^2}\left(\frac{2m_{c,w}}{\hbar^2}\right)^{3/2}\sqrt{\mathcal{E}}, \tag{6.35}$$

where we have assumed as a reference for the energy axis the bottom of the conduction band in the bulk case. Indeed:

$$\rho^{bulk}(\mathcal{E}_{nc}) = \rho^{bulk}(n^2\mathcal{E}_{1c}) = n\frac{m_{c,w}}{\pi\hbar^2 L_z} = \frac{\rho^{QW}}{L_z}. \tag{6.36}$$

6.4 Electron Density

At thermodynamic equilibrium, the probability of occupation of a particular state is given by the Fermi–Dirac distribution function. When electrons are excited in the conduction band, it is possible to assume that the energy relaxation between the conduction subbands and between the valence subbands is very fast (in the picosecond or subpicosecond time domain). As we have already seen, to calculate the electron and hole densities, it is still possible to make use of the Fermi–Dirac distribution function by substituting the Fermi energy with the quasi-Fermi energies in the conduction and valence bands:

$$n = \int \frac{\rho_c(\mathcal{E})}{L_z} f_c(\mathcal{E}) \, d\mathcal{E} \tag{6.37}$$

$$p = \int \frac{\rho_v(\mathcal{E})}{L_z} \bar{f}_v(\mathcal{E}) \, d\mathcal{E}. \tag{6.38}$$

Since ρ is constant within each subband, the previous integrals can be easily calculated. Let us consider the n-subband in the conduction band:

$$
\begin{aligned}
n_n &= \frac{m_{c,w}}{\pi \hbar^2 L_z} \int_{\mathcal{E}_{nc}}^{\infty} \frac{d\mathcal{E}}{1 + \exp\left(\frac{\mathcal{E} - \mathcal{E}_{Fc}}{k_B T}\right)} = \\
&= \frac{m_{c,w} k_B T}{\pi \hbar^2 L_z} \log\left[1 + \exp\left(\frac{\mathcal{E}_{Fc} - \mathcal{E}_{nc}}{k_B T}\right)\right],
\end{aligned}
\tag{6.39}
$$

and the total electron density is given by:

$$n = \sum_n n_n. \tag{6.40}$$

Exercise 6.1 Demonstrate that the density of electrons in the nth subband of a quantum well is given by Eq. 6.39.

The integral in Eq. 6.39 can be easily calculated by using the variable x defined as $x = \exp\left[(\mathcal{E} - \mathcal{E}_{Fc})/k_B T\right]$, so that:

$$dx = \frac{1}{k_B T} \exp\frac{\mathcal{E} - \mathcal{E}_{Fc}}{k_B T} \, d\mathcal{E} = \frac{x}{k_B T} \, d\mathcal{E}$$

$$n = \frac{m_{c,w} k_B T}{\pi \hbar^2 L_z} \int_{x_n}^{\infty} \frac{dx}{x(1+x)},$$

where $x_n = \exp\left[(\mathcal{E}_{nc} - \mathcal{E}_{Fc})/k_B T\right]$. Therefore:

$$
\begin{aligned}
n &= \frac{m_{c,w} k_B T}{\pi \hbar^2 L_z} \int_{x_n}^{\infty} \left(\frac{1}{x} - \frac{1}{1+x}\right) dx = \frac{m_{c,w} k_B T}{\pi \hbar^2 L_z} \log\left(1 + \frac{1}{x_n}\right) \\
&= \frac{m_{c,w} k_B T}{\pi \hbar^2 L_z} \log\left[1 + \exp\left(\frac{\mathcal{E}_{Fc} - \mathcal{E}_{nc}}{k_B T}\right)\right].
\end{aligned}
$$

∎

6.5 Transition Selection Rules

We will consider interband transitions from the valence band to the conduction band in a QW. The Fermi's Golden Rule is:

$$W_{12} = W_{vc} = \frac{2\pi}{\hbar}|H'_{cv}|^2\,\delta(\mathcal{E}_2 - \mathcal{E}_1 - \hbar\omega), \tag{6.41}$$

where the matrix element $|H'_{cv}|$ of the interaction Hamiltonian is given by:

$$H'_{cv} = \langle\psi_c|H'(\mathbf{r})|\psi_v\rangle = \int_V \psi_c^* H'(\mathbf{r})\psi_v\,d\tau, \tag{6.42}$$

and the interaction Hamiltonian is:

$$H'(\mathbf{r}) = -\frac{q}{m_0}\mathbf{A}\cdot\mathbf{p} = \frac{e}{m_0}\mathbf{A}(\mathbf{r})\cdot\mathbf{p}. \tag{6.43}$$

Considering a harmonic perturbation, the vector potential can be written as:

$$\mathbf{A}(\mathbf{r},t) = \hat{a}\frac{1}{2}\left[A(\mathbf{r})e^{-i\omega t} + A^*(\mathbf{r})e^{i\omega t}\right]. \tag{6.44}$$

where $A(\mathbf{r}) = \bar{A}(\mathbf{r})e^{i\mathbf{k}_{opt}\cdot\mathbf{r}}$. Due to the term $\delta(\mathcal{E}_2 - \mathcal{E}_1 - \hbar\omega)$ in Eq. 6.41 in the transition, the energy is conserved:

$$\mathcal{E}_1 + \hbar\omega = \mathcal{E}_2. \tag{6.45}$$

The wavefunctions $\psi_{v,c}$ in a QW can be written as in Eq. 6.2, within the envelope function approximation, so that:

$$\begin{aligned}
H'_{cv} &= \langle\psi_c|H'|\psi_v\rangle = \frac{e}{2m_0}\int_V F_c^* u_{c0}^* (A(\mathbf{r})\hat{a}\cdot\mathbf{p})F_v u_{v0}\,d\tau = \\
&= \frac{e}{2m_0}\int_V F_c^* u_{co}^* A(\mathbf{r})\hat{a}u_{v0}\mathbf{p}F_v\,d\tau + \frac{e}{2m_0}\int_V F_c^* u_{c0}^* A(\mathbf{r})\hat{a}F_v\mathbf{p}u_{v0}\,d\tau,
\end{aligned} \tag{6.46}$$

where we have used the following property ($\mathbf{p} = -i\hbar\nabla$):

$$\mathbf{p}(F_v u_{v0}) = u_{v0}\mathbf{p}F_v + F_v\mathbf{p}u_{v0}. \tag{6.47}$$

Equation 6.46 can be written as:

$$\begin{aligned}
H'_{cv} &= \frac{e}{2m_0}\int u_{c0}^* u_{v0}F_c^*[A(\mathbf{r})\hat{a}\cdot\mathbf{p}]F_v\,d\tau + \\
&\quad + \frac{e}{2m_0}\int [F_c^* A(\mathbf{r})F_v]u_{c0}^*\,\hat{a}\cdot\mathbf{p}\,u_{v0}\,d\tau.
\end{aligned} \tag{6.48}$$

Starting from the first integral, we can first calculate the integral over a single unit cell of the crystal and then sum the result considering all the unit cells in the crystal volume. It is reasonable to assume that the term $F_c^*[A(\mathbf{r})\hat{a}\cdot\mathbf{p}]F_v$ is approximately constant in each unit cell. In this case, the first integral in Eq. 6.48 can be written as:

$$\frac{e}{2m_0}\sum_j \left[F_c^*(A(\mathbf{r}_j)\hat{a}\cdot\mathbf{p})F_v\right]\int_{\text{u.c.}} u_{c0}^* u_{v0}\,d\tau, \tag{6.49}$$

where the integral is calculated over a single unit cell (u.c.), j is an index which refers to a given unit cell and $\mathbf{r}_j$ is the position vector of the jth cell. Since $\langle u_{c0}|u_{v0}\rangle = 0$ (see

Eq. 6.5), the integral calculated over the unit cell is zero, so that the first integral in Eq. 6.48 is zero.

We can treat the second integral in Eq. 6.48 in the same way, so that it can be rewritten as:

$$H'_{cv} = \frac{e}{2m_0} \sum_j \left[F_c^* A(\mathbf{r}_j) F_v \right] \int_{u.c.} u_{c0}^* \, \hat{a} \cdot \mathbf{p} \, u_{v0} \, d\tau. \tag{6.50}$$

Since the integral over the unit cell is the same for each cell, it can be extracted from the sum:

$$H'_{cv} = \frac{e}{2m_0} \left(\frac{1}{V_u} \int_{u.c.} u_{c0}^* \, \hat{a} \cdot \mathbf{p} \, u_{v0} \, d\tau \right) \sum_j \left[F_c^* A(\mathbf{r}_j) F_v \right] V_u. \tag{6.51}$$

Assuming that the volume of the unit cell is very small, the sum in the previous expression can be written as an integral, so that:

$$H'_{cv} = \frac{e}{2m_0} \langle u_{c0} | \hat{a} \cdot \mathbf{p} | u_{v0} \rangle \int F_c^* A(\mathbf{r}) F_v \, d\tau. \tag{6.52}$$

The envelope function overlap integral can be written as follows:

$$\int_V F_c^* A(\mathbf{r}) F_v \, d\tau = \int_V F_c^*(z) e^{-i\mathbf{K}_c \cdot \mathbf{r}} \bar{A}(\mathbf{r}) e^{i\mathbf{k}_{opt} \cdot \mathbf{r}} F_v(z) e^{i\mathbf{K}_v \cdot \mathbf{r}} \, d\tau, \tag{6.53}$$

this integral does not vanish if:

$$\mathbf{K}_c = \mathbf{K}_v + \mathbf{k}_{opt}. \tag{6.54}$$

Since $|\mathbf{k}_{opt}|$ is typically much smaller than $|\mathbf{K}_{v,c}|$, we can write the momentum selection rule 6.54 as:

$$\mathbf{K}_v = \mathbf{K}_c. \tag{6.55}$$

Since the well thickness is much smaller than the wavelength of the electromagnetic radiation, the vector potential can be assumed constant in the well region, so that the envelope function overlap integral can be written as:

$$\int_V F_c^* \bar{A}(\mathbf{r}) F_v \, d\tau \simeq A_0 \int_c F_c^*(z) F_v(z) \, dz = A_0 \langle F_c | F_v \rangle. \tag{6.56}$$

Therefore, the squared modulus of the matrix element H'_{cv} is given by:

$$|H'_{cv}|^2 = \left(\frac{eA_0}{2m_0} \right)^2 |\langle u_{c0} | \hat{a} \cdot \mathbf{p} | u_{v0} \rangle|^2 |\langle F_c | F_v \rangle|^2 = \left(\frac{eA_0}{2m_0} \right)^2 |M_T|^2. \tag{6.57}$$

The term $|\langle F_v | F_c \rangle|^2$ leads to another transition rule. In the case of a well with infinite potential barriers, we have calculated the functions $F_v(z)$ and $F_c(z)$:

$$F_{v,c} = \sqrt{\frac{2}{L_z}} \sin \left(n_{v,c} \frac{\pi}{L_z} z \right), \tag{6.58}$$

which are independent of the effective mass of electrons and holes. When $n_c = n_v$, the envelope functions $F(z)$ in the conduction and valence bands are equal, so that $|\langle F_c | F_v \rangle|^2 = 1$: These transitions are allowed. On the contrary, when $n_c \neq n_v$, the envelope functions in the conduction and valence bands are orthogonal to each other so that $|\langle F_c|F_v\rangle|^2 = 0$: These transitions are therefore strictly forbidden.

In the case of QWs with finite barriers, the wavefunction penetrates in the barrier regions so that the overall wavefunction weakly depends on the electron and hole effective masses. In this case when $n_c = n_v$, we still have $|\langle F_c|F_v\rangle|^2 \simeq 0.95 - 1$ but now if $n_c \neq n_v$, the envelope functions are not perfectly orthogonal, so that $|\langle F_c|F_v\rangle|^2 \simeq 0 - 0.1$ and the transition is not strictly forbidden although the probability of such transition is typically very small. Note that, also in the case of finite barriers, the transitions between a state with even envelope function and a state with odd envelope function are still strictly forbidden since in this case $|\langle F_c|F_v\rangle|^2 = 0$. As a final remark we add that, far from the band edge, band mixing in the valence band becomes rather strong and the envelope functions $F_v(z)$ are not purely even or odd functions, so that most transitions are weakly allowed.

Note that the first integral in Eq. 6.48 can be written in another way (in analogy to what has been done for the second integral):

$$\frac{e}{2m_0} \int u_{c0}^* u_{v0} F_c^* [A(\mathbf{r})\hat{a} \cdot \mathbf{p}] F_v \, d\tau = \frac{eA_0}{2m_0} \langle u_{c0}|u_{v0}\rangle \langle F_c|\hat{a} \cdot \mathbf{p}|F_v\rangle. \tag{6.59}$$

6.5.1 Intersubband Transitions

In a QW, we can have *intraband* (or *intersubband*) transitions, that is, transitions between different subbands either within the conduction band or the valence band. These intraband transitions have important applications in many photonic devices, ranging from far-infrared detectors to quantum cascade lasers. Let us consider intraband transitions in the conduction band of a QW grown, as always assumed in this chapter, along the z-direction. Let us consider two subbands, $n = 1, 2$. The corresponding wavefunctions can be written as:

$$\psi_{1c} = F_1(\mathbf{r})u_{c0}(\mathbf{r}) = F_1(z)e^{i\mathbf{K}\cdot\mathbf{r}}u_{c0}(\mathbf{r}) \tag{6.60}$$

$$\psi_{2c} = F_2(\mathbf{r})u_{c0}(\mathbf{r}) = F_2(z)e^{i\mathbf{K}\cdot\mathbf{r}}u_{c0}(\mathbf{r}), \tag{6.61}$$

where the central cell functions $u_{c0}(\mathbf{r})$ are the same for different subbands (this is valid in particular for the conduction band). We can use Eqs. 6.59 and 6.52 to write the matrix element H_{21}':

$$H_{21}' = \frac{eA_0}{2m_0} \langle u_{c0}|u_{c0}\rangle \langle F_2|\hat{a} \cdot \mathbf{p}|F_1\rangle + \frac{eA_0}{2m_0} \langle u_{c0}|\hat{a} \cdot \mathbf{p}|u_{c0}\rangle \langle F_2|F_1\rangle. \tag{6.62}$$

As we have pointed out before, in the case of interband transitions, the first term vanishes ($\langle u_{v0}|u_{c0}\rangle = 0$). In Eq. 6.62, the first term corresponds to the intersubband transition in the conduction band, while the second term vanishes due to the term $\langle F_2|F_1\rangle$, since $F_1(z)$ and $F_2(z)$ have even and odd parity, respectively (see Fig. 6.3(b)). Let us now analyze the first term in Eq. 6.62, which depends on the matrix element $\langle F_2|\hat{a} \cdot \mathbf{p}|F_1\rangle$, where $\mathbf{p} = -i\hbar\nabla$:

$$\langle F_2|\hat{a} \cdot \mathbf{p}|F_1\rangle = -i\hbar \int F_2^* e^{-i\mathbf{K}\cdot\mathbf{r}} \hat{a} \cdot \nabla F_1(z)e^{i\mathbf{K}\cdot\mathbf{r}} \, d\tau. \tag{6.63}$$

If the polarization of the electromagnetic field is in the $x - y$ plane, the matrix element in Eq. 6.63 is zero: That is, for light polarized in the $x - y$ plane, the intersubband transition rate is zero. Also in this case in the valence band, this condition does not strictly hold as a consequence of the strong mixing of the central cell function. On the contrary, if the electromagnetic field has a z-polarized component, the matrix element in Eq. 6.63 does not vanish. Indeed, if the radiation is z-polarized, Eq. 6.63 can be written as:

$$\langle F_2 | \mathbf{u}_z \cdot \mathbf{p} | F_1 \rangle = -i\hbar \int F_2^*(z) \mathbf{u}_z \cdot \left(\frac{dF_1(z)}{dz} \mathbf{u}_z \right) dz. \tag{6.64}$$

Since $F_1(z)$ and $F_2(z)$ have even and odd parity; respectively, $F_2(z)$ and $dF_1(z)/dz$ both have odd parity; therefore, the integral in Eq. 6.64 does not vanish.

6.6 Absorption and Gain in a Quantum Well

We will first calculate the joint density of states $\rho_j^{QW}(\mathcal{E}_0)$, where $\rho_j^{QW}(\mathcal{E}_0) \, d\mathcal{E}_0$ gives the number of allowed transitions per unit area with transition energy in the range between $\mathcal{E}_0$ and $\mathcal{E}_0 + d\mathcal{E}_0$. We will first consider interband transitions from the valence to the conduction band. Due to the transition selection rules derived in the previous section, and considering that the spin does not change in a transition, the number of transitions, $\rho_j^{QW}(\mathcal{E}_0) \, d\mathcal{E}_0$, is equal to the number of states (either in the valence or in the conduction band), $\rho^{QW}(K) \, dK$, corresponding to the same energy range. Let us assume a transition from the subband $n = 1$ in the valence band to the subband $n = 1$ in the conduction band:

$$\mathcal{E}_1 = -\mathcal{E}_{1v} - \frac{\hbar^2 K^2}{2m_v} \tag{6.65}$$

$$\mathcal{E}_2 = \mathcal{E}_g + \mathcal{E}_{1c} + \frac{\hbar^2 K^2}{2m_c}, \tag{6.66}$$

where we have assumed an energy axis with the zero at the top of the valence band. The transition energy $\mathcal{E}_0$ is given by:

$$\mathcal{E}_0 = \mathcal{E}_2 - \mathcal{E}_1 = \mathcal{E}_g + \mathcal{E}_{1c} + \mathcal{E}_{1v} + \frac{\hbar^2 K^2}{2m_r}, \tag{6.67}$$

where m_r is the reduced mass. From Eq. 6.67, we observe that the minimum (or threshold) transition energy is given by $\hbar\omega = \mathcal{E}_g + \mathcal{E}_{1c} + \mathcal{E}_{1v} > \mathcal{E}_g$, so that the absorption in a QW is blue-shifted with respect to the absorption in the case of a bulk semiconductor: The energy difference is given by the sum of the confinement energies in the valence and the conduction band and of the energy gap. The density of states in a QW is given by $\rho^{QW}(K) = K/\pi$, so that:

$$\rho_j^{QW}(\mathcal{E}_0) d\mathcal{E}_0 = \rho^{QW}(K) dK, \tag{6.68}$$

where $d\mathcal{E}_0 = \hbar^2 K dK / m_r$, thus obtaining:

$$\rho_j^{QW}(\mathcal{E}_0) = \frac{m_r}{\pi \hbar^2}, \qquad \rho_j^{QW}(\omega) = \frac{m_r}{\pi \hbar}. \tag{6.69}$$

The absorption coefficient from the subband $n = 1$ in the valence band to the subband $n = 1$ in the conduction band can be written by using the expression of $\alpha(\omega)$ obtained for bulk semiconductors (see Eq. 5.30) by replacing $\rho_j(\omega)$ with $\rho_j^{QW}(\omega)/L_z$ and by using Eq. 5.44

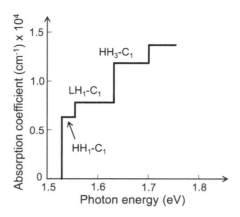

Calculated absorption coefficient in a 10-nm-thick GaAs/Al$_{0.3}$Ga$_{0.7}$As QW for in-plane polarized light. Clearly visible the transitions from the heavy-hole (HH) subbands 1 and 3 to the conduction band ($n = 1$) and from the $n = 1$ light-hole subband to the $n = 1$ conduction subband. Note the presence of the HH$_3$-C$_1$ transition, which is not truly forbidden in the case of wells with finite potential barriers (Adapted from [7]).

to write the squared modulus of the matrix element of the dipole moment in terms of the matrix element of the interaction Hamiltonian ($|\mu_{21}|^2 = 4|H'_{21}|^2/E_0^2 = 4|H'_{21}|^2/\omega^2 A_0^2$):

$$\alpha(\omega) \quad = \quad \frac{\omega\pi}{cn\epsilon_0\hbar}|\mu_{21}|^2\frac{\rho_j^{QW}(\omega)}{L_z}[f_v(\hbar\omega) - f_c(\hbar\omega)]. \tag{6.70}$$

By using Eq. 6.57, we can write:

$$\alpha(\omega) = \frac{\pi e^2}{cn\epsilon_0 m_0^2\omega\hbar}|M_T|^2\frac{m_r}{\pi\hbar L_z}[f_v(\hbar\omega) - f_c(\hbar\omega)], \tag{6.71}$$

or, in a more compact form:

$$\alpha(\omega) = \alpha_0(\omega)[f_v(\hbar\omega) - f_c(\hbar\omega)]. \tag{6.72}$$

If n_v and n_c are the quantum numbers identifying a given subband in the valence and conduction band, respectively, the absorption coefficient related to a given subband transition can be written as $\alpha_{sub}(\omega, n_v, n_c)$. The total absorption coefficient at a particular frequency ω is obtained by summing over all allowed subband transitions:

$$\alpha(\omega) = \sum_{n_v}\sum_{n_c}\alpha_{sub}(\omega, n_v, n_c). \tag{6.73}$$

Figure 6.7 shows the calculated absorption coefficient in a 10-nm GaAs/Al$_{0.3}$Ga$_{0.7}$As QW for in-plane polarized light.

The gain coefficient is directly obtained from Eq. 6.72:

$$g(\omega) = \alpha_0(\omega)[f_c(\hbar\omega) - f_v(\hbar\omega)], \tag{6.74}$$

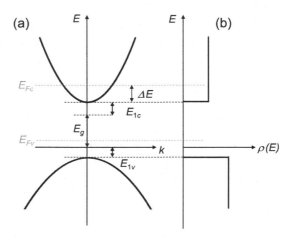

Figure 6.8 (a) Dispersion curves of the $n = 1$ subbands in the valence and conduction band; (b) corresponding density of states at transparency.

for a given interband transition. In order to have a net gain, it is required that:

$$f_c(\hbar\omega) - f_v(\hbar\omega) > 0, \tag{6.75}$$

which implies the following relation:

$$\hbar\omega = \mathcal{E}_2 - \mathcal{E}_1 < \mathcal{E}_{Fc} - \mathcal{E}_{Fv}, \tag{6.76}$$

which is the Bernard–Duraffourg condition we have already obtained in the case of bulk semiconductors. On the other hand, we also have:

$$\hbar\omega = \mathcal{E}_2 - \mathcal{E}_1 > \mathcal{E}_g + \mathcal{E}_{1c} + \mathcal{E}_{1v}, \tag{6.77}$$

thus, in the case of QWs the transparency condition is defined as follows:

$$\mathcal{E}_{Fc} - \mathcal{E}_{Fv} = \mathcal{E}_g + \mathcal{E}_{1c} + \mathcal{E}_{1v}. \tag{6.78}$$

Exercise 6.2   Consider a GaAs well ($m_c = 0.067\ m_0$, $m_v = 0.45\ m_0$) with thickness $L_z = 10$ nm. Calculate the energy difference between the quasi-Fermi level in the conduction band and the bottom of the conduction band at transparency. Calculate the transparency density. Finally, assuming that $m_c = m_v = 0.067\ m_0$, calculate the corresponding transparency density.

At transparency $\mathcal{E}_{Fc} - \mathcal{E}_{Fv} = \mathcal{E}_g + \mathcal{E}_{1c} + \mathcal{E}_{1v}$, which can be written as $\mathcal{E}_{Fc} - \mathcal{E}_g - \mathcal{E}_{1c} = \mathcal{E}_{Fv} + \mathcal{E}_{1v} = \Delta\mathcal{E}$ (see Fig. 6.8). The electron density in the conduction band can be calculated by using Eq. 6.39:

$$\begin{aligned} n &= \frac{m_c k_B T}{\pi \hbar^2 L_z} \log\left[1 + \exp\frac{\mathcal{E}_{Fc} - (\mathcal{E}_g + \mathcal{E}_{1c})}{k_B T}\right] = \\ &= \frac{m_c k_B T}{\pi \hbar^2 L_z} \log\left(1 + \exp\frac{\Delta\mathcal{E}}{k_B T}\right). \end{aligned} \tag{6.79}$$

If $\Delta\mathcal{E} > k_B T$, we can use the following approximated expression:

$$n \simeq \frac{m_c k_B T}{\pi \hbar^2 L_z} \frac{\Delta\mathcal{E}}{k_B T} = \frac{m_c}{\pi \hbar^2 L_z} \Delta\mathcal{E}. \tag{6.80}$$

The hole density in the valence band can be calculated by using the same procedure used to obtain Eq. 6.39. We have:

$$
\begin{aligned}
p &= \frac{m_v k_B T}{\pi \hbar^2 L_z} \log\left[1 + \exp\left(-\frac{\mathcal{E}_{1v} + \mathcal{E}_{Fv})}{k_B T}\right)\right] = \\
&= \frac{m_v k_B T}{\pi \hbar^2 L_z} \log\left[1 + \exp\left(-\frac{\Delta\mathcal{E}}{k_B T}\right)\right].
\end{aligned} \tag{6.81}
$$

Again, if $\Delta\mathcal{E} > k_B T$, we can write:

$$p \simeq \frac{m_v k_B T}{\pi \hbar^2 L_z} \exp\left(-\frac{\Delta\mathcal{E}}{k_B T}\right). \tag{6.82}$$

Since, for charge neutrality, we must have $n = p$, from Eqs. 6.80 and 6.82, we obtain:

$$\frac{\Delta\mathcal{E}}{k_B T} = \left(\frac{m_v}{m_c}\right) \exp\left(-\frac{\Delta\mathcal{E}}{k_B T}\right) = 6.7 \exp\left(-\frac{\Delta\mathcal{E}}{k_B T}\right), \tag{6.83}$$

which can be graphically solved. The result is $\Delta\mathcal{E} = 1.5\, k_B T = 39$ meV (at $T = 300$ K). Without approximations, we would have obtained $\Delta\mathcal{E} = 1.33\, k_B T = 34.6$ meV (at $T = 300$ K). The corresponding electron density at transparency is $N_{tr} \simeq 1.1 \times 10^{18}$ cm^{-3}. Note that if $m_c = m_v$, $\Delta\mathcal{E} = 0$ and the transparency density is given by:

$$N_{tr} = \frac{m_c k_B T}{\pi \hbar^2 L_z} \log 2 \simeq 5 \times 10^{17} \text{ cm}^{-3}. \tag{6.84}$$

      ■

A typical evolution of the gain coefficient versus photon energy for different values of the injected carrier density is shown in Fig. 6.9, which refers to an unstrained GaAs/Al$_{0.2}$Ga$_{0.8}$As QW, assuming parabolic subbands and taking into account the spectral broadening of each transition, due to scattering processes. The net effect of the lineshape broadening can be obtained by convolving a Lorentzian function with the calculated gain function obtained in this section. Better results can be achieved by replacing the Lorentzian function with another lineshape function obtained from a more exact calculation of the induced dipole phase damping.

Also in the case of QWs, above the transparency condition, the peak of the gain increases with carrier injection. It is possible to write an approximated linear dependence of the gain peak versus carrier injection, as we have done for bulk semiconductors:

$$g_P = \sigma^{QW}(N - N_{tr}), \tag{6.85}$$

where σ^{QW} is the differential gain (of the order of $\sim 7 \times 10^{-16}$ cm^2 in the case of GaAs/Al$_{0.2}$Ga$_{0.8}$As QWs). We note that in the case of QWs, the linear approximation between g_P and N is not so good as in the case of bulk semiconductors.

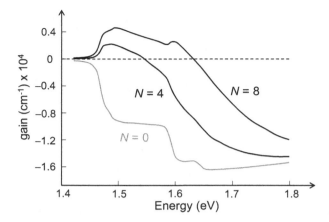

Figure 6.9 Gain coefficient as a function of the photon energy for different values of the injected carrier density, N (in units of 10^{18} cm^{-3}) for a 8-nm GaAs/Al$_{0.2}$Ga$_{0.8}$As QW, in the parabolic band approximation, considering the spectral broadening of each transition. The gray curve with $N = 0$ represents the absorption coefficient.

6.7 Intersubband Absorption

In the case of intersubband transitions, since the electronic subbands are parallel, all transitions between two subbands occur at the same energy (i.e., do not depend on $\mathbf{K}$); therefore, these interactions are resonant. The corresponding absorption coefficient can be calculated by using the same procedure used for bulk semiconductors. In the bulk case, the absorption coefficient associated to transitions with transition frequency around ω_0 is given by Eq. 5.28:

$$d\alpha(\omega) = \frac{\omega\pi}{nc\epsilon_0\hbar}|\mu_{21}|^2 g(\omega - \omega_0)\{\rho_j(\omega_0)[f_v(\hbar\omega_0) - f_c(\hbar\omega_0)]d\omega_0\}. \qquad (6.86)$$

In the case of intersubband transitions between two subbands in the conduction band, the terms in the curly braces give the difference between the electronic density in the n-th (N_n) and the m-th (N_m) subbands, so that the absorption coefficient can be written as:

$$\alpha(\omega) = \frac{\omega\pi}{nc\epsilon_0\hbar}|\mu_{mn}|^2 g(\omega - \omega_{mn})(N_n - N_m), \qquad (6.87)$$

where $\hbar\omega_{mn} = \mathcal{E}_m - \mathcal{E}_n$ and $g(\omega - \omega_{mn})$ are the Lorentzian lineshape of the transition. Here, we have assumed an electromagnetic field with a z-component. Let us consider a n-type semiconductor with $\mathcal{E}_n < \mathcal{E}_F < \mathcal{E}_m$ (where $\mathcal{E}_F$ is the Fermi energy). In this case, we have $N_m \simeq 0$ and the absorption coefficient is given by:

$$\alpha(\omega) = \frac{\omega\pi}{nc\epsilon_0\hbar}|\mu_{mn}|^2 g(\omega - \omega_{mn})N_n, \qquad (6.88)$$

which turns out to be proportional to the dopant concentration. If $\mathcal{E}_F > \mathcal{E}_m$, we can write (from Eq. 6.39):

$$N_n = \frac{m_c k_B T}{\pi \hbar^2 L_z} \log \left[1 + \exp \left(\frac{\mathcal{E}_F - \mathcal{E}_n}{k_B T} \right) \right] \simeq \frac{m_c}{\pi \hbar^2 L_z} (\mathcal{E}_F - \mathcal{E}_n)$$

$$N_m \simeq \frac{m_c}{\pi \hbar^2 L_z} (\mathcal{E}_F - \mathcal{E}_m), \tag{6.89}$$

and the absorption coefficient turns out to be independent of dopant concentration:

$$\alpha(\omega) = \frac{\omega \pi}{n c \epsilon_0 \hbar} |\mu_{mn}|^2 g(\omega - \omega_{mn}) \frac{m_c}{\pi \hbar^2 L_z} (\mathcal{E}_m - \mathcal{E}_n). \tag{6.90}$$

Exercise 6.3 Calculate the intersubband dipole moment in an infinite QW associated to the transition between the subband $n = 1$ and $n = 2$.

The envelope functions $F_{1,2}(z)$ are given by (see Eq. 6.17):

$$F_1(z) = \sqrt{\frac{2}{L_z}} \sin \left(\frac{\pi}{L_z} z \right), \tag{6.91}$$

$$F_2(z) = \sqrt{\frac{2}{L_z}} \sin \left(2 \frac{\pi}{L_z} z \right). \tag{6.92}$$

The dipole matrix element μ_{21} can be calculated as follows:

$$\begin{aligned}
\mu_{21} &= -e \int_0^{L_z} F_2(z) z F_1(z) \, dz = -\frac{2e}{L_z} \int_0^{L_z} \sin \left(\frac{\pi}{L_z} z \right) z \sin \left(2 \frac{\pi}{L_z} z \right) dz = \\
&= -\frac{2eL_z}{\pi^2} \int_0^{\pi} x \sin(x) \sin(2x) \, dx = \\
&= -\frac{2eL_z}{\pi^2} \frac{1}{18} [12x \sin^3(x) + 9\cos(x) - \cos(3x)]_0^{\pi} = \frac{16}{9} \frac{eL_z}{\pi^2}.
\end{aligned} \tag{6.93}$$

For example, assuming a well thickness $L_z = 10$ nm one obtains $\mu_{21} = 2.88 \times 10^{-28}$ C·m. The dipolar matrix element $|\langle F_2 | z | F_1 \rangle| = 1.8$ nm. ∎

Exercise 6.4 Calculate and plot the absorption spectrum corresponding to the intersubband transition from subband $n = 1$ to subband $n = 2$ in the conduction band of a GaAs QW with well thickness $L_z = 10$ nm for an injected electron density $N = 10^{18}$ cm^{-3}. Assume a dephasing time $T_2 = 110$ fs; the refractive index at the transition frequency is 3.3.

We first calculate the energy of the first two subbands in the conduction band (see Fig. 6.10):

$$\mathcal{E}_{1c} = \frac{\hbar^2 \pi^2}{2m_c L_z^2} = 55.8 \text{ meV}$$

$$\mathcal{E}_{2c} = 4\mathcal{E}_{1c} = 223.1 \text{ meV}.$$

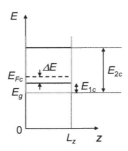

Figure 6.10 Schematic structure of the conduction band of the QW considered in Exercise 6.4.

We can now calculate the position of the Fermi level in the conduction band. From Eq. 6.39, we have:

$$N = \frac{m_c k_B T}{\pi \hbar^2 L_z} \log \left[1 + \exp \left(\frac{\mathcal{E}_{Fc} - (\mathcal{E}_g + \mathcal{E}_{1c})}{k_B T} \right) \right].$$ (6.94)

The energy difference $\Delta \mathcal{E}$ between the Fermi level and the first state in the conduction band is given by $\Delta \mathcal{E} = \mathcal{E}_{Fc} - (\mathcal{E}_g + \mathcal{E}_{c1})$. From Eq. 6.94, we obtain:

$$\frac{\Delta \mathcal{E}}{k_B T} = \log \left[\exp \left(\frac{\pi \hbar^2 L_z}{m_c k_B T} N \right) - 1 \right] = 1.08,$$

So that $\Delta \mathcal{E} = 1.08 k_B T = 28$ meV at $T = 300$ K. Therefore, the Fermi energy is well below the subband $n = 2$, and the absorption coefficient is given by Eq. 6.88, with $N_n = N$:

$$\alpha(\omega) = \frac{\omega \pi}{n c \epsilon_0 \hbar} |\mu_{21}|^2 g(\omega - \omega_{21}) N,$$ (6.95)

where the matrix element μ_{21} is given by Eq. 6.93. Assuming a Lorentzian lineshape, the absorption peak is given by:

$$\alpha_P(\omega_0) = \frac{\omega_0 \pi}{n c \epsilon_0 \hbar} \left(\frac{16}{9} \frac{e L_z}{\pi^2} \right)^2 \frac{T_2}{\pi} N,$$ (6.96)

where T_2 is the dephasing time and ω_0 is the transition frequency $\omega_0 = (\mathcal{E}_{2c} - \mathcal{E}_{1c})/\hbar$. The wavelength corresponding to the resonant frequency is:

$$\lambda_0 = \frac{2\pi c}{\omega_0} = \frac{2\pi \hbar c}{\mathcal{E}_{c2} - \mathcal{E}_{c1}} = 7.4 \ \mu\text{m}.$$

The absorption spectrum is shown in Fig. 6.11. Assuming a dephasing time $T_2 = 110$ fs, the peak absorption is $\alpha_P \simeq 2.5 \times 10^4$ cm^{-1}. ∎

6.8 Strained Quantum Wells

At the beginning of this chapter, we have mentioned that in the case of bulk heterostructures obtained by growing a thick semiconductor layer above a different bulk semiconductor, an

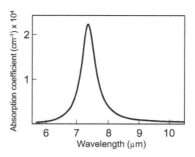

Figure 6.11 Absorption spectrum corresponding to the intersubband transition from subband $n = 1$ to subband $n = 2$ in the conduction band of the GaAs QW analyzed in Exercise 6.4.

almost perfect lattice matching is required in order to prevent the formation of misfit dislocations. One of the most notable characteristics of the largely used GaAs/Al$_x$Ga$_{1-x}$As heterostructures is that the lattice periods of GaAs and AlAs are almost equal, with a relative lattice period variation $\Delta a/a < 0.12\%$. In this case, different heterostructures can be used, with any combination of layer compositions without formation of dislocations. GaAs/AlGaAs heterostructures are used for the construction of double-heterostructure lasers and QW heterostructure lasers with emission in the wavelength range between 650 nm and 800 nm, where the long wavelength limit is related to the band gap of GaAs. Another largely used heterostructure is based on InP/In$_{1-x}$Ga$_x$As$_y$P$_{1-y}$. The quaternary alloy In$_{1-x}$Ga$_x$As$_y$P$_{1-y}$ can be lattice matched to InP for a specific ratio $y/x \simeq 2.2$. Upon changing x, while keeping fixed the y/x ratio to the lattice-matching value, the semiconductor energy gap and therefore the emission wavelength can be changed between 1.15 μm and 1.67 μm for cw room temperature operation. A very good material for emission in the range between 880 nm and 1.1 μm is In$_x$Ga$_{1-x}$As. In this case, there are no suitable binary materials allowing lattice-matched growth of In$_x$Ga$_{1-x}$As. The unit cell of In$_x$Ga$_{1-x}$As can be as much as 3.6% larger (for $x = 0.5$) than that of GaAs. Therefore, considering a thick GaAs substrate, when In$_x$Ga$_{1-x}$As layers are deposited on the top of GaAs the In$_x$Ga$_{1-x}$As cell is shortened in both directions parallel to the interface, thus giving a biaxial compression and it is elongated in the direction perpendicular to the interface, thus giving an uniaxial tension in this direction. Upon increasing the In$_x$Ga$_{1-x}$As layer thickness, the resulting strain energy increases, thus generating a force at the interface. When this force becomes larger than the tension in a dislocation line, a misfit dislocation is produced. In the case of In$_x$Ga$_{1-x}$As deposited on GaAs, for $x < 0.3$ the critical layer thickness, below which the strain produced by lattice mismatch does not lead to the formation of misfit dislocations, is greater than ~ 10 nm, so that it is possible to design QW devices properly working in a wide range of wavelengths.

In the QW configuration, both interfaces experience biaxial compression, as shown in Fig. 6.12. The biaxial compression in the directions parallel to the interface destroys the cubic symmetry of the semiconductor. Since the energy gap of a semiconductor is related to its lattice spacing, we expect that a change of the band gap of the strained layer is

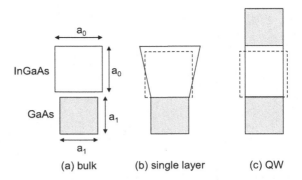

Figure 6.12 Crystal lattice deformation resulting from the epitaxial growth of a thin quantum-well layer of III–V material with original lattice constant a_0 (e.g., InGaAs, see panel (a)) between two thick layers of a material with a lattice constant $a_1 < a_0$ (e.g., GaAs, see panel (c)). (b) InGaAs deformation when a single layer is deposited on GaAs.

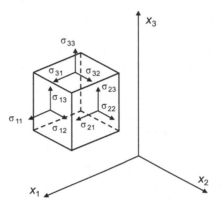

Figure 6.13 Components of the stress tensor in three dimensions.

produced by the lattice distortion. The valence and conduction bands of strained semiconductor layers are modified by the presence of the strain, with a corresponding variation of the effective masses and the generation of energy shifts of the band edges.

Soon after the development of the first strained QW heterostructures, many favorable characteristics of these structures were evidenced, ranging from lower threshold current density, higher differential gain and efficiency, increased range of available emission wavelengths. As a consequence, most of the commercially available semiconductor lasers are based on the use of strained semiconductors. Due to the remarkable practical importance of these structures, in this section we will analyze how strain affects the optical and electronic characteristics of III–V semiconductors.

The stress, or force per unit area, in a given crystal lattice is typically analyzed in terms of a *stress tensor*, with nine components σ_{ij}, where the index i indicates that the stress is acting on a plane which is perpendicular to the x_i axis, and the index j indicates the direction where the stress is acting. σ_{ii} are *normal stresses* and σ_{ij} are *shear stresses*. Figure 6.13 shows the

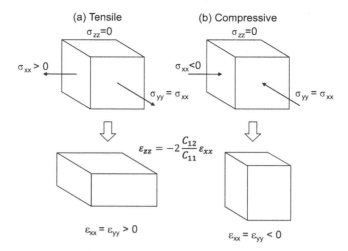

Figure 6.14   (a) Biaxial tensile strain and (b) biaxial compressive strain.

components of the stress tensor in three dimensions. In semiconductor lattices employed in practical devices, it is generally possible to assume that only normal stresses are present, that is, $\sigma_{ij} = 0$ (for $i \neq j$). A stress component σ_{ii} is positive if it acts in the direction indicated by the outward unit vector parallel to the x_i axis. When a stress is applied to a lattice, strain is produced, which is also a second-rank tensor as the stress. Stress and strain are related by the following expression:

$$\sigma_{ij} = C_{ijkl}\, \epsilon_{kl}. \tag{6.97}$$

C_{ijkl} is called the elastic stiffness tensor, and it is a fourth-rank tensor. Since it is generally sufficient to consider only normal stresses, we have to consider only the three diagonal components of the strain tensor, ϵ_{ii}, since the crystal axis, under the action of normal forces, remains orthogonal to each other. In the case of a cubic crystal, the diagonal elements of the stiffness tensor are all equal and also the off-diagonal elements are all equal, and Eq. 6.97 can be written as:

$$\begin{bmatrix} \sigma_{xx} \\ \sigma_{yy} \\ \sigma_{zz} \end{bmatrix} = \begin{bmatrix} C_{11} & C_{12} & C_{12} \\ C_{12} & C_{11} & C_{12} \\ C_{12} & C_{12} & C_{11} \end{bmatrix} + \begin{bmatrix} \epsilon_{xx} \\ \epsilon_{yy} \\ \epsilon_{zz} \end{bmatrix} \tag{6.98}$$

so that we have only two elastic moduli, C_{11} and C_{12}. In typical semiconductors $C_{11}, C_{12} > 0$ and $C_{11} > C_{12}$. In the case of a biaxial strain (as in the case of InGaAs/GaAs heterostructures), the source stress is applied on the four faces perpendicular to the x and y axis, so that $\sigma_{xx} = \sigma_{yy}$, while no forces act on the two faces perpendicular to the z axis, so that $\sigma_{zz} = 0$. When the forces are directed outward (so that $\sigma_{xx} > 0$), we have a tensile strain (see Fig. 6.14(a)), while when the applied forces are directed inward ($\sigma_{xx} < 0$), we have a compressive strain (as shown in Fig. 6.14(b)). In Eq. 6.98, for symmetry reasons, since $\sigma_{xx} = \sigma_{yy}$, we also have $\epsilon_{xx} = \epsilon_{yy}$; $\sigma_{zz} = 0$ so that:

$$\sigma_{xx} = C_{11}\epsilon_{xx} + C_{12}\epsilon_{xx} + C_{12}\epsilon_{zz} \tag{6.99}$$

$$0 = C_{12}\epsilon_{xx} + C_{12}\epsilon_{xx} + C_{11}\epsilon_{zz}, \tag{6.100}$$

so that, from Eq. 6.100:

$$\epsilon_{zz} = -2\frac{C_{12}}{C_{11}}\epsilon_{xx}. \tag{6.101}$$

Therefore, since both C_{11} and C_{12} are positive, in the case of biaxial strain the lattice deformation along the z axis is opposite to the deformation along x and y axes. We recall that the sum of the diagonal elements of the strain tensor gives the volumetric strain, or fractional change of the volume of the crystal lattice (if we neglect the cross-terms between the ϵ_i)

$$\frac{dV}{V} \simeq \epsilon_{xx} + \epsilon_{yy} + \epsilon_{zz}. \tag{6.102}$$

In the following, we will consider a biaxial strain. The strain $\epsilon = \epsilon_{xx} = \epsilon_{yy}$ is given by:

$$\epsilon = \frac{\Delta a}{a_0}, \tag{6.103}$$

where a_0 is the lattice constant of the substrate and Δa the variation of the lattice constant of the strained semiconductor. In the case of compressive strain $\Delta a < 0$ so that $\epsilon < 0$, while for tensile strain, Δa and ϵ are both positive. It is possible to demonstrate that the variation of the band gap induced by strain can be calculated by considering a suitable strain Hamiltonian. In particular, the band gap variation can be expressed in terms of two energies, P and Q, which are related to the strain tensor through the deformation potentials a and b:

$$P = a(\epsilon_{xx} + \epsilon_{yy} + \epsilon_{zz}) \tag{6.104}$$

$$Q = b\left[\frac{1}{2}(\epsilon_{xx} + \epsilon_{yy}) - \epsilon_{zz}\right]. \tag{6.105}$$

By considering Eqs. 6.102 and 6.104, it is evident that P is proportional to the fractional change of the volume of the crystal lattice induced by the strain. This energy term is called the hydrostatic component of the strain and a is also called hydrostatic deformation potential. In the case of biaxial strain:

$$P = 2a\epsilon\left(1 - \frac{C_{12}}{C_{11}}\right). \tag{6.106}$$

Q is called the shear component of the strain and gives the asymmetry in the strain parallel $((\epsilon_{xx} + \epsilon_{yy})/2)$ and perpendicular (ϵ_{zz}) to the stress plane (do not confuse the shear strain with the shear stress, which is assumed to be zero in our discussion). b is also called tetragonal shear deformation potential. In the case of biaxial strain:

$$Q = b\epsilon\left(1 + 2\frac{C_{12}}{C_{11}}\right). \tag{6.107}$$

Assuming a biaxial strain in the $x - y$ plane, the Hamiltonian can be simplified and the eigenvalues can be calculated. The variation in band gap is given by:

$$\Delta\mathcal{E}_g = P, \tag{6.108}$$

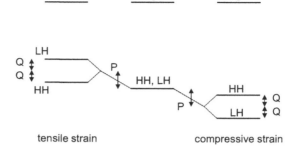

tensile strain   compressive strain

Figure 6.15 Schematic representation of the band energy shift for biaxial tensile and compressive strain. Although in this very schematic plot, the conduction band edge is not changed, this does not mean that the strain affects only the valence band. From the experimental point of view, it is difficult to determine separately the shifts of the conduction and valence band edges.

while the splitting between the HH and LH bands is given by:

$$\Delta \mathcal{E}_{HH-LH} = 2Q. \tag{6.109}$$

Therefore, as shown schematically in Fig. 6.15, the variation of the energy difference between the conduction band and valence band at $k = 0$ is given by:

$$\Delta \mathcal{E}_{HH} = P - Q = 2a\epsilon \left(1 - \frac{C_{12}}{C_{11}} \right) - b\epsilon \left(1 + 2\frac{C_{12}}{C_{11}} \right) \tag{6.110}$$

$$\Delta \mathcal{E}_{LH} = P + Q = 2a\epsilon \left(1 - \frac{C_{12}}{C_{11}} \right) + b\epsilon \left(1 + 2\frac{C_{12}}{C_{11}} \right), \tag{6.111}$$

where $\mathcal{E}_{HH}$ represents the shift of the HH valence band edge with respect to the conduction band edge and $\mathcal{E}_{LH}$ the corresponding shift of the LH band edge. In the case of biaxial compressive strain, P and Q are both positive. For biaxial tensile strain, P and Q are both negative.

It is instructive to analyze a particular example: $In_xGa_{1-x}As/GaAs$ (where the ternary alloy is the strained well semiconductor). The strain ϵ, given by Eq. 6.103, changes with In fraction x as shown in Fig. 6.16. The parameters used in Eqs. 6.110 and 6.111 for the calculation of the shifts in the energy of the band edges can be interpolated from the corresponding values of the binary semiconductors GaAs ($x = 0$) and InAs ($x = 1$), which are given in Table 6.2.

By using Eqs. 6.110 and 6.111 and the numerical values, which can be obtained from Table 6.2, it is possible to calculate the increase of the energy separation between the conduction band edge and the HH and LH valence bad edges at $k = 0$, which are reported in Fig. 6.17 as a function of x. Figure 6.17 shows that the shift of the LH valence band edge is much larger than the corresponding shift for the HHs; therefore, the optical transitions in

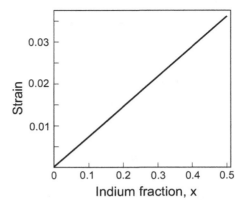

Figure 6.16   Variation of strain ϵ as a function of In fraction, x for an $In_xGa_{1-x}As/GaAs$ strained layer system.

Table 6.2 List of specific values of material parameters used in the calculations.

		GaAs	InAs	AlAs
Lattice constant (Å)	a_0	5.6535	6.0585	5.6600
Elastic coefficient (10^{12} dyne/cm^2)	C_{11}	1.188	0.865	1.25
	C_{12}	0.538	0.485	0.534
Hydrostatic deformation potential (eV)	a	−8.2	−6	−8.11
Shear deformation potential (eV)	b	−2	−1.8	−1.5

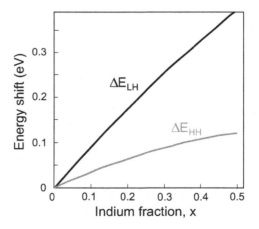

Figure 6.17   Shift of the heavy-hole valence band edge, $\Delta\mathcal{E}_{HH}$, and of the light-hole valence band edge, $\Delta\mathcal{E}_{LH}$, with respect to the conduction band edge calculated by using Eqs. 6.110 and 6.111 versus indium fraction, x, for an $In_xGa_{1-x}As/GaAs$ strained layer system.

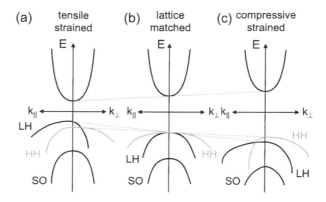

Figure 6.18 Electronic band structure of strained InGaAsP on InP in the case of: (a) tensile strain, (b) lattice matching, and (c) compressive strain (Adapted with permission from [8]).

strained layer $In_xGa_{1-x}As$ (e.g., recombination or absorption) are dominated by transitions between the conduction band and the HH valence band.

In a strained semiconductor, the strain does not just affect the positions of the band edges (in the conduction and valence bands) at $k = 0$. Figure 6.18 clearly shows the effect of strain on the energy band structure of InGaAsP-material on InP in the case of tensile and compressive strain. The strain removes the energy degeneracy of HH and LH valence bands at $k = 0$. In the case of compressive strain, the in-plane HH effective mass is reduced, thus leading to a decrease of the density of states in the HH valence band and, consequently, to an increase of the band filling at the same level of electrical pumping, thus producing a lower threshold current density and a higher differential gain.

6.9 Transparency Density and Differential Gain

We have seen that in order to achieve a net gain, the Bernard–Duraffourg condition must be satisfied. It is obvious that if the semiconductor is in thermal equilibrium $\mathcal{E}_{Fc} = \mathcal{E}_{Fv}$ and it is not possible to achieve a net gain. In the case of nondegenerate semiconductors $f_c(\mathcal{E}_2) \simeq 0$ and $f_v(\mathcal{E}_1) \simeq 1$, so that photons with energy $\hbar\omega = \mathcal{E}_2 - \mathcal{E}_1$ are strongly absorbed. In the case of a degenerate n^+−type semiconductor (still in thermal equilibrium), we have $f_c(\mathcal{E}_2) \simeq 1$ and $f_v(\mathcal{E}_1) \simeq 1$, so that the material is almost transparent for photons with energy $\hbar\omega = \mathcal{E}_2 - \mathcal{E}_1$. Therefore, we conclude that net gain requires a nonequilibrium condition in the semiconductor in order to produce simultaneously a high concentration of electrons in the conduction band and a high concentration of holes in the valence band. This condition can be achieved for example in the depletion region of a $p - n$ junction under forward bias or by optical pumping, which generates a high density of electron–hole pairs.

The minimum carrier density required to have a positive gain is the transparency density, N_{tr}, which is related to the density of states in the valence and conduction bands. In Exercise 6.2, the transparency carrier density has been calculated in the case of a GaAs QW with $m_v/m_c \simeq 7.5$: In this case, the density of states in the valence and conduction bands is strongly asymmetric. We have seen that in this case, the quasi-Fermi level $\mathcal{E}_{Fc}$ moves into the conduction band and $\mathcal{E}_{Fv}$ moves outside the valence band in order to preserve charge neutrality. In general, the effective mass asymmetry shifts both quasi-Fermi levels toward the band characterized by the lighter effective mass. The asymmetry leads to an increase of the transparency density. Indeed, as calculated in Exercise 6.2, in the case of a symmetric QW ($m_c = m_v$), the transparency density decreases by a factor of ~ 2.4. As we have discussed in Section 6.8, the use of strained QWs allows one to significantly decrease the effective mass in the valence band, so that symmetric structures can be achieved. Note that, in terms of transparency density, the best situation is obtained in the case of a symmetric bulk semiconductor, since in this case the density of states smoothly increases from zero both in the valence and in the conduction bands.

Exercise 6.5   Consider a symmetric bulk semiconductor with effective masses $m_c = m_v$ equal to the conduction effective mass of GaAs ($m_c = 0.067 \, m_0$). Calculate the transparency carrier density.

Since the semiconductor is symmetric, the position of the quasi-Fermi levels at transparency coincides with the band edges. The density of electrons in the conduction band at transparency can be calculates as follows ($\mathcal{E}_{Fc} = \mathcal{E}_c$ and $\mathcal{E}_{Fv} = \mathcal{E}_v$):

$$N_{tr} = \int_{\mathcal{E}_c}^{\infty} \rho_c(\mathcal{E}) f(\mathcal{E}) \, d\mathcal{E}.$$

Therefore, we have:

$$N_{tr} = \frac{1}{2\pi^2} \left(\frac{2m_c}{\hbar^2} \right)^{3/2} \int_{\mathcal{E}_c}^{\infty} \frac{\sqrt{\mathcal{E} - \mathcal{E}_c}}{1 + \exp(\mathcal{E} - \mathcal{E}_c)/k_B T} \, d\mathcal{E}. \tag{6.112}$$

If, as usual, the parameter $x = (\mathcal{E} - \mathcal{E}_c)/k_B T$ is used in Eq. 6.112, the following result is obtained:

$$N_{tr} = N_c F_{1/2}(0), \tag{6.113}$$

where $F_{1/2}(y)$ is the Fermi function introduced in Section 2.4 and $F_{1/2}(0) = 0.765$. Since $N_c = 4.4 \times 10^{17} \, \text{cm}^{-3}$ (see Exercise 2.3), we obtain $N_{tr} = 3.37 \times 10^{17} \, \text{cm}^{-3}$, which is about 1.5 times smaller than the transparency density calculated in the case of a symmetric QW with the same effective mass in the conduction and valence bands. ∎

We can conclude that, in order to minimize the transparency carrier density, it is useful to use symmetric semiconductors with small density of states and a "soft" band edge.

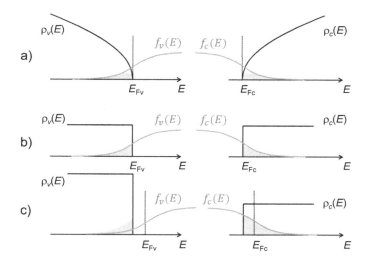

Figure 6.19 Calculation of the transparency carrier density (represented by the shaded areas) in the case of (a) a symmetric bulk semiconductor; (b) a symmetric QW, and (c) an asymmetric QW with $m_c < m_v$.

6.9.1 Differential Gain

Another very important parameter is the differential gain, $\sigma = dg/dN$, which determines how the gain of a given semiconductor device reacts to a variation of the injected carrier density. The gain coefficient g is proportional to $\rho_j(\mathcal{E})[f_c(\mathcal{E}) - f_v(\mathcal{E})]$, which can be written as:

$$g \propto \rho_j(\mathcal{E})f_c(\mathcal{E}) + \rho_j(\mathcal{E})[1 - f_v(\mathcal{E})] - \rho_j(\mathcal{E}). \tag{6.114}$$

In symmetric structures, the joint density of states is almost equal to the density of states (in the conduction or valence band) since the reduced mass m_r contained in the expression of $\rho_j(\mathcal{E})$ can be replaced by $m_c/2$ (or $m_v/2$). Therefore, in Eq. 6.114, the first term on the right side, $\rho_j(\mathcal{E})f_c(\mathcal{E})$, is directly related to the electron density in the conduction band, while the second term, $\rho_j(\mathcal{E})[1 - f_v(\mathcal{E})]$, is directly related to the hole density in the valence band. A high differential gain corresponds to a high variation of the band edge carrier density in response to a given increase in the separation between the quasi-Fermi levels, starting from the transparency condition. Figures 6.19(a) and 6.19(b) show the position of quasi-Fermi levels at transparency and the corresponding carrier density (displayed by the shaded areas) in the case of two symmetric structures (bulk and QW). It is evident that an increase of the separation between the two quasi-Fermi levels is more effective in the case of the symmetric QW (6.19(b)) due to the step-like character of the corresponding density of states. Therefore, it is reasonable to conclude that a symmetric QW is characterized by a higher differential gain in comparison to a symmetric bulk structure. On the other hand, it is also evident that the differential gain of a symmetric QW is higher than the differential gain of an asymmetric QW (shown in Fig. 6.19(c)). Indeed, in the symmetric case at transparency, the maximum slope of the Fermi–Dirac distribution function coincides with the band edges, so that the band edge carrier density is very sensitive to a variation of the position of the quasi-Fermi level with respect to the transparency condition. The situation is completely different in the case of an asymmetric structure, since in this case the band edges do not correspond to the maximum slope of the Fermi–Dirac function, so that the band edge carrier density turns out to be much less sensitive to a variation of the position of the quasi-Fermi levels, thus resulting in a smaller differential gain. In conclusion, in terms of differential gain, it is useful to use a symmetric band structure with step edges. Also in this case the use of strained QWs turns out to be beneficial to improve the performances of semiconductor devices.

6.10 Excitons

Before closing this chapter, it is useful to briefly discuss important effects on the optical properties of a semiconductor related to electron–electron interactions. So far we have always adopted the single-electron approximation, which is very useful since it allowed us to investigate several electronic and optical properties of a semiconductor by using a single-particle Schrödinger equation. In the following, we will consider only the Coulomb component of the electron–electron interaction and we will neglect both exchange and correlation terms.

When an electron is excited to the conduction band, a hole is generated in the valence band; the attraction between the electron and the hole gives rise to a correlated motion of the electron–hole pair, which behaves as a single particle called *exciton*. There are two types of excitons: the *Frenkel excitons* and the *Wannier–Mott excitons*. The Frenkel exciton is localized to a few unit cells; therefore, it is delocalized in the k-space and its analysis requires a full-band treatment. This type of exciton is important in materials with low dielectric constant, where screening effects are weak, for example in the case of organic molecules. A Frenkel exciton generally has a binding energy of the order of 0.1 to 1 eV. In semiconductors, where the dielectric constant is typically large, the Coulomb interaction between the electron and hole in the exciton can be strongly screened by the valence electrons, so that the exciton binding energy is rather small (typically of the order of 0.01 eV) and the exciton dimension is much larger than the lattice spacing, with a typical dimension of a few hundreds Angstroms. For this reason, near band edge electron and hole states can be used to describe these Wannier–Mott excitons and the effective mass approximation can be used. Within this approximation, the electron and hole that form the exciton are described as particles moving with their effective masses, in the conduction and valence bands, respectively.

The simplest approach for the investigation of excitonic effects in semiconductors is based on the analogy with a hydrogen atom, which we have already adopted in Section 2.6 for the description of the donor and acceptor states, which can be seen as particular "excitons," where one of the two coupled particles has an infinite mass. In analogy with the hydrogen atom problem, we can assume, as a first order approximation, that the presence of excitons leads to the formation of discrete states within the gap of the semiconductor, with energies given by the following expression:

$$\mathcal{E} = \mathcal{E}_g - \mathcal{E}_{Bn}, \tag{6.115}$$

where $\mathcal{E}_g$ is the energy gap and $\mathcal{E}_{Bn}$ is the binding energy of the n-th state of the exciton, given by:

$$\mathcal{E}_{Bn} = \frac{m_r e^4}{2(4\pi\epsilon\hbar)^2} \frac{1}{n^2}, \tag{6.116}$$

where we have used the reduced mass of the electron–hole pair and the dielectric constant of the semiconductor ($\epsilon = \epsilon_0 \epsilon_r$). The exciton radius is given by:

$$R_B = \frac{4\pi\epsilon\hbar^2}{m_r e^2}. \tag{6.117}$$

By using Eqs. 6.116 and 6.117, we can calculate a first-order approximation for the energy of the excitonic state $n = 1$ in a semiconductor and the corresponding radius. In the case of GaAs: $m_c = 0.067\ m_0$, $m_{hh} = 0.45\ m_0$, $\epsilon_r = 12.9$ so that $m_r = 0.058\ m_0$, $\mathcal{E}_{B1} = 4.7$ meV and $R_B = 117$ Å. Note that the lattice constant of GaAs is $a_0 = 5.6$ Å, so that the exciton Bohr radius is more than 20 times larger, thus justifying the use of the effective mass approximation for the description of the Wannier exciton.

We will now study the exciton motion in terms of the motion of its center of mass (CM) and of the relative motion of the electron and hole about the exciton CM. We anticipate that, within the effective mass approximation, the exciton CM moves as a free particle with mass $M = m_e + m_h$, where m_e and m_h are the electron and hole masses, respectively, while the relative motion of the electron and hole is similar to the relative motion of an electron and proton in an hydrogen atom. In the effective mass approximation, we have to solve the following two-particle Schrödinger equation, which contains the electron–hole Coulomb interaction:

$$\left(-\frac{\hbar^2}{2m_e}\nabla^2_{\mathbf{r}_e} - \frac{\hbar^2}{2m_h}\nabla^2_{\mathbf{r}_h} - \frac{e^2}{4\pi\epsilon|\mathbf{r}_e - \mathbf{r}_h|} \right) \psi(\mathbf{r}_e, \mathbf{r}_h) = \mathcal{E}\psi(\mathbf{r}_e, \mathbf{r}_h). \tag{6.118}$$

In Eq. 6.118, $\mathbf{r}_e$ and $\mathbf{r}_h$ are the position vectors of the electron and hole, so that $|\mathbf{r}_e - \mathbf{r}_h|$ represents their distance. The effective mass equations for a free electron and a free hole are:

$$-\frac{\hbar^2}{2m_e}\nabla^2_{\mathbf{r}_e}\psi_c(\mathbf{r}_e) = \mathcal{E}_c\psi_c(\mathbf{r}_e) \tag{6.119}$$

$$-\frac{\hbar^2}{2m_h}\nabla^2_{\mathbf{r}_h}\psi_v(\mathbf{r}_h) = \mathcal{E}_v\psi_v(\mathbf{r}_h). \tag{6.120}$$

Without the Coulomb interaction term in Eq. 6.118, we would have: $\psi(\mathbf{r}_e, \mathbf{r}_h) = \psi_c(\mathbf{r}_e)\psi_v(\mathbf{r}_h)$. In the following, we will always assume a weak Coulomb interaction due to the screening effect of the valence electrons, so that the effective mass approximation is valid. Moreover, we will assume:

$$\mathcal{E}_c(\mathbf{k}_e) = \mathcal{E}_g + \frac{\hbar^2 k_e^2}{2m_e} \tag{6.121}$$

$$\mathcal{E}_v(\mathbf{k}_h) = -\frac{\hbar^2 k_h^2}{2m_h}. \tag{6.122}$$

If we now consider the effects of the Coulomb interaction between the electron and hole, it is convenient to use the following transformations:

$$\mathbf{R} = \frac{m_e \mathbf{r}_e + m_h \mathbf{r}_h}{M}, \tag{6.123}$$

which gives the position of the exciton CM, and:

$$\mathbf{r} = \mathbf{r}_e - \mathbf{r}_h, \tag{6.124}$$

which gives the relative position of the electron with respect to the hole. Moreover, we can introduce the CM wavevector, $\mathbf{K}$, as follows:

$$\mathbf{K} = \mathbf{k}_e + \mathbf{k}_h. \tag{6.125}$$

By using Eqs. 6.123 and 6.124 in the Schrödinger equation (Eq. 6.118), we obtain:

$$\left(-\frac{\hbar^2}{2M} \nabla_{\mathbf{R}}^2 - \frac{\hbar^2}{2m_r} \nabla_{\mathbf{r}}^2 - \frac{e^2}{4\pi\epsilon r} \right) \psi(\mathbf{R}, \mathbf{r}) = \mathcal{E}\psi(\mathbf{R}, \mathbf{r}). \tag{6.126}$$

Exercise 6.6 Show that the Schrödinger equation (Eq. 6.118) can be written as in Eq. 6.126.

Since $\mathbf{r} = \mathbf{r}_e - \mathbf{r}_h$, the components of $\mathbf{r}$ along x, y, and z axes are: $x = x_e - x_h$, $y = y_e - y_h$, and $z = z_e - z_h$, while the components of $\mathbf{R}$ are:

$$X = \frac{m_e x_e + m_h x_h}{M}$$
$$Y = \frac{m_e y_e + m_h y_h}{M}$$
$$Z = \frac{m_e z_e + m_h z_h}{M}.$$

The derivative with respect to x_e can be written as:

$$\frac{\partial}{\partial x_e} = \frac{\partial}{\partial x}\frac{\partial x}{\partial x_e} + \frac{\partial}{\partial X}\frac{\partial X}{\partial x_e} = \frac{\partial}{\partial x} + \frac{m_e}{M}\frac{\partial}{\partial X}.$$

In the same way:

$$\begin{aligned}
\frac{\partial^2}{\partial x_e^2} &= \frac{\partial}{\partial x_e}\left(\frac{\partial}{\partial x_e}\right) = \frac{\partial}{\partial x}\left(\frac{\partial}{\partial x_e}\right)\frac{\partial x}{\partial x_e} + \frac{\partial}{\partial X}\left(\frac{\partial}{\partial x_e}\right)\frac{\partial X}{\partial x_e} = \\
&= \frac{\partial}{\partial x}\left(\frac{\partial}{\partial x} + \frac{m_e}{M}\frac{\partial}{\partial X}\right) + \frac{\partial}{\partial X}\left(\frac{\partial}{\partial x} + \frac{m_e}{M}\frac{\partial}{\partial X}\right)\frac{m_e}{M} = \\
&= \frac{\partial^2}{\partial x^2} + \left(\frac{m_e}{M}\right)^2\frac{\partial^2}{\partial X^2} + 2\frac{m_e}{M}\frac{\partial^2}{\partial x \partial X}.
\end{aligned}$$

In the same way, we can write:

$$\frac{\partial^2}{\partial x_h^2} = \frac{\partial^2}{\partial x^2} + \left(\frac{m_h}{M}\right)^2\frac{\partial^2}{\partial X^2} - 2\frac{m_h}{M}\frac{\partial^2}{\partial x \partial X}.$$

By repeating the same procedure for $\partial^2/\partial y_e^2$, $\partial^2/\partial y_h^2$, $\partial^2/\partial z_e^2$ and $\partial^2/\partial z_h^2$ we obtain:

$$-\frac{\hbar^2}{2m_c}\nabla_{\mathbf{r}_e}^2 - \frac{\hbar^2}{2m_v}\nabla_{\mathbf{r}_h}^2 = -\frac{\hbar^2}{2M}\nabla_{\mathbf{R}}^2 - \frac{\hbar^2}{2m_r}\nabla_{\mathbf{r}}^2,$$

so that the complete Schrödinger equation can be written as in Eq. 6.126.   ∎

In this way, we have written an equation where the coordinates of the exciton CM and the relative coordinates are decoupled, so that we can write the wavefunction $\psi(\mathbf{R}, \mathbf{r})$ as a product of two wavefunctions:

$$\psi(\mathbf{R}, \mathbf{r}) = F(\mathbf{R})g(\mathbf{r}). \tag{6.127}$$

In this way from Eq. 6.126, we can write two equations:

$$-\frac{\hbar^2}{2M}\nabla_{\mathbf{R}}^2 F(\mathbf{R}) = \mathcal{E}_R F(\mathbf{R}), \tag{6.128}$$

which describes the motion of the exciton CM and:

$$\left(-\frac{\hbar^2}{2m_r}\nabla_{\mathbf{r}}^2 - \frac{e^2}{4\pi\epsilon r}\right) g(\mathbf{r}) = \mathcal{E}_r g(\mathbf{r}), \tag{6.129}$$

which describes the relative motion of the electron about the hole. The total energy of the exciton is $\mathcal{E}_R + \mathcal{E}_r$. We note that Eq. 6.128 is the Schrödinger equation of a free particle with mass M, so that:

$$F(\mathbf{R}) = \frac{1}{\sqrt{V}}\exp{(i\mathbf{K} \cdot \mathbf{R})} \tag{6.130}$$

$$\mathcal{E}_R = \frac{\hbar^2 K^2}{2M}, \tag{6.131}$$

this last equation represents the kinetic energy due to the motion of the exciton CM. Equation 6.129 is the same equation which describes the hydrogen atom, so that the wavefunction $g(\mathbf{r})$ can be written, in polar coordinate (r, θ, ϕ), as:

$$g_{n\ell m}(\mathbf{r}) = R_{n\ell}(r)Y_{\ell m}(\theta, \phi), \tag{6.132}$$

where n is the principal quantum number, ℓ is the angular momentum quantum number, and m is the magnetic quantum number. $R_{n\ell}(r)$ can be written in terms of Laguerre polynomials, and $Y_{\ell m}(\theta, \phi)$ are the spherical harmonic functions. In the case of isotropic effective masses, the eigenvalues $\mathcal{E}_r$ are given by:

$$\mathcal{E}_r = \mathcal{E}_g - \frac{\text{Ry}^*}{n^2}, \tag{6.133}$$

where the energy gap $\mathcal{E}_g$ represents the minimum energy of the continuum states and Ry^* is the Rydberg constant for the exciton defined as:

$$\text{Ry}^* = \frac{m_r e^4}{2(4\pi\epsilon\hbar)^2} = 13.6\left(\frac{m_r}{m_0\epsilon_r^2}\right) \text{ eV}. \tag{6.134}$$

Therefore, the wavefunctions and energy values for an exciton are given by:

$$\psi_{n\ell m}(\mathbf{R}, \mathbf{r}) = \frac{1}{\sqrt{V}}\exp{(i\mathbf{K} \cdot \mathbf{R})}R_{n\ell}(r)Y_{\ell m}(\theta, \phi) \tag{6.135}$$

$$\mathcal{E}_{n\ell m} = \mathcal{E}_g + \frac{\hbar^2 K^2}{2M} - \frac{\text{Ry}^*}{n^2}. \tag{6.136}$$

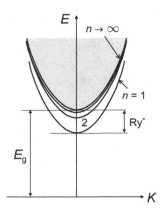

Figure 6.20 Energy states of a Wannier excitons. The bound states from $n = 1$ to $n = 3$ are clearly visible; the shaded area represents the continuum states.

The dispersion curves of a Wannier exciton are shown in Fig. 6.20. The exciton envelope functions $g_{n\ell m}(r)$ are the hydrogen-like functions. Therefore, the ground-state exciton envelope function is given by:

$$g_{100}(r) = \frac{1}{\sqrt{\pi a_B^{*3}}} \exp\left(-\frac{r}{a_B^*}\right) \qquad (6.137)$$

with

$$a_B^* = \left(\epsilon_r \frac{m_0}{m_r}\right) a_B \qquad (6.138)$$

and a_B is the Bohr radius ($a_B = 0.53$ Å). Therefore, the interaction between the electron and hole in an exciton leads to the formation of discrete states below the conduction band edge. The energy term $\hbar^2 K^2 / 2M$ corresponds to the kinetic energy of the free motion of the exciton CM. Above $\mathcal{E}_g + \frac{\hbar^2 K^2}{2M}$, there is a continuum of allowed energy states.

6.10.1 Absorption Spectrum

We will now calculate how the presence of excitons modifies the absorption spectrum of a semiconductor. From a general point of view, the absorption spectrum can be calculated starting from the eigenfunctions and eigenenergies given in Eqs. 6.135 and 6.136 and considering a suitable Hamiltonian describing the interaction between excitons and photons. The procedure is rather complex; therefore, we will limit to mention the main final results.

It is possible to demonstrate that the imaginary part of the optical susceptibility can be written as:

$$\text{Im}\chi(\omega) = \frac{e^2 |p_{vc}|^2 \sqrt{\text{Ry}^*}}{2\epsilon_0 m^2 \omega^2} \left(\frac{2m_r}{\hbar^2}\right)^{3/2} \left[\sum_n \frac{4}{n^3} \delta\left(\Delta + \frac{1}{n^2}\right) + \frac{H(\Delta)e^{\pi/\sqrt{\Delta}}}{\sinh\left(\pi/\sqrt{\Delta}\right)}\right], \quad (6.139)$$

which is called *3D Elliott formula*. In Eq. 6.139:

$$\Delta = \frac{\hbar\omega - \mathcal{E}_g}{\text{Ry}^*} \tag{6.140}$$

and $H(\Delta)$ is the Heaviside step function. By using Eq. 5.51, $|p_{vc}|^2 = m^2\omega^2|\mu_{vc}|^2/e^2$, the 3D Elliott formula can be rewritten as:

$$\text{Im}\chi(\omega) = \frac{|\mu_{vc}|^2\sqrt{\text{Ry}^*}}{2\epsilon_0}\left(\frac{2m_r}{\hbar^2}\right)^{3/2}\left[\sum_n \frac{4}{n^3}\delta\left(\Delta + \frac{1}{n^2}\right) + \frac{H(\Delta)e^{\pi/\sqrt{\Delta}}}{\sinh\left(\pi/\sqrt{\Delta}\right)}\right]. \tag{6.141}$$

The absorption coefficient is given by:

$$\alpha = \frac{\omega}{nc}\text{Im}\chi(\omega). \tag{6.142}$$

As we have already observed, the absorption spectrum is composed by a series of sharp lines with an amplitude rapidly decreasing upon increasing the quantum number n (the amplitude is proportional to n^{-3}) and by a continuum related to the ionized electron and hole particles.

It is instructive to consider the 3D Elliott formula when Coulomb effects become negligible, that is, when excitonic effects are not present. In this situation, the relative dielectric constant ϵ_r becomes very large so that Ry* becomes very small. We will assume Ry$^* \to 0$, so that $\Delta \to \infty$:

$$\lim_{\Delta\to\infty}\frac{e^{\pi/\sqrt{\Delta}}}{\sinh\left(\pi/\sqrt{\Delta}\right)} = \frac{\sqrt{\Delta}}{\pi}. \tag{6.143}$$

Moreover, we do not have to consider the contribution from bound exciton, so that the absorption spectrum given by Eqs. 6.142 and 6.141 is given by:

$$\begin{aligned}\alpha_{free}(\omega) &= \frac{\omega|\mu_{vc}|^2\sqrt{\text{Ry}^*}}{2\epsilon_0 nc}\left(\frac{2m_r}{\hbar^2}\right)^{3/2}\frac{\sqrt{\Delta}}{\pi} = \\ &= \frac{\omega|\mu_{vc}|^2}{2\pi\epsilon_0 nc}\left(\frac{2m_r}{\hbar^2}\right)^{3/2}\sqrt{\hbar\omega - \mathcal{E}_g} = \\ &= \frac{\omega\pi}{nc\epsilon_0\hbar}|\mu_{vc}|^2\rho_j(\omega),\end{aligned} \tag{6.144}$$

which is the expression we have obtained in the case of bulk semiconductors assuming $f_v(\hbar\omega) - f_c(\hbar\omega) = 1$.

From Eqs. 6.141 and 6.142, the contribution to the absorption coefficient from the continuum states is given by:

$$\alpha_{cont} = \frac{\omega|\mu_{vc}|^2\sqrt{\text{Ry}^*}}{2\epsilon_0 nc}\left(\frac{2m_r}{\hbar^2}\right)^{3/2}\frac{H(\Delta)e^{\pi/\sqrt{\Delta}}}{\sqrt{\Delta}\sinh(\pi/\sqrt{\Delta})}, \tag{6.145}$$

which can be written in terms of $\alpha_{free}(\omega)$ as:

$$\alpha_{cont}(\omega) = \alpha_{free}(\omega)\frac{\pi}{\sqrt{\Delta}}\frac{e^{\pi/\sqrt{\Delta}}}{\sinh\left(\pi/\sqrt{\Delta}\right)} = \alpha_{free}(\omega)S(\omega), \tag{6.146}$$

where $S(\omega)$ is called *Sommerfeld or Coulomb enhancement factor*. At the band edge $\Delta \to 0$

$$\lim_{\Delta\to 0}\frac{e^{\pi/\sqrt{\Delta}}}{\sinh\left(\pi/\sqrt{\Delta}\right)} = 2, \tag{6.147}$$

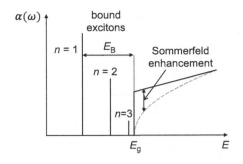

Very schematic plot of the absorption spectrum of a bulk semiconductor with (solid line) and without (dashed gray line) excitonic effects.

so that the Sommerfeld factor is approximately given by:

$$S(\omega) \to 2\frac{\pi}{\sqrt{\Delta}}. \tag{6.148}$$

Therefore, at the band edge, the absorption does not go to zero following a square-root law as in the bulk case without excitonic effects, but it assumes a constant value given by:

$$\alpha(\Delta \to 0) = \frac{\omega|\mu_{vc}|^2\sqrt{\mathrm{Ry}^*}}{\epsilon_0 nc}\left(\frac{2m_r}{\hbar^2}\right)^{3/2}. \tag{6.149}$$

Figure 6.21 shows a schematic band edge absorption spectrum of a bulk semiconductor. It is evident that the presence of excitons does not only introduce discrete states in the gap, but significantly changes also the absorption in the continuum. Moreover, if the spectral broadening is taken into account (induced mainly by scattering of the electron–hole pairs with phonons), only a few bound states can be spectrally resolved, since the higher bound states merge together with the absorption of the ionized states. In real bulk semiconductors, the absorption features related to the bound excitons can be experimentally observed at low temperatures in good-quality semiconductors, as shown in Fig. 6.22, which displays the absorption spectra of GaAs near the band gap for several sample temperatures, in the range between 21 K and 294 K. Since the binding energy of the exciton in GaAs is about 5 meV, only a broadened n = 1 bound state is visible at 21 K. We note that at room temperature, the GaAs absorption edge is still modified by excitonic effects.

6.10.2 Excitons in Quantum Wells

In the case of QWs, the well thickness is typically smaller than the exciton Bohr radius in three-dimensional semiconductors; moreover, the Coulomb potential in the direction perpendicular to the well does not play a significant role compared to the confinement potential. We have a kind of "two-dimensional" hydrogen atom, which introduces discrete energy levels in the gap, whose dispersion relations are given by the following expression:

$$\mathcal{E}_{nK} = \mathcal{E}_g + \mathcal{E}_{1c} + \mathcal{E}_{1v} + \frac{\hbar^2 K^2}{2M} - \frac{\mathrm{Ry}^*}{\left(n - \frac{1}{2}\right)^2}. \tag{6.150}$$

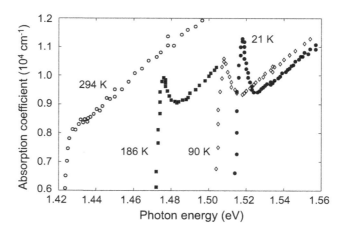

Figure 6.22 Excitonic absorption spectra of GaAs near band gap for several sample temperatures: 294 K (open circles), 186 K (solid square), 90 K (open diamond), and 21 K (solid circles). Adapted with permission from [9].

The fundamental binding energy of excitons in QWs is four times the corresponding binding energy in the three-dimensional case, indeed $\mathcal{E}_B = 4\text{Ry}^*$.

It is possible to demonstrate that the imaginary part of the optical susceptibility describing the excitonic absorption in a QW with thickness L is given by the *2D Elliott formula*

$$\text{Im}\chi(\omega) = \frac{e^2|p_{vc}|^2 m_r}{\epsilon_0 m^2 \hbar^2 \omega^2 L}\left\{\sum_n \frac{4}{\left(n+\frac{1}{2}\right)^3}\delta\left[\Delta + \frac{1}{\left(n+\frac{1}{2}\right)^2}\right] + \frac{H(\Delta)e^{\pi/\sqrt{\Delta}}}{\cosh\left(\pi/\sqrt{\Delta}\right)}\right\} \quad (6.151)$$

where:

$$\Delta = \frac{\hbar\omega - (\mathcal{E}_g + \mathcal{E}_{1c} + \mathcal{E}_{1v})}{\text{Ry}^*}. \quad (6.152)$$

Equation 6.151 can be also written as:

$$\text{Im}\chi(\omega) = \frac{|\mu_{vc}|^2 m_r}{\epsilon_0 \hbar^2 L}\left\{\sum_n \frac{4}{\left(n+\frac{1}{2}\right)^3}\delta\left[\Delta + \frac{1}{\left(n+\frac{1}{2}\right)^2}\right] + \frac{H(\Delta)e^{\pi/\sqrt{\Delta}}}{\cosh\left(\pi/\sqrt{\Delta}\right)}\right\} \quad (6.153)$$

and the absorption spectrum can be calculated by using Eq. 6.142. Also in the case of QWs, in the limit of completely negligible Coulomb interaction (limit of independent electrons, $\Delta \to \infty$), the absorption coefficient given by Eq. 6.153 turns out to be given by what we have already calculated without excitonic effects. Indeed, in this case:

$$\lim_{\Delta \to \infty} \frac{e^{\pi/\sqrt{\Delta}}}{\cosh\left(\pi/\sqrt{\Delta}\right)} = 1, \quad (6.154)$$

so that the absorption spectrum is given by:

$$\alpha_{free}(\omega) = \frac{\omega|\mu_{vc}|^2 m_r}{cn\epsilon_0 \hbar^2 L}H[\hbar\omega - (\mathcal{E}_g + \mathcal{E}_{1c} + \mathcal{E}_{1v})], \quad (6.155)$$

which is the expression we have obtained in the case of free charges assuming $f_v(\hbar\omega) - f_c(\hbar\omega) = 1$.

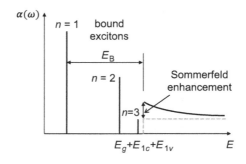

Very schematic plot of the absorption spectrum of a QW with (solid line) and without (dashed gray line) excitonic effects.

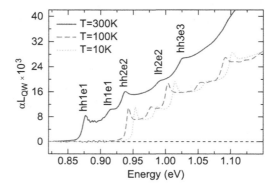

Linear absorption spectrum of a Ge multiple quantum well structure at several temperatures (adapted from [10]). Note: N.S. Köster et al., New J. Phys. 15, 075004 (2013), CC BY 3.0.

Also in this case we can write the continuum contribution as:

$$\alpha_{cont}(\omega) = \alpha_{free}(\omega)\frac{e^{\pi/\sqrt{\Delta}}}{\cosh\left(\pi/\sqrt{\Delta}\right)} = \alpha_{free}(\omega)S_{free}(\omega). \tag{6.156}$$

Since at the band edge ($\Delta \to 0$), the Sommerfeld factor is 2, the absorption at the band edge is twice the free-carrier absorption. Figure 6.23 shows a schematic band edge absorption spectrum in a QW. As in the bulk case, considering the spectral broadening of the sharp excitonic states in the gap, the absorption in the higher band states merges with the absorption of the ionized states. In contrast to the bulk case, the fundamental $1s$ bound excitonic peak is much better spectrally resolved due to the strong increase (factor of 4) of the binding energy in the case of QWs. This is clearly visible in Fig. 6.24, which shows the linear absorption spectra of a Ge QW structure at 10 K, 100 K, and 300 K.

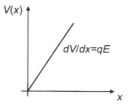

Figure 6.25 Triangular quantum well.

6.11 Exercises

Exercise 6.1   Consider a triangular quantum well, as shown in Fig. 6.25, characterized by an infinite potential barrier at $x = 0$ and a slope $dV/dx = qE$ for $x > 0$.
a) Write the Schrödinger equation for a charge q in this potential well.
b) Demonstrate that this Schrödinger equation can be written in the form of the Airy equation:

$$\frac{d^2\psi}{d\xi^2} - \xi\psi(\xi) = 0.$$

Hint: you have to make a substitution into the original equation using a change of variables

$$\xi = \left(\frac{2meE}{\hbar^2}\right)^{1/3}\left(x - \frac{\mathcal{E}_n}{eE}\right).$$

c) The general solution of the previous Airy equation is:

$$\psi(\xi) = C \cdot \mathrm{Ai}(\xi),$$

where $\mathrm{Ai}(\xi)$ is the Airy function and C is a constant, which can be determined by normalization. Demonstrate that the energy eigenvalues, $\mathcal{E}_n$, are given by:

$$\mathcal{E}_n = -\left(\frac{q^2E^2\hbar^2}{2m}\right)^{1/3} a_n,$$

where a_n are the nth zeros of the Airy function, which can be approximated by:

$$a_n = -\left[\frac{3\pi}{2}\left(n - \frac{1}{4}\right)\right]^{2/3} \qquad n = 1, 2, \cdots$$

d) Plot the Airy function normalized to $\mathrm{Ai}(0)=1$.
e) Describe a real physical situation where a triangular quantum well comes into play.

Exercise 6.2   Assume an electric field $E = 2.3 \times 10^7$ V/m and consider an electron with effective mass $m = 0.91\, m_0$. By using the results of Exercise 6.1, calculate the first four energy levels and plot the corresponding wavefunctions.

Exercise 6.3   The normalized stationary states of an infinite potential wells are:

$$\psi_n = \sqrt{\frac{2}{L}} \sin \left(n\frac{\pi}{L}x \right)$$

demonstrate that they are orthogonal.

Exercise 6.4   Consider a particle in the second state $\psi(x)$ ($n = 2$) of an infinite potential well, with thickness L. Calculate:
a) the expectation value of the position, $\langle x \rangle$;
b) the probability of finding the particle on the region $0.4L \leq x \leq 0.6L$.

Exercise 6.5   Consider a 10-nm-thick GaAs quantum well (GaAs well/AlGaAs) at $T = 0$ K. Assume $\mathcal{E}_g = 1.519$ eV, $m_c = 0.067\ m_0$, $m_{hh} = 0.5\ m_0$, $m_{lh} = 0.087\ m_0$, refractive index $n = 3.5$ and Kane energy $\mathcal{E}_P = 25.7$ eV.
a) Assuming the infinite potential well approximation, calculate the confinement energies of the first subband in the conduction band and of the first heavy-hole and light-hole subbands in the valence band.
b) Assuming that only the heavy-hole band is populated, determine the difference between the quasi-Fermi levels as a function of the carrier density at $T = 0$ K.

Exercise 6.6   Same as Exercise 6.5. Plot the gain coefficient for the quantum well as a function of the photon energy for two different values of the splitting between the quasi-Fermi levels: $\mathcal{E}_{Fc} - \mathcal{E}_{Fv} = 0$ and $\mathcal{E}_{Fc} - \mathcal{E}_{Fv} = 1.05(\mathcal{E}_g + |\mathcal{E}_{c1}| + |\mathcal{E}_{hh1}|)$. Consider only transitions between the first conduction, heavy-hole and light-hole subbands.

Exercise 6.7   Determine the density of states in the subband $n = 1$, $n/\rho^{QW}k_BT$, as a function of the position of the Fermi level, $(\mathcal{E}_F - \mathcal{E}_{c1})/k_BT$ and draw the corresponding logarithmic plot.

Exercise 6.8   Determine the transparency carrier concentration in a symmetric quantum well and in an asymmetric quantum well with $m_v^* = 6m_c^*$.

Exercise 6.9   Determine and plot the normalized difference between the quasi-Fermi energy, $\mathcal{E}_F$, and the energy of the $n = 1$ subband as a function of injected carriers, for both electrons and holes, in a 10-nm GaAs/AlGaAs quantum well.

Exercise 6.10   Consider a particle confined in a well with infinite barriers, extending from $x = 0$ to $x = L$.
a) Calculate the probability that the particle in the eigenstate $\psi_n(x)$ is located in the region between $x = 0$ and $x = L/4$.
b) Determine the value of n corresponding to the largest probability to find the particle in the region $0 \leq x \leq L/4$.

Exercise 6.11   Consider a particle confined in a well with infinite barriers in the region $0 \leq x \leq L$. Assume that the system at $t = 0$ is in a superposition of two eigenstates $\psi = a\psi_n + b\psi_m$

a) write the wavefunction ψ at time t;

b) determine the expectation value of the energy at time t.

Exercise 6.12   Same as Exercise 6.11. Assume that the system has been prepared in a nonstationary state so that at the time of the measurement of the energy of the particle, the state is described by the following normalized wavefunction

$$\psi(x) = \sqrt{\frac{30}{L^5}} x(L - x) \qquad 0 \le x \le L$$

and $\psi(x) = 0$ elsewhere.

a) Calculate the probability that a measurement of the energy of the particle will give the value $\mathcal{E}_n = (n\pi\hbar)^2/(2mL^2)$ for any given value of n.

b) Determine the expectation value of H (average energy of the system) for the system described by the wavefunction $\psi(x)$.

Exercise 6.13   Calculate the binding energy and the Bohr radius of an exciton in InN ($m_c = 0.11\,m_0$, $m_{hh} = 1.63\,m_0$, $\epsilon_r = 8.4$).

Exercise 6.14   Consider a double heterostructure composed by the quaternary alloy $Ga_xIn_{1-x}As_yP_{1-y}$ grown on InP substrate. Assume that for the quaternary alloy, the lattice constant, a_0 in Å, and the room-temperature band gap, $\mathcal{E}_g$ in eV, are given by:

$$a_0 = 5.87 + 0.18y - 0.42x + 0.02xy$$
$$\mathcal{E}_g(\Gamma) = 1.35 + 0.668x - 1.068y + 0.758x^2 + 0.078y^2 +$$
$$- 0.069xy - 0.322x^2y + 0.03xy^2.$$

The lattice constant of InP is $a_0 = 5.869$ Å.

a) Calculate the composition of the quaternary alloy lattice matched to InP and with an energy gap at 1,550 nm.

b) Same as point a) but at 1,330 nm.

c) Determine the smallest and the widest band gap (in nm) that can be obtained with the quaternary alloy lattice matched to InP and the corresponding composition.

Exercise 6.15   Consider a GaAs quantum well with thickness $L = 10$ nm ($m_c = 0.067\,m_0$). Determine:

a) the energy difference between the first two subbands in the conduction band;

b) the dipole matrix element for the intersubband transition from $n = 1$ to $n = 2$;

c) the intersubband absorption at resonance for a wave incident at 45°, assuming a lineshape broadening of 10 meV.

Exercise 6.16   Linearly polarized light is incident on a quantum well at an angle θ to the normal direction, z. The polarization direction is within the plane of incidence. Determine the maximum fraction of the power in the input beam, which can be absorbed by intersubband transitions, assuming a refractive index $n = 3.3$.

7 Light Emitting Diodes

7.1 Basic Concepts

In this chapter, the generation of light by the process of electroluminescence in a semiconductor will be considered. This phenomenon was first reported in 1907 by Henry Joseph Round, who observed the emission of visible light from carborundum (i.e., an impure form of polycristalline SiC) when a potential was applied between two points of the crystal. In 1951, Lehovec and coworkers explained the electroluminescence in SiC as the result of injection of carriers across a junction followed by radiative recombination of the electron–hole pairs. The first demonstration of a direct band gap visible (GaAsP) light-emitting diode (LED) was reported by Holonyak and Bevaqua in 1962. Toward the end of the 1960s, the mass production of low-cost GaAsP red LEDs began (first in Monsanto and then at Hewlett-Packard).

A LED is essentially a $p-n$ junction diode under forward bias, as shown in Fig. 7.1 in the case of a homojunction. We will first describe this simple case although homojunctions are not used in real devices (as we will discuss in this chapter). Under forward bias electrons from the n-side of the junction and holes from the p-side cross the depletion layer thus becoming minority carriers. These excess minority carriers diffuse in the semiconductor (minority electrons diffuse in the p-type semiconductor, while minority holes diffuse in the n-type semiconductor) and recombine with majority carriers. For example, the excess minority electron concentration in the p-type portion varies along the semiconductor as:

$$\Delta n(x) = \Delta n(0) \exp(-x/L_n), \tag{7.1}$$

where L_n is the diffusion length of electrons, defined as the average distance a minority carrier diffuses before recombination. The diffusion length is given by:

$$L_n = \sqrt{D_n \tau_n}, \qquad L_p = \sqrt{D_p \tau_p}, \tag{7.2}$$

where D_n and D_p are the diffusion constants for electrons and holes, respectively, τ_n and τ_p are the electron and hole carrier lifetimes. In general, the diffusion length in a semiconductor is of the order of a few micrometers. As we have discussed in Chapter 5, electron–hole recombination can occur by means of radiative and nonradiative processes: A LED is designed to maximize the fraction of radiative events. Of course, if the $p-n$ junction is under reverse bias, no carriers are injected across the junction and no photons can be generated. The current-to-voltage characteristic for a diode can be written as follows:

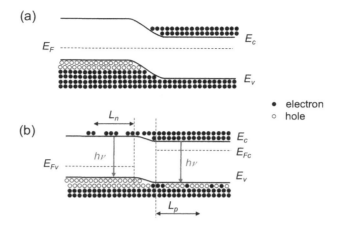

Figure 7.1 $p - n$ homojunction under zero bias (a) and under forward bias (b).

$$I = I_s \left[\exp\left(\frac{eV}{k_B T} \right) - 1 \right] \simeq I_s \exp\left(\frac{eV}{k_B T} \right), \tag{7.3}$$

where I_s is the reverse saturation current and V is the voltage applied to the diode.

In the following, we will recall a few important concepts about radiative and nonradiative recombination processes we have already discussed in Chapter 5. Equation 5.69 gives the spontaneous emission rate, $R_{sp}(\hbar\omega)$ per unit time, volume, and energy (for photons with energy $\hbar\omega$):

$$R_{sp}(\hbar\omega) = \frac{1}{\tau_r} \rho_j(\hbar\omega) f_c(\hbar\omega)[1 - f_v(\hbar\omega)], \tag{7.4}$$

where τ_r is the spontaneous emission lifetime. In general, in the case of LEDs, the excess carrier concentration is rather small so that the quasi-Fermi levels are in the gap and quite far from the band edges, as shown in Fig. 7.1(b). Therefore, the Fermi–Dirac distribution function can be approximated by the Boltzmann distribution function. With this approximation, the spontaneous emission rate, which gives the spontaneous emission lineshape, can be written as (see Eq. 5.79):

$$R_{sp}(\hbar\omega) = \Re_{sp} (\hbar\omega - \mathcal{E}_g)^{1/2} \exp\left(-\frac{\hbar\omega - \mathcal{E}_g}{k_B T} \right), \tag{7.5}$$

where $\Re_{sp}$ is given by the following expression (see Eq. 5.78):

$$\Re_{sp} = \frac{1}{2\pi^2 \tau_r} \left(\frac{2m_r}{\hbar^2} \right)^{3/2} \exp\left(\frac{\Delta\mathcal{E}_F - \mathcal{E}_g}{k_B T} \right), \tag{7.6}$$

where $\Delta\mathcal{E}_F$ is the difference between the quasi-Fermi energies: $\Delta\mathcal{E}_F = \mathcal{E}_{Fc} - \mathcal{E}_{Fv}$. In Chapter 5, it has been demonstrated that the photon energy corresponding to the peak value of the spontaneous emission lineshape is $\hbar\omega = \mathcal{E}_g + k_B T/2$ with a full-width at half-maximum $\Delta\mathcal{E} = 1.8 k_B T$, which can be written in terms of the emission wavelength as follows (see Exercise 5.4):

$$\Delta\lambda = 1.8 k_B T \frac{\lambda^2}{hc}. \tag{7.7}$$

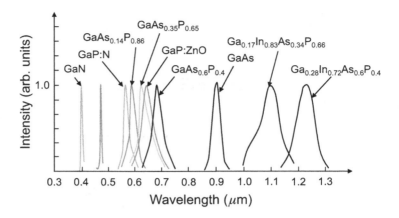

Emission spectra of various LEDs from violet to near-infrared.

For example, an AlGaAs/GaAs LED emitting at 625 nm has a (theoretical) linewidth $\Delta\mathcal{E} = 46$ meV (at $T = 300$ K), corresponding to $\Delta\lambda \simeq 15$ nm. Figure 7.2 shows the spectral distribution of various LEDs with emission from the violet to the near-infrared.

We have then calculated the total radiative recombination rate, $R_{sp,T}$, for photons of all energies (see Eq. 5.93) in the case of nondegenerate semiconductors, so that Maxwell–Boltzmann statistics can be applied. This is the typical situation in the case of LEDs, so that the radiative recombination rate can be written as:

$$R_{sp,T} = Bnp, \tag{7.8}$$

where B is the bimolecular recombination coefficient, given by Eq. 5.94. The net recombination rate is given by (see Eq. 5.105):

$$R_r = \frac{\Delta n}{\tau_{rad}}, \tag{7.9}$$

where Δn is the nonequilibrium carrier density in the junction region of the LED and τ_{rad} is the radiative recombination time. In the following, we will assume $\Delta n = \Delta p = n$. We can calculate the total photon flux, Φ, per second as:

$$\Phi = R_r \frac{V_{jun}}{S} = \frac{n}{\tau_{rad}}d, \tag{7.10}$$

where V_{jun} is the volume of the junction ($V_{jun} = Sd$, with d the effective thickness of the junction, where electron–hole recombination takes place).

The nonradiative recombination processes include the Shockley–Read–Hall (SRH) process and the Auger recombination, which depend on the minority carrier concentration, n (electrons in a p-type layer, holes in a n-type layer). We have seen that the SRH process increases linearly with n, $A_{nr}n$ (see Eq. 5.137), whereas the Auger recombination process in the high-excitation limit increases as the cube of n, Cn^3 (or, more generally, $f(n)$), since it is a three-particle process. A_{nr} and C are constants. Therefore, the nonradiative lifetime, τ_{nr}, can be written as:

$$\frac{1}{\tau_{nr}} = A_{nr} + Cn^2. \tag{7.11}$$

The total recombination time, τ (see Eq. 5.191), is given by:

$$\frac{1}{\tau} = \frac{1}{\tau_{rad}} + \frac{1}{\tau_{nr}} = A_{nr} + Bn + Cn^2. \tag{7.12}$$

The probability of radiative recombination, called *radiative efficiency*, η_r (Eq. 5.192), is given by:

$$\eta_r = \frac{1/\tau_{rad}}{1/\tau_{rad} + 1/\tau_{nr}} = \frac{\tau}{\tau_{rad}}, \tag{7.13}$$

that is, it is defined as the ratio between the probability of radiative recombination and the total recombination probability. If J_e and J_h are the current densities for electrons and holes, the electron and hole fluxes into the junction region are given by J_e/e and J_h/e. To preserve charge neutrality, we must have $J_e = J_h = J$. The injected carriers per unit volume can be written as:

$$\Psi_c = \eta_i \frac{J}{ed}, \tag{7.14}$$

where η_i, called *injection efficiency*, gives the fraction of current which generates carriers in the active region. In stationary state conditions, the injected flux of carriers which are flowing through the surface S of the junction equals the total recombination rate, R_t, in the volume V_{jun}, so that:

$$\eta_i \frac{J}{ed} = A_{nr}n + Bn^2 + Cn^3 = \frac{n}{\tau}. \tag{7.15}$$

By using this last expression in Eq. 7.10, the total photon flux can be written as:

$$\Phi = \eta_i \eta_r \frac{J}{e} = \eta_{int} \frac{J}{e}, \tag{7.16}$$

where $\eta_{int} = \eta_i \eta_r$ is called *internal quantum efficiency*. From this relation, the physical meaning of η_{int} is clear: It represents the ratio between the number of photons generated inside the LED and the total number of injected electrons.

7.2 Double-Heterostructure LEDs

At low carrier densities, the nonradiative SRH term (A_{nr}) becomes the dominant term, thus leading to a low internal quantum efficiency. This is the situation in the case of a $p-n$ homojunction LED, shown in Fig. 7.1. In this case, the electrons diffuse into the p-type layer and the holes diffuse into the n-type layer. The effective thickness d of the junction is given by $L_n + L_p$, and the diffusion length for minority carriers is typically in a range between 1 μm and 10 μm; therefore, the minority carriers diffuse over a large region thus reducing their density. In order to significantly decrease the effective thickness d thus increasing the minority carrier density, LEDs are typically based on double heterostructures, in which the thickness of the active layer (i.e., the layer where radiative recombinations occur) is typically around 0.2–0.3 μm. Therefore, at the same value of the current density, the carrier concentration increases, thus leading to a significant increase of the probability of radiative recombination (Bn) and of the LED efficiency.

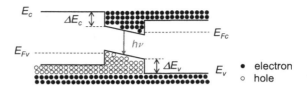

Figure 7.3   Schematic energy bands of a double-heterostructure LED under forward bias.

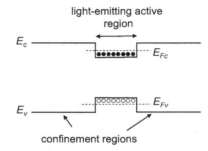

Figure 7.4   Qualitative interpretation of the Burstein–Moss shift.

A double heterostructure is formed by an active region, where radiative recombination occurs, and two confinement layers cladding the active region, characterized by a larger energy gap, as shown schematically in Fig. 7.3. In order to obtain an effective carrier confinement, the band discontinuities $\Delta\mathcal{E}_c$ and $\Delta\mathcal{E}_v$ in the conduction and valence band, respectively, must be larger than the thermal energy $k_B T$, to block carrier leakage into the cladding layers. Double heterostructures can be employed both in the case of bulk semiconductors and in the case of quantum wells, where additional carrier localization is achieved in the well region.

An additional important advantage of double heterostructures with respect to homojunctions is that the photons emitted in the active region are not absorbed by the confinement semiconductors, which have a larger energy gap. Since the confinement regions are rather thin, they are almost totally transparent to the emitted radiation. Moreover, under typical carrier injection levels, also the active region is almost transparent to the emitted radiation. Indeed, as shown in Fig. 7.4, due to carrier injection into the active layer, the quasi-Fermi levels in the conduction and valence bands are within the bands, so that all states close to the conduction band edge are populated by electrons and transitions from the valence band to these states cannot occur, thus leading to a shift of the absorption edge to higher energies: This effect is called *Burstein–Moss shift*. It is therefore evident why homojunctions are not used in LEDs.

7.2.1 Burstein–Moss shift

If we assume that all levels in the conduction band with energy below $\mathcal{E}_{Fc} - 4k_B T$ are populated, the minimum energy of an absorbed photon is given by (see Fig. 7.5):

$$(\hbar\omega)_{min} = \mathcal{E}_g + \Delta\mathcal{E}_1 + \Delta\mathcal{E}_2. \tag{7.17}$$

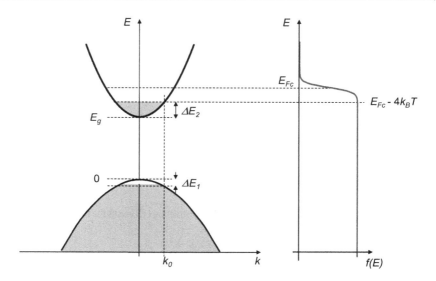

Figure 7.5 Dispersion curves in the conduction and valence band of a bulk semiconductor; the shaded areas represent occupied states. On the right the Fermi–Dirac distribution function is reported. It is assumed that all levels with energy below $\mathcal{E}_{Fc} - 4k_BT$ are populated.

The corresponding k-value is k_0 as displayed in Fig. 7.5. At $k = k_0$, the energy $\mathcal{E}_2$ of the corresponding state in the conduction band is:

$$\mathcal{E}_2 = \mathcal{E}_g + \frac{\hbar^2 k_0^2}{2m_c} = \mathcal{E}_{Fc} - 4k_BT, \tag{7.18}$$

so that

$$\frac{\hbar^2 k_0^2}{2m_c} = \mathcal{E}_{Fc} - 4k_BT - \mathcal{E}_g = \Delta\mathcal{E}_2. \tag{7.19}$$

The energy $\mathcal{E}_1$ of the corresponding state in the valence band is given by:

$$\mathcal{E}_1 = -\frac{\hbar^2 k_0^2}{2m_v} = -\frac{m_c}{m_v}(\mathcal{E}_{Fc} - 4k_BT - \mathcal{E}_g) = -\Delta\mathcal{E}_1. \tag{7.20}$$

The Burnstein–Moss shift of the absorption edge is given by:

$$\Delta\mathcal{E} = (\hbar\omega)_{min} - \mathcal{E}_g = \Delta\mathcal{E}_1 + \Delta\mathcal{E}_2. \tag{7.21}$$

From Eqs. 7.19–7.21, we have:

$$\Delta\mathcal{E} = (\mathcal{E}_{Fc} - 4k_BT - \mathcal{E}_g)\left(1 + \frac{m_c}{m_v}\right). \tag{7.22}$$

As we have seen in Section 2.4, the relationship between the electron density and Fermi energy is given by:

$$n = N_c F_{1/2}(y), \tag{7.23}$$

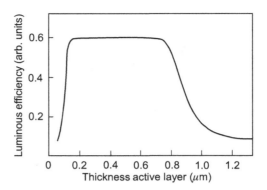

where:

$$y = \frac{\mathcal{E}_{Fc} - \mathcal{E}_c}{k_B T} = \frac{\mathcal{E}_{Fc} - \mathcal{E}_g}{k_B T}. \qquad (7.24)$$

In the case we are considering y is positive and can be assumed $\gg 1$, so that the Fermi function can be approximated as in Eq. 2.44, so that:

$$n = \frac{4 N_c}{3 \sqrt{\pi}} \left(\frac{\mathcal{E}_{Fc} - \mathcal{E}_g}{k_B T} \right)^{3/2}. \qquad (7.25)$$

From this expression, it is possible to calculate $\mathcal{E}_{Fc} - \mathcal{E}_g$:

$$\mathcal{E}_{Fc} - \mathcal{E}_g = k_B T \left(\frac{3 \sqrt{\pi}}{4} \frac{n}{N_c} \right)^{2/3}. \qquad (7.26)$$

A very important device parameter which strongly influences the LED efficiency is the thickness of the active region. Typical thickness is of the order of a few tenth of micrometer in the case of bulk active regions, as shown in Fig. 7.6 in the case of an AlGaInP double-heterostructure LED. The decrease of LED efficiency upon increasing the thickness of the active region above $\sim 1~\mu$m can be easily understood: Carrier confinement induced by the double heterostructure is less effective and when the thickness becomes comparable or larger than the carrier diffusion length, double heterostructures do not present any advantage compared to homojunctions. If the thickness of the active region is decreased too much at moderately high levels of carrier injection, there is a significant carrier overflow, thus leading to an increase of the losses. An estimation of the carrier concentration required to have carrier overflow can be obtained as follows.

7.2.2 Carrier Overflow

We will consider the conduction band, but the same considerations can be repeated also for the valence band. Under steady-state conditions, the injection of carriers equals the rate of carrier recombination. Therefore, assuming only radiative recombination and $\eta_i = 1$:

$$\frac{J}{ed} = Bnp. \tag{7.27}$$

For high injection density $n = p$ and from 7.27:

$$n = \sqrt{\frac{J}{eBd}}. \tag{7.28}$$

The condition for carrier overflow is reached when the separation between the quasi-Fermi level and the bottom of the conduction band, $\mathcal{E}_{Fc} - \mathcal{E}_c$, equals the height of the potential barrier between the active region and the cladding layers:

$$\mathcal{E}_{Fc} - \mathcal{E}_c = \mathcal{E}_{Fc} - \mathcal{E}_g = \mathcal{E}_B. \tag{7.29}$$

The net barrier height includes residual band bending under forward bias and the conduction band discontinuity $\Delta \mathcal{E}_c$. By using Eq. 7.26, we can write:

$$k_B T \left(\frac{3\sqrt{\pi}}{4} \frac{n}{N_c} \right)^{2/3} = \mathcal{E}_B, \tag{7.30}$$

so that:

$$n = \frac{4}{3\sqrt{\pi}} N_c \left(\frac{\mathcal{E}_B}{k_B T} \right)^{3/2}. \tag{7.31}$$

The corresponding driving current density required to generate an electron density giving carrier overflow can be written as:

$$J = eBd \left(\frac{4}{3\sqrt{\pi}} N_c \right)^2 \left(\frac{\mathcal{E}_B}{k_B T} \right)^3. \tag{7.32}$$

∎

Exercise 7.1   Calculate the current density required for carrier overflow at room temperature in a GaAs double heterostructure assuming a barrier height $\Delta \mathcal{E}_c = 180$ meV and a thickness $d = 0.1$ μm. The bimolecular recombination coefficient is $B = 1.7 \times 10^{-10}$ cm^{3}s^{-1}.

In Exercise 2.3, the effective density of states in the conduction band of GaAs has been calculated: $N_c = 4.4 \times 10^{17}$ cm^{-3}. From Eq. 7.33, one obtains:

$$J = eBd \left(\frac{4}{3\sqrt{\pi}} N_c \right)^2 \left(\frac{\mathcal{E}_B}{k_B T} \right)^3 = 9.9 \times 10^3 \text{ A/cm}^2. \tag{7.33}$$

∎

The current density for carrier overflow decreases with the thickness, d, of the active region since less carriers are required to completely fill the active region. When carrier overflow is activated, an increase of carrier injection does not lead to an increase of the emitted radiation, thus leading to a saturation of the emitted light intensity upon increasing the diode current (i.e., it is an effective loss process). This effect is important also in the case of heterostructures with quantum wells.

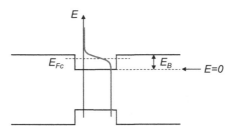

Figure 7.7   Carrier leakage over barrier: due to the shape of the Fermi–Dirac distribution function, the high-energy electrons can escape from the active region and diffuse in the barrier cladding region.

7.3 Carrier Leakage over Barrier

At lower levels of carrier injection, far from carrier overflow, another loss mechanism is active, related to the energy distribution of the carriers in the conduction and valence bands. Typical heights of the potential barriers confining carriers in the active region ($\Delta\mathcal{E}_c$ and $\Delta\mathcal{E}_v$) are of the order of a few hundreds of meV, much larger than the thermal energy $k_B T$. Due to the shape of the Fermi–Dirac distribution function, a few carriers have energy larger than the height of the confining barriers: These carriers can escape from the active region and diffuse in the barrier cladding region. This situation is schematically displayed in Fig. 7.7, in the case of electrons in the conduction band. The density of electrons in the conduction band with energy larger than the effective barrier height $\mathcal{E}_B$ is given by:

$$n_B = \int_{\mathcal{E}_B}^{\infty} \rho_c(\mathcal{E}) f_c(\mathcal{E}) d\mathcal{E}. \tag{7.34}$$

Since we are at moderate charge injection and we are considering electrons with energy much larger than the Fermi energy, the Fermi–Dirac distribution function can be well approximated by the Boltzmann distribution:

$$f(\mathcal{E}) \simeq \exp\left(-\frac{\mathcal{E} - \mathcal{E}_{Fc}}{k_B T}\right), \tag{7.35}$$

so that the electron density (Eq. 7.34) is given by (see Eq. 2.47):

$$n_B = N_c \exp\left(-\frac{\mathcal{E}_B - \mathcal{E}_{Fc}}{k_B T}\right). \tag{7.36}$$

A fraction of the electron population with energy above the barrier is reflected at the interface between the active region and the cladding region. Another fraction diffuse into the cladding layer. Note that, since the velocities of the high-energy electrons have random orientations, only a portion of these electrons are moving toward the cladding layer. In addition, a fraction of the electrons that enter the cladding material can diffuse back toward the active region. Therefore, it is not trivial to calculate the effective carrier leakage over the barrier.

To calculate the electron leakage current, we can assume that the electrons diffuse into the p-type cladding layer as minority carriers, so that the variation of the electron density in this region can be written as:

$$n_c(x) = n_c(0) \exp\left(-\frac{x}{L_n}\right) = n_B \exp\left(-\frac{x}{L_n}\right), \tag{7.37}$$

where $L_n = \sqrt{D_n \tau_n}$ is the diffusion length, D_n is the diffusion constant, and τ_n is the minority carrier lifetime in the p-cladding medium. We have assumed an x-axis with origin at the active region-cladding interface and $n_c(0) = n_B$, given by Eq. 7.36. The corresponding diffusion current is given by:

$$J_n|_{x=0} = -eD_n\frac{dn}{dx}\bigg|_{x=0} = eD_n\frac{n_B}{L_n} = e\frac{L_n}{\tau_n}n_B. \tag{7.38}$$

Exercise 7.2 Consider a GaAs/Al$_{0.3}$Ga$_{0.7}$As double heterostructure with an injected electron density in the GaAs active region $n = 2.86 \times 10^{18}$ cm^{-3} ($T = 300$ K). The barrier height between the active and the cladding regions is 200 meV. The electron mobility in the cladding region is $\mu_n = 2,300$ cm^2/Vs and the minority carrier lifetime is 14 ns. Calculate the leakage current density.

In Exercise 2.3, the position of the quasi-Fermi level in the conduction band of GaAs with an electron density $n = 2.86 \times 10^{18}$ has been calculated:

$$\mathcal{E}_{Fc} - \mathcal{E}_c = 109 \text{ meV}.$$

The density of electrons with energy above the barrier can be calculated by using Eq. 7.36:

$$n_B = N_c \exp\left(-\frac{\mathcal{E}_B - \mathcal{E}_{Fc}}{k_B T}\right) = 1.3 \times 10^{16} \text{ cm}^{-3}.$$

The diffusion constant is given by

$$D_n = \mu_n \frac{k_B T}{e} = 59.5 \text{ cm}^2/\text{s},$$

and the diffusion length is:

$$L_n = \sqrt{D_n \tau_n} = 9.1 \ \mu\text{m}.$$

The leakage current density is given by Eq. 7.38:

$$J_n = e\frac{D_n}{L_n}n_B = 136 \text{ A/cm}^2.$$

This result shows that the leakage current represents a significant loss mechanism. ∎

Note that, as a consequence of the exponential dependence of n_B on temperature, the leakage current increases rapidly with temperature. Moreover, a reduced minority carrier lifetime caused by high defect density determines an increase of the leakage current.

To reduce the carrier leakage from the active region, which can be an important loss mechanism, in particular in the case of double heterostructures with low barrier heights at the interface between the active region and the confinement materials, carrier-blocking layers can be used. In particular, electron blocking layers are employed, due to the typically

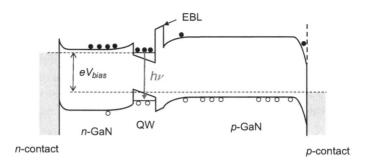

Energy-level diagram of a GaN quantum well (QW) LED showing an electron blocking layer (EBL). V_{bias} is the forward bias voltage applied to the device.

larger diffusion constant of electrons in comparison with holes in III–V semiconductors. A blocking layer is formed by a thin semiconductor layer with high gap, placed at the interface between active and confining regions, as schematically shown in Fig. 7.8.

Figure 7.8 also shows the band bending at the interfaces between different semiconductors, whose origin can be easily explained. Let us consider the interface between two semiconductors with different energy gaps. Carriers in the semiconductor with larger energy gap diffuse toward the semiconductor with smaller gap, filling states with lower energy. Let us consider the electron transfer in the conduction band: A charge separation is created at the interface with a depletion region with positive charge in the large-gap semiconductor and a negatively charged electron accumulation layer in the small-gap semiconductor. This charge separation leads to the generation of an electric field, which causes a bending of the band, as shown in Fig. 7.8. The potential energy discontinuities at the interfaces, which can cause an increase of the resistance to current flow, can be greatly reduced by using graded heterostructures, that is, heterostructures where the semiconductor composition does not change abruptly. For example, in a $p-n$ junction $n-Al_{0.3}Ga_{0.7}As/p-GaAs$ graded heterostructure, $Al_xGa_{1-x}As$ is graded from $x = 0$ to $x = 0.3$ over a short distance.

7.4 External Efficiency of a LED

So far we have discussed the mechanisms leading to an increase or a decrease of the internal quantum efficiency of a LED. In a real device not all the photons generated in the active region are emitted. There are a number of different processes leading to a reduction of the number of photons at the output of the LED. Photons can be reabsorbed during their travel inside the semiconductor materials (this loss process can be minimized by using properly designed heterostructures). Photons incident on metallic electrodes are absorbed by the metal. Another (large) fraction of photons cannot exit from the LED due to total internal reflection. For this reason, it is important to define a transmission efficiency, η_t, as the ratio between the number of photons emitted into the free space per second and the number of photons emitted from the active region per second. The product of the transmission efficiency and the internal quantum efficiency gives the ratio between the number of photons emitted into the free space per second and the number of electrons injected into the LED per second, which is called *external quantum efficiency*, η_{ext}:

$$\eta_{ext} = \eta_t \eta_{int}. \tag{7.39}$$

With this definition, the external photon flux is given by (see Eq. 7.16 for the photon flux inside the LED):

$$\Phi_{ext} = \eta_{ext}\frac{J}{e}. \tag{7.40}$$

The output power, P_{ext}, can be calculated by multiplying the external photon flux by the surface S and the energy of the emitted photons:

$$P_{ext} = \Phi_{ext}Sh\nu = \eta_{ext}\frac{I}{e}h\nu. \tag{7.41}$$

The *responsivity* is defined as the ratio between the emitted optical power and the injected current:

$$R = \frac{P_{ext}}{I} = \eta_{ext}\frac{h\nu}{e}. \tag{7.42}$$

The responsivity in W/A can be written as follows, when the emission wavelength is expressed in μm:

$$R = \eta_{ext}\frac{1.24}{\lambda_0}. \tag{7.43}$$

The linear dependence of the output power versus input current is valid for relatively small values of current (typically a few tens of mA). Upon increasing the current, efficiency starts to decrease, as previously mentioned (current droop).

The power efficiency, or *wall-plug efficiency*, is defined as the ratio between the emitted optical power and the electrical power provided to the LED:

$$\eta_{wp} = \frac{P_{ext}}{IV}, \tag{7.44}$$

here V is the voltage drop across the device.

As already mentioned, a major mechanism that strongly reduces the number of photons emitted into the free space is due to total internal reflection at the semiconductor–air interface. It is possible to calculate an approximated value of the fraction of the radiation which can exit from the interface between two materials with refractive index n_1 and n_2 ($n_1 > n_2$). We will assume an isotropic radiation source in the medium with refractive index n_1 and, as a first-order approximation, we will consider a constant transmission at the interface between the two media: $T(\theta) = T(0)$, where $T(0)$ is the normal incidence transmission. The light rays outside a cone defined by the critical angle θ_c corresponding to the condition of total internal reflection are completely reflected. θ_c can be easily calculated by using the Snell's law:

$$n_1 \sin \theta_c = n_2, \qquad \theta_c = \arcsin \frac{n_2}{n_1}. \tag{7.45}$$

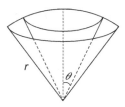

Figure 7.9   Calculation of the solid angle corresponding to the light escape cone.

Since the refractive index of a semiconductor is typically quite high ($n_1 \gg n_2$), we can use the following approximation:

$$\sin \theta_c \simeq \theta_c, \qquad \theta_c = \frac{n_2}{n_1}. \tag{7.46}$$

Indeed, if $n_1 = 3.4$ and $n_2 = 1$, $\theta_c = 17°$. The light rays which can escape from the semiconductor are those inside the light escape cone with semi-aperture θ_c, as shown in Fig. 7.9. The corresponding solid angle Ω_c is given by:

$$\Omega_c = \int_0^{\theta_c} \frac{2\pi r \sin \theta \; rd\theta}{r^2} = 2\pi \int_0^{\theta_c} \sin \theta \; d\theta = 2\pi(1 - \cos \theta_c). \tag{7.47}$$

For small angles θ_c, $\cos \theta_c \simeq 1 - \theta_c^2/2$, so that:

$$\Omega_c \simeq \pi \theta_c^2 = \pi \left(\frac{n_2}{n_1} \right)^2. \tag{7.48}$$

The power fraction emitted inside the light escape cone assuming an isotropic source is given by $\Omega_c/4\pi$, so that the power that can escape from the semiconductor is given by:

$$P_{ext} = P_{source} T(0) \frac{\Omega_c}{4\pi}, \tag{7.49}$$

where:

$$T(0) = 1 - \left(\frac{n_1 - n_2}{n_1 + n_2} \right)^2 \tag{7.50}$$

and the extraction efficiency related to total internal reflection is:

$$F = \frac{P_{ext}}{P_{source}} \approx \frac{1}{4} \left[1 - \left(\frac{n_1 - n_2}{n_1 + n_2} \right)^2 \right] \left(\frac{n_2}{n_1} \right)^2. \tag{7.51}$$

A simple way to increase the critical angle and therefore also the fraction of emitted power is to encapsulate the LED in a plastic material with a refractive index larger than 1.

The light rays emitted outside the light escape cone are trapped inside the semiconductor, and most of them are absorbed by the substrate. In order to avoid this problem, a LED should have hemispherical shape with a point-like source in the center so that all the emitted light rays are normal to the semiconductor–air interface and can escape from the device (only Fresnel reflection must be considered at the interface, if the surface of the sphere does not have an antireflection coating). The problem of this structure is the fabrication cost, related to the nonplanar geometry. A shaping technology more suitable for mass production,

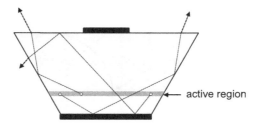

LED with truncated inverted pyramid structure.

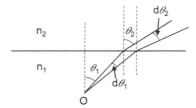

An isotropic point source placed in O in the medium with refractive index n_1 forms a Lambertian source in the medium with refractive index n_2.

which has proven to increase LED efficiency, consists in the fabrication of LEDs in the shape of truncated inverted pyramid, which can redirect toward the exit axis side-emitted rays, as schematically shown in Fig. 7.10. An increase in efficiency of a factor of two is possible by using this particular geometry.

7.5 Emission Pattern of a LED

The emission at the output of a LED is in general nonisotropic even though light inside the semiconductor can be assumed as isotropic. Let us consider a planar LED with high refractive index n_1 placed in air (refractive index n_2). As in the previous section, we will further assume an isotropic point source placed in the semiconductor, quite close to the semiconductor–air interface, as shown in Fig. 7.11. Moreover, a constant transmission is assumed, independent of the incidence angle. The photon flux emitted from the point source with directions between θ_1 and $\theta_1 + d\theta_1$ (i.e., within the solid angle $d\Theta_1$ delimited by two conical surfaces with semi-aperture θ_1 and $\theta_1 + d\theta_1$) is given by:

$$d\Phi_1(\theta_1) = I_1(\theta_1)2\pi \sin\theta_1 \, d\theta_1 = I_1(0)2\pi \sin\theta_1 \, d\theta_1, \qquad (7.52)$$

where I is the luminous intensity (i.e., the luminous power per unit solid angle), which does not depend on θ_1 since the source is isotropic. The photon flux emerging in the second medium in the solid angle corresponding to $d\theta_2$ is:

$$d\Phi_2(\theta_2) = T(\theta_1)d\Phi_1(\theta_1). \qquad (7.53)$$

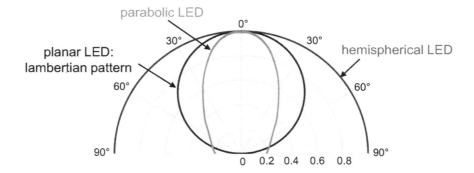

Figure 7.12 Polar plot of the far-field pattern of LEDs with planar, hemispherical, and parabolic surfaces.

We will assume $T(\theta_1) = T(\theta = 0) = T_0$. The photon flux $d\Phi_2$ can be written in terms of the luminous intensity in the region outside the LED:

$$d\Phi_2(\theta_2) = I_2(\theta_2) 2\pi \sin\theta_2 \, d\theta_2, \qquad (7.54)$$

so that:

$$I_2(\theta_2) = T_0 I_1(0) \frac{\sin\theta_1 d\theta_1}{\sin\theta_2 d\theta_2}. \qquad (7.55)$$

From Snell's law:

$$n_1 \sin\theta_1 = n_2 \sin\theta_2, \qquad n_1 \cos\theta_1 d\theta_1 = n_2 \cos\theta_2 d\theta_2, \qquad (7.56)$$

so that Eq. 7.55 can be written as:

$$I_2(\theta_2) = T_0 I_1(0) \left(\frac{n_2}{n_1}\right)^2 \frac{\cos\theta_2}{\cos\theta_1}. \qquad (7.57)$$

Since typically $n_2 \ll n_1$, $\cos\theta_1 \simeq 1$ and Eq. 7.57 can be written as:

$$I_2(\theta_2) = T_0 I_1(0) \left(\frac{n_2}{n_1}\right)^2 \cos\theta_2. \qquad (7.58)$$

Therefore, the light intensity at the output of a planar LED is characterized by a lambertian emission pattern, since it is proportional to $\cos\theta_2$. From Eq. 7.58, it turns out that the intensity is maximum in the direction perpendicular to the semiconductor surface and decreases upon increasing the emission angle θ_2. The polar plot of the lambertian emission is shown in Fig. 7.12. The emission pattern can be changed by using nonplanar LED surfaces. Indeed, in the case of a hemispherical LED with light-emitting region in the center of the hemisphere, an isotropic emission pattern is obtained. A highly directional emission can be achieved by using a LED with a parabolic shape as shown in Fig. 7.12.

7.6 Luminous Efficiency

An important parameter, which characterizes the performances of a LED, is the *luminous efficiency*, whose definition requires the use of photometric units, which are related to the

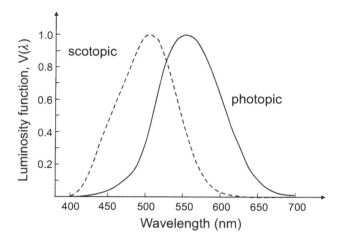

Figure 7.13   Spectral luminous efficiency function, $V(\lambda)$, for photopic (solid line) and scotopic (dashed line) vision regimes.

characteristics of the human eye. The basic photometric unit is the luminous flux, Φ_l, which represents the light power emitted by a source spectrally weighted with the human eye's spectral luminous efficiency function $V(\lambda)$, also called luminosity function. The unit of luminous flux is lumen (lm), and it is defined as follows: At 555 nm, where the sensitivity of human eye is maximum (for photopic vision regime), an optical power of 1 W corresponds to a luminous flux of 683 lm. Therefore, in the case of monochromatic radiation at the wavelength λ, the luminous flux, Φ_l, in lumen is given by:

$$\Phi_l = 683 \, V(\lambda)P, \tag{7.59}$$

where P is the optical power (in W). The function $V(\lambda)$ is shown in Fig. 7.13 for photopic and scotopic vision regimes. Photopic vision is the vision of human eye under well-lit conditions, while scotopic vision is the vision of the eye under low light conditions. If the light is not monochromatic, the luminous flux can be calculated by using the following expression:

$$\Phi_l = 683 \int V(\lambda)P(\lambda) \, d\lambda, \tag{7.60}$$

where $P(\lambda)$ is the power spectral density. The total optical power emitted by a non-monochromatic source is given by:

$$P = \int_\lambda P(\lambda) \, d\lambda. \tag{7.61}$$

The luminous efficiency of a LED (measured in units of lm/W) is the ratio between the luminous flux of the LED and the electrical input power:

$$\eta_l = \frac{\Phi_l}{IV}, \tag{7.62}$$

where I and V are current and voltage applied to the LED.

While the first red GaAsP LEDs with emission at 650 nm had a luminous efficiency of less than 0.1 lm/W (for comparison, an incandescent lamp has an efficiency of $\sim$ 15 lm/W), current white light LEDs show efficiencies well above 100 lm/W.

7.7 Blue LED

In this last section, we well discuss a particularly important class of LEDs: the blue LEDs. Their development has led to the generation of highly efficient white light sources, with important technological applications. Indeed, blue LEDs can be used to excite a phosphor material (like Ce:YAG), which emits in the green and red spectral regions. The combination of the residual blue light with green and red leads to the generation of white light. Another technique for the generation of white light is based on the combined use of three LEDs of complementary colors (blue, green, and red). These white-light LED sources are now progressively replacing the fluorescence lamps for lighting applications, with enormous reduction of electrical energy consumption, since LED-based white light sources require about 10 times less energy than typical light bulbs. The invention of efficient blue LEDs was awarded with a Nobel prize in Physics in 2014 to the three Japanese researchers I. Akasaki, H. Amano, and S. Nakamura, with the following motivation: "for the invention of efficient blue LEDs which has enabled bright and energy-saving white light sources."

The material of choice for blue LEDs is GaN (gallium nitride), which belongs to the III–V class, with a wurtzite crystal structure. It has a direct band gap of 3.4 eV (corresponding to an emission wavelength of 365 nm, in the violet spectral region). GaN can be grown on a substrate of sapphire (Al_2O_3) or SiC: The problem is the large mismatch in the lattice constants ($\sim$ 16%). Sapphire substrate is used since it is stable even under the harsh MOVPE (metalorganic vapor-phase epitaxy), namely, a temperature above 1,000 °C and an ammonia (NH_3) atmosphere. Moreover, it is similar to GaN in terms of crystallographic symmetry. GaN can be doped, for example, with Si for n-type doping and with Mg for p-type doping. The growth of high-quality GaN layers on sapphire substrates and the p-doping have been the two major problems, which delayed considerably the development of efficient blue LEDs.

The main idea at the basis of the novel growing technique, which allowed a high-quality deposition of GaN on sapphire, was the use of a soft or flexible thin buffer layer between the substrate and the GaN film. This was obtained by developing in 1985 a low-temperature buffer layer technology. First a thin layer ($\sim$ 30 − 50 nm) of polycrystalline AlN was deposited on a sapphire substrate at low temperature (500 °C), and then it was heated up to the temperature required for epitaxial growth of GaN single crystal ($\sim$ 1,000 °C). The AlN layer is composed of columnar fine crystals, with a diameter of the order of 10 nm: This structure is reached during the heating process, starting from an initial amorphous-like structure (when deposited at low temperature). GaN is subsequently grown on this buffer layer. As shown in Fig. 7.14, the density of dislocations of the GaN crystal is very high in contact to the buffer layer, but then it rapidly decreases after a few-micrometer growth, thus leading to a very high-quality surface. Figure 7.15 shows scanning electron micrographs

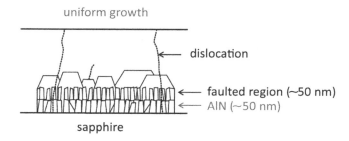

Figure 7.14 Low-temperature buffer layer technology for the growth of high-quality GaN on sapphire (adapted with permission from [11]).

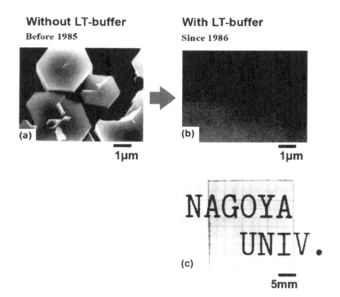

Figure 7.15 Scanning electron micrographs of GaN on sapphire (a) without the low temperature (LT) buffer layer and (b) with the LT-buffer layer. (c) A photograph of specular and transparent GaN film grown on sapphire with the LT-buffer layer (reproduced with permission from [11]).

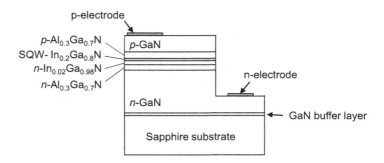

p-electrode

p-Al$_{0.3}$Ga$_{0.7}$N

SQW- In$_{0.2}$Ga$_{0.8}$N

n-In$_{0.02}$Ga$_{0.98}$N

n-Al$_{0.3}$Ga$_{0.7}$N

p-GaN

n-electrode

n-GaN

GaN buffer layer

Sapphire substrate

Figure 7.16   Structure of the single-quantum well (SQW) blue LED described in the text.

of surfaces of GaN films grown on a sapphire substrate. The surface morphology of the films is crucially improved by the use of low-temperature buffer layers. The GaN film displays a specular surface with no pits or cracks and is so transparent that letters written on an underlying paper can be clearly seen, as shown in Fig. 7.15(c). In 1991, by using a similar method, the AlN buffer layer was replaced by a thin GaN buffer layer grown at low temperature.

The second major problem was the p-doping of GaN. In 1988, it was observed that the luminescence of a high-quality GaN film with Zn-doping was greatly increased when it was irradiated by a low-energy electron beam. In the same way, the blue luminescence of Mg-doped GaN irradiated by low-energy electrons was significantly increased and the material was converted to a low-resistivity p-type crystal. Later it was found that p-type GaN can be obtained by thermal annealing above 400 °C in a H$_2$-free atmosphere and it was shown that Mg is easily passivated by forming Mg-H complexes, so that in order to obtain p-doping, it was required to activate the Mg acceptor by releasing the hydrogen. Another very important progress was the growth and p-doping of high crystal quality alloys (AlGaN, InGaN) for the development of heterostructure. InGaN is an ideal choice for the active layer in a double heterostructure. Upon changing the amount of In, it is possible to tune the wavelength of the emission peak from 400 nm to 445 nm. A challenging aspect is that the introduction of In atoms in a GaN lattice gives rise to a large strain since the radius of In atoms is $\sim$ 20% bigger than that of Ga atoms.

Figure 7.16 shows the structure of a single-quantum well (SQW) blue LED, developed by S. Nakamura and coworkers in 1995 [12]. III–V nitride films were grown by the two-flow MOCVD method, using a sapphire substrate. The LED structure consists of a 30-nm-thick GaN buffer layer grown at a low temperature (550 °C), a 4 μm-thick layer of n-type GaN, a 100-nm-thick layer of n-type Al$_{0.3}$Ga$_{0.7}$N (Si doped), a 50-nm-thick layer of n-type In$_{0.02}$Ga$_{0.98}$N, a 2-nm-thick active layer of undoped In$_{0.2}$Ga$_{0.8}$N, a 100-nm-thick layer of p-type Al$_{0.3}$Ga$_{0.7}$N (doped with Mg), and a 500-nm-thick layer of p-type GaN. The active region forms a SQW structure consisting of a 2-nm In$_{0.2}$Ga$_{0.8}$N well layer sandwiched by 50-nm n-type In$_{0.02}$Ga$_{0.98}$N and 100-nm p-type Al$_{0.3}$Ga$_{0.7}$N barrier layers. After deposition, the surface of the p-type GaN layer was partially etched to expose the n-type GaN layer. A semitransparent Ni/Au contact was then deposited onto the p-type GaN layer and a Ti/Al contact onto the n-type GaN layer. A portion of the top p-contact was covered by a nontransparent bonding pad. The wafer was cut into a square shape (350 μm $\times$ 350 μm). The LEDs produced 4.8 mW at 20 mA and sharply peaked at 450 nm, corresponding to an external quantum efficiency of 8.7 %.

An obvious way to increase the emitted radiation is to increase the input current density: This is particularly effective in high-efficiency devices since less input energy is converted in thermal energy by nonradiative recombination processes. On the other hand, upon increasing the input current density, the LED efficiency first increases, reaching a maximum value, and then it decreases if the current density is further increased. This phenomenon is called *efficiency droop*, and it is common in all LED devices. For this reason, the operational current densities of LEDs are smaller than the maximum current densities that would lead to excessive heating of the structure. A particularly effective physical process leading to efficiency droop is the Auger recombination process, whose efficiency scales with the cube of the carrier concentration (Cn^3) at high injection levels. Due to the high density of carriers in highly efficient LEDs, this term starts to be the dominant (nonradiative) recombination process, thus reducing the internal quantum efficiency of the device.

Blue LEDs, in combination with red and green LEDs, allow the development of completely solid-state full-color displays, traffic lights, and other specialized lighting applications. Efficient blue LEDs in combination with phosphors to obtain white light are characterized by very long lifetimes ($\sim$ 100,000 hours) and are two times more efficient than fluorescent lamps: For this reasons, white and UV LEDs are now largely used in TVs, mobile phones, and computer displays.

7.8 Exercises

Exercise 7.1   The following table lists the main characteristics of a few LEDs. Calculate the emitted power (in W) and the wall-plug efficiency for the typical values of I and ΔV. Assume that the intensity does not depend on the vision angle.

λ_p (nm)	I_c (typ) (mA)	I_c (max) (mA)	ΔV (typ) (V)	ΔV_{max} (V)	I (typ) at 20 mA (mcd)	θ_v (°)
470	20	30	3.6	5	1000	15
525	20	30	3.5	5	3400	15
470	20	30	3.6	5	300	45
525	20	30	3.5	5	1050	45

Exercise 7.2   A LED emits at $\lambda = 550$ nm ($V(\lambda) = 0.88$) with wall-plug efficiency $\eta_{wp} = 0.1\%$. The operational current is $I = 50$ mA. Consider a point-like and isotropic source. Determine, in photometric units, the flux, the luminous intensity and the luminous efficiency.

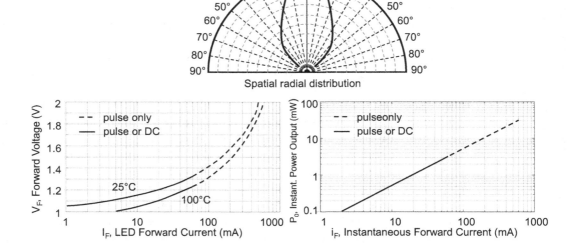

Spatial radial distribution

Figure 7.17   Main characteristics of the LED considered in Exercise 7.3.

Exercise 7.3   The main characteristics of a LED are reported in Fig. 7.17. The LED is polarized with $\Delta V = 1.2$ V, and it is at room temperature. Calculate how many of the emitted photons per unit time impinge on a circular target with a radius of 5 mm, placed at 1 m from the LED, whose axis forms an angle $\theta = 30°$ with the perpendicular to the emitting surface.

Exercise 7.4   Consider a LED with a carrier injection giving a Burstein–Moss shift of 110 meV. The LED is based on an GaAs heterostructure ($m_c = 0.067\, m_0$, $m_v = 0.47\, m_0$).
a) Calculate, at room temperature ($T = 300$ K), the electron density and the difference between the quasi-Fermi level in the conduction band and the bottom of the same band.
b) Calculate the change in the wavelength of the absorption edge and compare the shift to the typical bandwidth.

Exercise 7.5   Consider Exercise 7.4, where the claddings are made of $Al_{0.3}Ga_{0.7}As$ ($\mathcal{E}_g = 1.8$ eV). Do we have current overflow? Calculate the corresponding current density, considering an active region with a thickness $d = 0.1\ \mu m$. Consider a bimolecular recombination coefficient $B = 1.7 \times 10^{-10}$ cm^3/s.

Exercise 7.6   Consider two n^+p GaAs LEDs with nonradiative recombination time $\tau_{nr} = 100$ ns and p-type doping of 10^{15} cm^{-3} and 10^{18} cm^{-3}. Calculate the radiative efficiency, η_r, of the two LEDs.

Exercise 7.7   Consider an $Al_xGa_{1-x}As$ double-heterostructure LED. The energy gap at room temperature of the ternary alloy is ($x < 0.45$:)

$$\mathcal{E}_g(eV) = 1.424 + 1.266x + 0.266x^2.$$

The carrier mobility is $\mu_n = 1,300$ cm^{2}Vs^{-1}, and the carrier lifetime is $\tau_n = 5$ ns. The LED has the following characteristics:

well region: $x = 0.2$, energy difference between quasi-Fermi level in the conduction band and conduction band edge, $\mathcal{E}_{Fc} - \mathcal{E}_c = 20$ meV, effective density of states $N_c = 6 \times 10^{17}$ cm^{-3}.

barrier region: $x = 0.4$.

a) Estimate the leakage current density at $T = 300$ K.

b) Assuming a forward current density of 100 A/cm^2, determine the fraction of the total LED current represented by the leakage current.

Exercise 7.8 Consider the double heterostructure of Exercise 7.7. The well thickness is $d = 20$ nm, and the bimolecular recombination coefficient is $B = 1.9 \times 10^{-10}$ cm^3/s. Calculate the current level at which the well overflows.

Exercise 7.9 Consider an AlGaAs LED emitting at 890 nm. The active region is p-doped with $N_a = 4 \times 10^{17}$ cm^{-3}, the nonradiative lifetime is $\tau_{nr} = 50$ ns, and the bimolecular recombination coefficient is $B = 2 \times 10^{-10}$ cm^3/s. At a forward current of 55 mA, the voltage is $\Delta V = 1.5$ V and the emitted power is 12 mW. Calculate the wall-plug efficiency, the internal quantum efficiency, and the external quantum efficiency.

Exercise 7.10 Consider an Al$_x$Ga$_{1-x}$As LED. The peak emission wavelength at $T = 300$ K is at 820 nm and the bandwidth is $\Delta\lambda = 40$ nm. Determine the badgap of the semiconductor and its composition, assuming that the band gap of Al$_x$Ga$_{1-x}$As is given by the following empirical expression:

$$\mathcal{E}_g(\text{eV}) = 1.424 + 1.266x + 0.266x^2.$$

Exercise 7.11 Demonstrate that in the case of a Lambertian source, half of the total emitted power is contained in a cone of angle $2\pi/3$ around the normal to the surface.

Exercise 7.12 A LED has an active area of 1 mm^2 and emits 100 μW in a Lambertian profile. Estimate the power received by a target with an area of 50 mm^2 normal to the emission direction at a distance of 0.2 m.

Exercise 7.13 When the temperature of a GaAs LED is changed from $T_0 = 300$ K to $T_1 = 310$ K, the variation of the wavelength of the emission peak is 3 nm. Assuming a linear variation with temperature, determine the change in the band gap energy of GaAs in the same temperature range (the energy gap at T_0 is 1.42 eV).

Exercise 7.14 Consider the quaternary alloy $(\text{Al}_x\text{Ga}_{1-x})_{0.5}\text{In}_{0.5}\text{P}$. At room temperature, the direct gap at the Γ point is given by:

$$\mathcal{E}_\Gamma(\text{eV}) = 1.91 + 0.61x,$$

while the indirect gap at the X point of the Brillouin zone is given by:

$$\mathcal{E}_X(\text{eV}) = 2.19 + 0.085x.$$

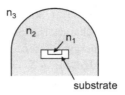

substrate

Dome-encapsulated LED.

Determine the range of Al mole fractions x in order to have a direct band gap semiconductor and determine the emission wavelength at the direct–indirect crossover point.

Exercise 7.15   Consider the dome-encapsulated LED shown in Fig. 7.18.
a) Write an approximated expression of the extraction efficiency.
b) Calculate the extraction efficiency of the dome-encapsulated LED assuming $n_1 = 3.5$, $n_2 = 1.6$, $n_3 = 1$ and compare the result with the extraction efficiency without dome encapsulation.

Exercise 7.16   Consider a LED formed by an asymmetric $n^+ - p$ junction.
a) Show that the injection efficiency, that is, the fraction of total diode current due to diffusion, can be written as:

$$\eta_{in} = \left(1 + \frac{\mu_p p_{n0} L_n}{\mu_n n_{p0} L_p}\right)^{-1}.$$

b) Calculate the injection efficiency in a GaAs LED with $N_A = 10^{17}$ cm^{-3} and $N_D = 10^{17}$ cm^{-3}, $\mu_n = 8500$ cm^{2}V^{-1}s^{-1}, $\mu_p = 400$ cm^{2}V^{-1}s^{-1}, and $L_n = L_p$.

Exercise 7.17   Consider a GaAs quantum well with thickness $d = 10$ nm and a barrier height $\Delta\mathcal{E}_c = 200$ meV. The bimolecular recombination coefficient for bulk GaAs is $B = 10^{-10}$ cm^{3}s^{-1}, and the effective mass in the conduction band is $m_c = 0.067\ m_0$. Calculate the energy of the first quantized state in the well (assuming infinite potential barriers) and the overflow current density.

Exercise 7.18   Consider a LED die with vertical sidewalls. If all six sidewalls of the die are considered, estimate the maximum outcoupling efficiency of two LEDs composed of GaAs and GaN in air. Repeat the same calculation if the die is embedded in an epoxy resin with index of refraction $n = 1.6$.

8 Semiconductor Lasers

8.1 Introduction

In Chapter 5, we have seen that a semiconductor far from thermodynamic equilibrium can present a net gain if the Bernard–Duraffourg condition is achieved by proper carrier injection:

$$\mathcal{E}_2 - \mathcal{E}_1 < \mathcal{E}_{Fc} - \mathcal{E}_{Fv}. \tag{8.1}$$

This condition is required in order to generate a carrier inversion and can be obtained, for example, by electrical pumping of a $p - n$ junction composed by degenerate semiconductors, so that both the quasi-Fermi levels, $\mathcal{E}_{Fc}$ and $\mathcal{E}_{Fv}$, fall in the conduction and valence bands, respectively (a very schematic view of a semiconductor laser is shown in Fig. 8.1). Laser action in a semiconductor was first reported in 1962 by using $p - n$ homojunctions. As discussed in Chapter 7 in the case of light-emitting diodes, homojunctions are not used any more since they present a number of severe drawbacks: They can operate CW only at cryogenic temperatures ($T = 77$ K) due to the high threshold current density at room temperature, of the order of 10^5 A/cm^2. Double heterostructures present various advantages which lead to a strong reduction of the threshold current density at room temperature ($\sim 10^3$ A/cm^2). Figure 8.2 shows schematically the energy band in the case of a forward-biased double-heterostructure laser. As in the case of LEDs, the active region, with a typical thickness of $0.1-0.2$ μm, is placed between the n-type and p-type cladding layers with a larger energy gap. Therefore, the generated photons are not absorbed outside the active region. The energy barriers between the active region and the cladding layers, due to the significantly different energy gaps, confine effectively the injected electrons and holes inside the active layer (*carrier confinement*), thus increasing their concentrations at a given current density. Moreover, the refractive index of the active layer is significantly larger than that of the p- and n-type cladding layers, thus giving rise to a waveguiding structure, which effectively confines the laser beam in the active layer (*photon confinement*). The number of modes in this waveguide depends on the thickness and on the refractive indexes of the semiconductors. By a proper design, a single transverse mode can be obtained. Quantum-well double heterostructures can be used, as will be discussed in this chapter.

In order to achieve laser action, a suitable positive optical feedback is required. In the case of semiconductor, lasers Fresnel reflection at the semiconductor–air interfaces are large enough to provide the required feedback for laser action: The parallel end faces are prepared by cleavage along crystal planes. This is due to the high refractive index of semiconductors (e.g., in the case of GaAs $n = 3.6$, thus leading to a reflectance $R = 0.32$)

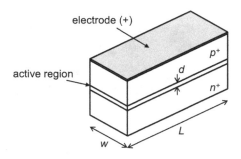

Figure 8.1   Schematic structure of a semiconductor diode laser.

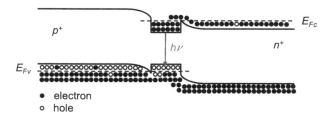

Figure 8.2   Energy band structure of a forward-biased double-heterostructure laser.

together with the high gain which can be obtained from semiconductors.

In this chapter, we will first write the rate equations of a semiconductor laser to deduce the threshold condition for laser action and the output power. Various device structures will be described, which have been introduced to optimize the laser performances. Finally, the main spectral and spatial characteristics of semiconductor laser emission will be discussed.

8.2 Rate Equations and Threshold Conditions for Laser Action

We will describe the properties of a semiconductor laser by using a set of rate equations. Space-independent rate equations will be written since we assume that the laser oscillates on a single mode; moreover, the pumping and mode energy are uniform within the active layer with volume V. Leakage currents will be completely neglected. The mode transverse profile is considered uniform. In the following rate equations, N is the carrier density (due to charge neutrality, the electron density equals the hole density). The rate equation for N takes into account the electron generation rate, due to electron injection, and the recombination rate per unit volume in the active region. The generation rate per unit volume (pumping rate) is given by:

$$R_p = \eta_i \frac{I}{eV}, \tag{8.2}$$

where I is the pumping current and η_i is the injection efficiency, which gives the fraction of injected current generating carriers in the active region. The recombination rate is related to all radiative and nonradiative recombination processes. The spontaneous radiative recombination rate can be written using Eq. 5.93:

$$R_{sp,T} = BN^2,\tag{8.3}$$

where B is the bimolecular recombination coefficient. The nonradiative recombination rate is given by Eq. 7.11:

$$R_{nr} = A_{nr}N + CN^3.\tag{8.4}$$

Therefore, the rate equation for N can be written as:

$$\frac{dN}{dt} = \eta_i \frac{I}{eV} - A_{nr}N - BN^2 - CN^3 - R_{st},\tag{8.5}$$

where R_{st} is the stimulated recombination rate per unit volume, which takes into account both stimulated emission and absorption. This rate equation can be written in a more compact form by introducing the carrier lifetime, τ, given by:

$$\frac{N}{\tau} = A_{nr}N + BN^2 + CN^3.\tag{8.6}$$

With this definition, the rate equation 8.5 can be written as:

$$\frac{dN}{dt} = \eta_i \frac{I}{eV} - \frac{N}{\tau} - R_{st}.\tag{8.7}$$

In the case of LEDs, the stimulated recombination rate can be neglected (when the photon density is not too large), so that the previous rate equation can be written as:

$$\frac{dN}{dt} = \eta_i \frac{I}{eV} - \frac{N}{\tau},\tag{8.8}$$

and, under steady-state conditions ($dN/dt = 0$) we have:

$$\eta_i \frac{I}{eV} = \frac{N}{\tau},\tag{8.9}$$

which coincides with Eq. 7.15.

The second relevant rate equation gives the rate of change of the photon density, ϕ, in the laser cavity. In the following, we will consider a resonant cavity oscillating on a single mode. The photon generation rate is related to the spontaneous emission, $R_{sp,T}$, and to the net stimulated recombination rate, R_{st}. Since the resonant cavity is oscillating on a single mode, only a fraction of the total spontaneous emission, $\beta_{sp}R_{sp,T}$, is coupled into the lasing mode, where β_{sp} is called spontaneous emission factor. Generation processes, relevant for laser action, must be considered inside the volume, V, of the laser active region. In general, the volume, V_p, occupied by photons is larger than the volume of the active region occupied by electrons. For this reason, it is useful to introduce a confinement factor, Γ, which represents the fraction of the laser beam power in the active region. A first and simple expression of Γ is:

$$\Gamma = \frac{V}{V_p}\tag{8.10}$$

(a formal definition of Γ will be given in the following). Therefore, the photon density generation rate can be written as:

$$\left(\frac{d\phi}{dt}\right)_{gen} = \Gamma R_{st} + \Gamma \beta_{sp} R_{sp,T}. \tag{8.11}$$

In a laser cavity, there are also photon losses generated by absorption, scattering in some optical elements inside the cavity and by mirror reflectivity, which is always smaller than unity. To quantify the rate of photon density decay in a given cavity mode, it is useful to introduce the photon lifetime, τ_c, which accounts for the decrease of photons in the cavity due to cavity losses. We assume that when no generating processes are present, the photon density in the cavity mode decays as:

$$\phi(t) = \phi(0) \exp\left(-\frac{t}{\tau_c}\right), \tag{8.12}$$

Therefore, the rate equation for ϕ can be written as:

$$\frac{d\phi}{dt} = \Gamma R_{st} + \Gamma \beta_{sp} R_{sp} - \frac{\phi}{\tau_c}, \tag{8.13}$$

In order to solve the two rate Eqs. 8.7 and 8.13 (in steady-state conditions), we have first to write the stimulated emission rate R_{st} in terms of the photon density and other characteristics of the laser cavity. We will assume, without loss of generality, that the confinement factor, Γ, is equal to 1; therefore, we assume a complete overlap between the active region and the optical mode. If g is the gain per unit length, the growth of photon density in the cavity due to stimulated processes and neglecting any losses can be written as:

$$\phi(z) = \phi(0) \exp(gz). \tag{8.14}$$

z is the cavity propagation coordinate, $z = v_g t$, where v_g is the group velocity in the cavity. Equation 8.14 can be rewritten as:

$$\phi(z) = \phi(0) \exp(gv_g t), \tag{8.15}$$

and the corresponding rate of change of the photon density due to stimulated emission is given by:

$$R_{st} = \left(\frac{d\phi}{dt}\right)_{st} = gv_g\phi(0) \exp(gv_g t) = gv_g\phi. \tag{8.16}$$

8.2.1   Photon Lifetime, τ_c.

The relation between the photon lifetime, τ_c, and cavity losses can be easily calculated. Let R_1 and R_2 be the power reflectivities of the two-end mirrors of the cavity and T_i the fractional internal loss per pass caused by diffraction and any other internal loss. If $I_0(x, y, 0)$ is the initial beam intensity inside the cavity at time $t = 0$ and at a given point with transverse coordinates x and y, the intensity at the same point after n round-trips (therefore at time $t = t_n = n2L/v_g$, where L is the cavity length) is given by:

$$I(x, y, t_n) = [R_1 R_2 (1 - T_i)^2]^n I_0(x, y, 0). \tag{8.17}$$

Since we are considering a mode of the cavity, it maintains its shape after each round trip, so that we can assume that the photon density is proportional to the beam intensity, so that:

$$\phi(t_n) = [R_1 R_2 (1 - T_i)^2]^n \phi(0), \tag{8.18}$$

where $\phi(0)$ is the initial photon density in the cavity. As already noted, the decrease of photons due to all loss processes can be described by using the photon lifetime, so that $\phi(t_n)$ can be also written as:

$$\phi(t_n) = \phi(0) \exp\left(-\frac{t_n}{\tau_c}\right). \tag{8.19}$$

We can assume that this exponential law holds at any time t ($t > 0$). From the last two equations, we get:

$$\tau_c = -\frac{2L}{v_g \ln\left[R_1 R_2 (1 - T_i)^2\right]}. \tag{8.20}$$

By using the logarithmic losses, defines as:

$$\gamma_1 = -\ln R_1 \tag{8.21}$$
$$\gamma_2 = -\ln R_2 \tag{8.22}$$
$$\gamma_i = -\ln(1 - T_i) \tag{8.23}$$

and the logarithmic loss per pass:

$$\gamma = \gamma_i + \frac{\gamma_1 + \gamma_2}{2} \tag{8.24}$$

the photon lifetime can be written as:

$$\tau_c = \frac{L}{v_g \gamma}. \tag{8.25}$$

Since the mode distribution in the laser cavity is space-dependent, the threshold condition for laser action can be calculated by imposing that the spatially averaged gain must equal the spatially averaged losses. This threshold condition can be written as:

$$\langle g_{th} \rangle L = \langle \alpha_a \rangle L + \langle \alpha_n \rangle L + \langle \alpha_p \rangle L + \gamma_m, \tag{8.26}$$

where α_a is the scattering loss in the active layer, α_n and α_p are the losses in the n-type and p-type cladding layers, respectively (in the following we will assume $\alpha_n \simeq \alpha_p = \alpha$), and $\gamma_m = \log(1/R)$ represents the mirror loss. In Eq. 8.26, the averaged values are calculated over the field intensity distribution in the cavity, $|u(x, y, z)|^2$:

$$\langle g \rangle = \frac{\int_a g|u|^2 dV}{\int |u|^2 dV}, \tag{8.27}$$

where the integral at the numerator is calculated over the volume of the active region and the integral at the denominator is calculated over the whole volume of the laser cavity. $\langle g \rangle$ is called modal gain coefficient and $\langle g \rangle L$ is the modal gain. In the following, we will neglect

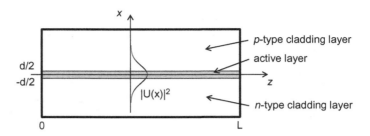

Figure 8.3 Schematic structure of a semiconductor laser showing the field intensity distribution, $|u(x)|^2$, in the cavity.

the spatial variation of the field intensity distribution along the longitudinal coordinate z and along the coordinate y parallel to the junction, so that $u = u(x)$ (see Fig. 8.3), where x is the coordinate in the direction perpendicular to the junction. Therefore, we can write:

$$\langle g \rangle = \frac{\int_{-d/2}^{d/2} g |u(x)|^2 dx}{\int_{-\infty}^{\infty} |u(x)|^2 dx} = g\Gamma,$$

(8.28)

where Γ is the beam confinement factor defined as:

$$\Gamma = \frac{\int_{-d/2}^{d/2} |u(x)|^2 dx}{\int_{-\infty}^{\infty} |u(x)|^2 dx}.$$

(8.29)

From the definition of Γ, it is clear that the confinement factor gives the fraction of the beam power contained in the active region. From the threshold condition 8.26, by using the definition of Γ, we can write:

$$g_{th}\Gamma = \alpha_a \Gamma + \alpha(1 - \Gamma) - \frac{1}{L} \log R.$$

(8.30)

If we define the average internal loss as:

$$\langle \alpha_i \rangle = \alpha_a \Gamma + \alpha(1 - \Gamma),$$

(8.31)

the previous equation can be written as:

$$g_{th}\Gamma = \langle \alpha_i \rangle - \frac{1}{L} \log R = \frac{\gamma}{L}.$$

(8.32)

If, as pointed out in Section 5.4, we approximate g as:

$$g = \sigma(N - N_{tr}),$$

(8.33)

we can obtain an expression for the threshold carrier density:

$$N_{th} = \frac{\gamma}{\sigma L \Gamma} + N_{tr}.$$

(8.34)

Above threshold, when the photon density increases from the initial value determined by spontaneous emission, the steady-state value of the carrier density can be calculated by using Eq. 8.13 by setting $d\phi/dt = 0$. Assuming that only a very small portion of the spontaneous emission is coupled into the oscillating mode, the term $\Gamma \beta_{sp} R_{sp,T}$ can be neglected and, by using Eq. 8.16, we obtain:

$$g\Gamma = \frac{1}{v_g \tau_c} = \frac{\gamma}{L},$$

(8.35)

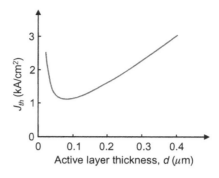

Evolution of the threshold current density, J_{th}, as a function of the thickness, d, of the active layer for a 300 μm-long GaAs double-heterostructure laser.

which is the same expression obtained at threshold (Eq. 8.32). Upon using Eq. 8.33, it is possible to calculate the carrier density above threshold (in steady-state conditions):

$$N = \frac{\gamma}{\sigma L \Gamma} + N_{tr}, \tag{8.36}$$

therefore, even above threshold $N = N_{th}$, thus meaning that the staedy-state carrier density remains damped to its value at threshold.

By using Eq. 8.7 in steady-state conditions, it is possible to calculate the current and the current density at threshold. At threshold $\phi = 0$, therefore since $dN/dt = 0$ and $N = N_{th}$, from Eq. 8.7, it is possible to calculate the threshold current (and the threshold current density):

$$I_{th} = \frac{eV}{\eta_i \tau} N_{th}, \qquad J_{th} = \frac{ed}{\eta_i \tau} N_{th}. \tag{8.37}$$

By using Eq. 8.37 and Eq. 8.34, the threshold current density can be written as:

$$J_{th} = \frac{ed}{\eta_i \tau} \left(\frac{\gamma}{\sigma L \Gamma} + N_{tr} \right). \tag{8.38}$$

It is worth to mention that a simple expression for the confinement factor is given by:

$$\Gamma \simeq \frac{V^2}{2 + V^2}, \tag{8.39}$$

where:

$$V = \frac{2\pi d}{\lambda_0} \sqrt{n_a^2 - n_c^2}, \tag{8.40}$$

where n_a and n_c are the refractive indexes of the active medium and of the cladding layers, respectively.

Figure 8.4 shows the evolution of the threshold current density as a function of the thickness, d, of the active layer for a double-heterostructure GaAs laser. Upon increasing the active layer thickness, the threshold current density first decreases and reaches a minimum value, then it starts to increase rapidly. This behavior can be understood on the basis of Eq. 8.38. For sufficiently large values of d ($d > 0.15$ μm), N_{tr} is the dominant term with respect to $\gamma/\sigma L \Gamma$. In this case, J_{th} turns out to be proportional to d. When d is very small

($d < 0.1$ μm) also the confinement factor becomes very small. Indeed, by using Eq. 8.39, for very small values of d, Γ becomes proportional to d^2. In this case, the first term in the parenthesis of Eq. 8.38 is the dominant one, so that the threshold current density becomes proportional to $1/d$.

Exercise 8.1   Let us consider a GaAs/Al$_{0.3}$Ga$_{0.7}$As double-heterostructure laser. Calculate the threshold current density assuming the following device parameters: refractive index of GaAs $n_1 = 3.6$, refractive index of Al$_{0.3}$Ga$_{0.7}$As $n_2 = 3.4$, thickness of the active medium $d = 0.2$ μm, cleaved an uncoated end faces, absorption coefficient $\alpha = 10$ cm^{-1}, cavity length $L = 250$ μm, differential gain coefficient $\sigma = 1.5 \times 10^{-16}$ cm^2, transparency carrier density $N_{tr} = 2 \times 10^{18}$ cm^{-3}, emission wavelength $\lambda = 850$ nm, injection efficiency $\eta_i = 1$, and recombination time constant $\tau = 4$ ns.

We first calculate the confinement factor Γ by using the approximation given in Eq. 8.39, where V is given by (see Eq. 8.40):

$$V = \frac{2\pi d}{\lambda}(n_1^2 - n_2^2) = 1.75$$

$$\Gamma \simeq \frac{V^2}{2 + V^2} = 0.60.$$

The threshold carrier density is given by Eq. 8.34:

$$N_{th} = \frac{\gamma}{\sigma L \Gamma} + N_{tr}.$$

$\gamma = \alpha L - \log R$, where $R = (n_1 - 1)^2/(n_1 + 1)^2 \simeq 0.32$ so that $\gamma = 1.39$.

$$N_{th} = \frac{\gamma}{\sigma L \Gamma} + N_{tr} = (0.61 \times 10^{18} + 2 \times 10^{18})\ \text{cm}^{-3} = 2.61 \times 10^{18}\ \text{cm}^{-3}.$$

The threshold current density is given by:

$$J_{th} = \frac{ed}{\eta_i \tau} N_{th} = 2.1 \times 10^3\ \text{A/cm}^2.$$

∎

The threshold current has been calculated by assuming a linear dependence of the gain with respect to the carrier density (see Eq. 8.33). In a more general case, the dependence of gain versus carrier density can be well approximated by a three-parameter logarithmic expression:

$$g = g_0' \ln \frac{N + N_s}{N_{tr} + N_s}. \tag{8.41}$$

Upon increasing N_s in the previous formula, the relationship between g and N becomes more linear. For positive gain and when g does not linearly increase with N, a two-parameter approximation gives a very good fitting of the experimental data:

$$g = g_0 \ln \frac{N}{N_{tr}}, \tag{8.42}$$

where g_0 is an empirical gain coefficient. The differential gain is given by $dg/dN = g_0/N$. At threshold, by using Eq. 8.32 and Eq. 8.42, it is possible to write:

$$N_{th} = N_{tr} \exp\left(\frac{g_{th}}{g_0}\right) = N_{tr} \exp\left(\frac{\gamma}{\Gamma g_0 L}\right). \tag{8.43}$$

In good laser materials, the recombination at threshold is dominated by the spontaneous recombination rate, BN_{th}^2, so that $1/\tau \simeq BN_{th}$ and the threshold current is given by:

$$I_{th} \simeq \frac{eV}{\eta_i} BN_{th}^2. \tag{8.44}$$

By using Eq. 8.43 in 8.44, an expression for the threshold current can be obtained:

$$I_{th} \simeq \frac{eV}{\eta_i} BN_{tr}^2 \exp\left(\frac{2\gamma}{\Gamma g_0 L}\right), \tag{8.45}$$

which clearly shows how I_{th} depends on transparency carrier density, gain, cavity volume, and confinement factor. It is evident that to have a small threshold current, it is required to reduce the transparency density and to increase the differential gain. In the case of a quantum well laser with a separate confinement waveguide, the small volume V and the small confinement factor Γ may lead to an increase of I_{th}. In this case, it can be useful to increase the number of quantum wells (multiple quantum well laser). In this case, Eq. 8.45 can still be used by multiplying the volume, V_1, and the confinement factor, Γ_1, of a single well by the number of wells, n_w:

$$I_{th} \simeq \frac{e n_w V_1}{\eta_i} BN_{tr}^2 \exp\left(\frac{2\gamma}{n_w \Gamma_1 g_0 L}\right). \tag{8.46}$$

The number of wells cannot be increased too much: It is limited by the spatial extension of the optical mode inside the laser cavity.

If the nonradiative recombination events related to Auger processes cannot be neglected, the threshold current increases. In this case:

$$\frac{1}{\tau} \simeq BN_{th} + CN_{th}^2, \tag{8.47}$$

and the threshold current is given by:

$$I_{th} \simeq \frac{eV}{\eta_i} (BN_{th}^2 + CN_{th}^3). \tag{8.48}$$

This is the case, for example, of long-wavelength InGaAsP/InP lasers, where nonradiative recombination events due to Auger processes cannot be neglected. The increase of the threshold current due to the Auger term in Eq. 8.48 is given by:

$$I_{th,nr} = \frac{eV}{\eta_i} CN_{tr}^3 \exp\left(\frac{3\gamma}{\Gamma g_0 L}\right). \tag{8.49}$$

Since this additional term is proportional to the cube of N_{tr}, particular attention has to be paid in these materials to reduce N_{tr}. For InGaAsP at room temperature $C = 2 - 3 \times 10^{-29}$ cm^6/s at 1.3 μm and $C = 7 - 9 \times 10^{-29}$ cm^6/s at 1.55 μm, so that the contribution to the threshold current from the nonradiative processes given by the previous equation becomes the dominant term in Eq. 8.48 for carrier concentrations $N_{th} \simeq 3 \times 10^{18}$ cm^{-3} and $N_{th} \simeq 1.5 \times 10^{18}$ cm^{-3} at 1.3 μm and 1.55 μm, respectively. For this reason, particular attention must be paid in these cases to reduce cavity losses and to maintain a large confinement factor. This can be achieved by using strained InGaAs/InGaAsP or InGaAs/InGaAlAs quantum wells on InP. Moreover, the splitting of the valence bands induced by the strain can contribute to reduce the Auger coefficient C.

8.3 Temperature Dependence

It is found that in typical double-heterostructure lasers, the threshold current changes with temperature as:

$$I_{th}(T) = I_0 \exp\left(\frac{T}{T_0}\right),$$

(8.50)

where T is the temperature in kelvin, I_0 is a constant, and T_0 is a characteristic temperature (measured in kelvin), which depends on the particular laser diode. T_0 can be used as a measurement of the temperature sensitivity of the diode. Indeed, the larger T_0 the less sensitive is I_{th} to temperature variations, since $\Delta I_{th}/I_{th} = \Delta T/T_0$.

In Eq. 8.45, the transparency density N_{tr}, the gain g_0, and the internal losses (γ factor) typically depend on temperature. In particular, N_{tr} increases with temperature while g_0 decreases with temperature, since at high temperature, the injected carriers can occupy a larger energy interval. The characteristic temperature T_0 is measured experimentally and depends on a large number of factors, including the physical properties of the semiconductors used in the laser diode and the particular laser structure. For example, in the case of GaAs/AlGaAs double-heterostructure laser, $T_0 \geq 120$ K around room temperature. In the case of GaAs/AlGaAs quantum well lasers, T_0 is in the range 150–180 K, while for strained InGaAs/AlGaAs quantum well lasers, $T_0 \geq 200$ K. The increase of the characteristic temperature in quantum well lasers is another advantage of the quantum well devices, and it is mainly a result of the weaker dependence of the differential gain on temperature. In the case of InGaAsP/InP double-heterostructure and quantum well lasers emitting between 1.3 and 1.55 μm, T_0 is in the range of 50–70 K at room temperature. These quite low values are mainly related to Auger recombination and to leakage current and limit the performances of these lasers under high-temperature operation (at 400 K T_0 is of the order of 30 K).

8.4 Output Power

In this section, a general expression for the output power of a laser diode as a function of the driving current will be calculated. Under steady-state conditions, above threshold, Eq. 8.7 can be rewritten as:

$$\eta_i \frac{I}{eV} - \frac{N_{th}}{\tau} - R_{st} = 0.$$

(8.51)

By using Eq. 8.16 and Eq. 8.37, the previous equation allows one to calculate the photon density ϕ corresponding to the injection current I:

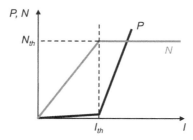

Figure 8.5   Qualitative evolution of the output power, P, and of the carrier density, N, as a function of the driving current.

$$\phi = \frac{\eta_i(I - I_{th})}{eVv_gg_{th}},\tag{8.52}$$

where g_{th} is given by Eq. 8.32. The optical energy stored in the laser cavity can be calculated as follows:

$$\mathcal{E}_{st} = \phi h\nu V_p,\tag{8.53}$$

where $h\nu$ is the energy of the emitted photons. The photon loss rate due to the mirror transmission is given by $-(v_g/L)\log R$, so that the output power is:

$$P = \mathcal{E}_{st}\left(-\frac{v_g}{L}\log R\right) = \eta_i\frac{h\nu}{e}\frac{-\log R}{\alpha L - \log R}(I - I_{th}).\tag{8.54}$$

It is useful to define the external quantum efficiency, η_{ext}, as the ration between the increase in emitted photons and the corresponding increase in injected carriers:

$$\eta_{ext} = \frac{d(P/h\nu)}{d(I/e)} = \frac{e}{h\nu}\frac{dP}{dI} = \eta_i\frac{-\log R}{\alpha L - \log R}.\tag{8.55}$$

By using this definition, the output power can be written as (when $I > I_{th}$):

$$P = \eta_{ext}\frac{h\nu}{e}(I - I_{th}).\tag{8.56}$$

The previous equation gives to total power out of both end faces of the diode. If one mirror has zero transmission, all power is emitted out of the other mirror.

Another useful parameter is the *slope efficiency* of the laser, defined as:

$$\eta_s = \frac{dP}{V_{el}dI},\tag{8.57}$$

where V_{el} is the applied voltage. From Eqs. 8.55 and 8.57, we see that the relation between the external quantum efficiency and the slope efficiency is:

$$\eta_{ext} = \eta_s\frac{eV_{el}}{h\nu}.\tag{8.58}$$

From Eq. 8.56, it is evident that the $P - I$ curve of a semiconductor laser diode above threshold is a straight line with a slope defined by η_s. Below the threshold current, the output power is due to spontaneous emission and depends on the portion of spontaneous emission in the lasing mode. Figure 8.5 shows schematically the evolution of the output power and of the carrier density as a function of the injected current.

Below threshold, the stimulated emission term, ΓR_{st}, can be neglected in Eq. 8.13, so that, in steady-state conditions, we can write:

$$\phi = \Gamma \beta_{sp} R_{sp} \tau_c. \tag{8.59}$$

If the stimulated emission rate, R_{st}, is neglected in Eq. 8.7, in steady-state conditions, we have:

$$\eta_i \frac{I}{eV} = \frac{N}{\tau}, \tag{8.60}$$

which can be rewritten in terms of R_{sp}:

$$\eta_i \frac{I}{eV} = \frac{R_{sp}}{\eta_{int}}. \tag{8.61}$$

where $\eta_{int} = \tau/\tau_{rad}$ is the internal quantum efficiency. The output power can be calculated as before:

$$
\begin{aligned}
P &= \phi h\nu V_p \left(-\frac{v_g}{L} \log R\right) = \frac{V}{V_p} \beta_{sp} \eta_{int} \eta_i \frac{I}{eV} \frac{L}{v_g(\alpha L - \log R)} h\nu V_p \left(-\frac{v_g}{L} \log R\right) \\
&= \eta_{int} \eta_i \frac{h\nu}{e} \frac{-\log R}{\alpha L - \log R} \beta_{sp} I.
\end{aligned}
\tag{8.62}
$$

At threshold, the spontaneous emission rate clamps to the value N_{th}/τ_{rad}.

8.5 Quantum Well Lasers

Quantum wells can be used as active materials in diode lasers to take advantage of the good optical properties of quantum wells compared to the corresponding bulk materials: in particular, the typically larger differential gain and weaker dependence of this gain on temperature and the possibility to use strained layers to improve the optical characteristics and the band gap tunability. Due to the very small thickness of the active material, the corresponding confinement factor, Γ is very small, thus giving rise to a large threshold current.

In order to increase the confinement factor separate confinement heterostructures (SCHs) are used, with various configurations, as shown in Fig. 8.6, which displays schematically the conduction band structure of a simple SCH and of a graded-index SCH. At the center of both structures, there is a thin quantum well, with a thickness of a few nanometers. The beam confinement is achieved by the barrier layers (with a thickness of the order of 0.1 μm), characterized by a higher refractive index compared to the cladding layers. The contribution of the quantum well to beam confinement is negligible, while the carriers are confined by the quantum well. In the case of the graded-index SCH (called GRIN-SCH, from GRaded-INdex SCH), the refractive index of the inner confinement structure is smoothly changed from a maximum value at the center of the structure to the value in the cladding region. For example, in a GaAs/AlGaAs quantum well laser, where GaAs is the well material, the composition of the inner barrier $Al_x Ga_{1-x}As$ changes from, for example, $x = 0.3$ at the center to $x = 0.7$ at the interface with the two cladding layers formed by $Al_{0.7}Ga_{0.3}As$.

Figure 8.6 Qualitative plot of the conduction energy band of (a) a step-index separate confinement quantum well heterostructure and of (b) a graded-index separate confinement quantum well heterostructure.

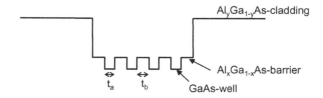

Figure 8.7 Qualitative plot of the conduction energy band of a multiple quantum well laser.

The active material can be a single quantum well or a multiple quantum well, which consists of a few alternating thin layers of wide and narrow energy gap semiconductors, as schematically shown in Fig. 8.7. By using a separate confinement structure, it is possible to significantly increase the confinement factor. A very good and simple approximation for the confinement parameter can be obtained as follows. The structure shown in Fig. 8.7 can be analyzed considering a three-region waveguide composed by the same cladding layers and by a central layer of thickness:

$$t = N_a t_a + N_b t_b,$$ (8.63)

where t_a and t_b are the thickness of the active (well) and barrier layers, respectively (see Fig. 8.7); N_a and N_b are the number of well and barrier layers, respectively. The central region is characterized by an average refractive index given by:

$$n = \frac{N_a t_a n_a + N_b t_b n_b}{t},$$ (8.64)

where n_a and n_b are the refractive indexes of the active and barrier layers, respectively. This equivalent three-region waveguide is characterized by a confinement factor Γ given by Eq. 8.39, where:

$$V = \frac{2\pi t}{\lambda_0}\sqrt{n^2 - n_c^2}.$$ (8.65)

The confinement factor of the real multiple quantum well structure must take into account the fraction of the modal energy in the active regions and can be calculated as:

$$\Gamma_{MQW} = \Gamma \frac{N_a t_a}{t}.$$ (8.66)

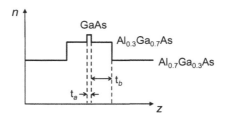

Figure 8.8 Variation of the refractive index of the quantum well diode laser with separate confinement heterostructure analyzed in Exercise 8.2.

Note that this simple three-region waveguide model is a fairly accurate model when the waveguide supports only the fundamental mode (the approximation is still relatively good also when higher-order modes can propagate).

Exercise 8.2   Consider a quantum well diode laser (with a single well) with a separate confinement heterostructure with the following characteristics. Active layer: GaAs, thickness $t_a = 10$ nm, refractive index $n_a = 3.6$; inner barrier layers: $Al_{0.3}Ga_{0.7}As$, thickness $t_b = 0.1$ μm, refractive index $n_b = 3.4$; cladding layer: $Al_{0.7}Ga_{0.3}As$, refractive index $n_c = 3.18$; emission wavelength $\lambda_0 = 850$ nm. Calculate the confinement factor.

The variation of the refractive index in the laser structure is shown in Fig. 8.8. The three-region equivalent waveguide has a central layer with thickness:

$$t = N_a t_a + N_b t_b = 0.21 \ \mu\text{m},$$

where $N_a = 1$ and $N_b = 2$. The average refractive index in this layer is given by Eq. 8.64:

$$n = \frac{N_a t_a n_a + N_b t_b n_b}{t} = 3.41.$$

The V-number is:

$$V = \frac{2\pi t}{\lambda_0}\sqrt{n^2 - n_c^2} = 1.91,$$

the confinement factor of the three-region waveguide is:

$$\Gamma = \frac{V^2}{2 + V^2} = 0.65,$$

and the corresponding confinement factor of the quantum well structure is:

$$\Gamma_{QW} = \Gamma\frac{t_a}{t} = 0.031.$$

∎

Exercise 8.3   Use the confinement factor obtained in the previous exercise to calculate the threshold current density assuming: $N_{tr} = 2 \times 10^{18}$ cm^{-3}, differential gain $\sigma = 6 \times 10^{-16}$ cm^2, absorption coefficient $\alpha = 10$ cm^{-1}, cavity length $L = 250$ μm, injection efficiency $\eta_i = 1$, recombination time constant $\tau = 4$ ns, and mirror reflectivity $R = 0.32$.

As we have calculated in Exercise 8.1 $\gamma = 1.39$. The threshold carrier density is given by:

$$N_{th} = \frac{\gamma}{\sigma L \Gamma} + N_{tr} = (3 \times 10^{18} + 2 \times 10^{18})\ \text{cm}^{-3} = 5 \times 10^{18}\ \text{cm}^{-3},$$

so that the threshold current density is given by:

$$J_{th} = \frac{e t_a}{\eta_i \tau} N_{th} = 200\ \text{A/cm}^2.$$

∎

8.6 Laser Structures

The simple laser structure shown in Fig. 8.1 cannot be used, also in the case of double-heterostructure configuration, due to the high driving current and the poor quality of the output beam. Lateral confinement of carriers and photons is required in order to achieve stable laser operation and good quality output beams. A lateral confinement of the injected carriers is required to avoid leakage currents, which may bypass the active region, thus strongly reducing the laser efficiency. Several laser configurations have been developed, which can be grouped in two main classes: gain-guided structures and index-guided structures. The *gain-guided* is the simplest structure: Fig. 8.9(a) shows a scheme of an edge-emitting gain-guided semiconductor laser. By adding an oxide layer, the current injected from the positive electrode is forced to flow in a stripe with narrow width ($w \simeq 5$ μm) thus strongly reducing the required current flow ($I = JLw$). Also the width of the pumped active layer is roughly equal to w thus leading to a (relatively weak) lateral waveguiding, which weakly confines the beam transverse dimension in the direction parallel to the junction. The structure is called gain-guided since the optical beam is mainly guided by the variation of the gain. If the stripe width is < 10 μm, the laser mode is confined to the fundamental transverse mode in the direction parallel to the junction, while in the direction perpendicular to the junction, the optical beam is confined to the fundamental mode by the double heterostructure or by the separate confinement structure. A major disadvantages of the gain-guided structure is that the un-pumped portions of the active layer are strongly absorbing thus introducing losses. Moreover, since the lateral waveguiding effect is rather weak, small variations of the refractive index, due for example to variations of the temperature or of the injected carriers, may lead to unstable operation. Threshold current is typically rather high of the order of 50–100 mA and the power-current curves generally present kinks, as shown in Fig. 8.9(b), associated to unstable lateral waveguiding.

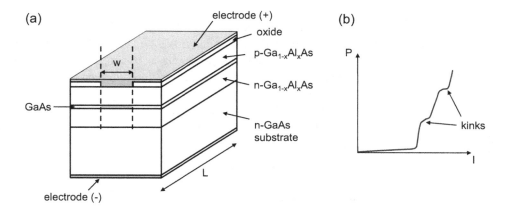

Figure 8.9 (a) Typical structure of gain-guided laser. (b) Power-current curve of a gain-guided laser with kinks associated to unstable lateral waveguiding

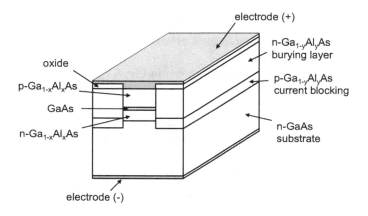

Figure 8.10 Typical structure of a buried heterostructure index-guided laser.

The optical spectrum of the output beam is multimodal due to the enhanced spontaneous emission.

A typical structure of a (strongly) index-guided edge-emitting laser is shown in Fig. 8.10. In this case, a small stripe of the active material is embedded into semiconductor layers with larger band gap (and lower refractive index). Due to reduced lateral width w of the active layer, lateral single-mode operation is possible (w must be smaller than a critical value for higher-order lateral modes). Figure 8.10 refers to a buried-heterostructure GaAs/AlGaAs index-guided laser. The n-type burying layer and the p-type current blocking layers form a reverse-biased diode, which does not allow the current to bypass the active region. By using strong index-guided laser structures, very stable laser performances can be achieved, with lateral single-mode operation and low threshold currents (<10 mA). For these reasons, they are largely used in commercial devices, despite the rather complex fabrication procedures required for the construction.

8.7 Spectral and Spatial Characteristics of Diode Laser Emission

The output spectrum of a diode laser depends on the structure of the optical resonator and on the optical gain curve of the active medium. In a Fabry–Perot (FP) cavity, the length L determines the separation between the longitudinal modes. In a typical semiconductor laser, while below threshold, a large number of FP cavity resonances are visible across a wide spectrum, above threshold only a few modes close to the peak of the gain curve oscillate, as shown in Fig. 8.11 in the case of a gain-guided laser emitting at ~ 779 nm. The frequency separation between adjacent modes is given by:

$$\Delta \nu = \frac{c}{2Ln_g},$$

(8.67)

where n_g is the group index, which takes into account the dispersion of the waveguide:

$$n_g = n_{eff} + \nu \frac{dn_{eff}}{d\nu} = n_{eff} - \lambda \frac{dn_{eff}}{d\lambda}.$$

(8.68)

n_{eff} is the effective refractive index experienced by an optical mode. For a typical FP laser diode, assuming $n_g = 3.5$ and a cavity length $L = 300$ μm, the mode separation is $\Delta \nu \simeq 140$ GHz, which is rather small in comparison to the typical linewidth of the gain curve of a semiconductor, which can be of the order of a few THz, so that longitudinal multimode operation is achieved above threshold in typical FP diode lasers. A possible way to obtain single-mode operation is to reduce the cavity length L so that the mode spacing exceeds the width of the gain curve: This requires cavity lengths of the order of a few micrometers. These very short cavities can be generated along the epitaxial deposition axis, and the corresponding devices are called vertical cavity surface-emitting lasers (see Chapter 11). Another technique used for single-mode laser operation is based on the use of distributed Bragg reflector lasers or of distributed feedback lasers, which will be analyzed in Chapter 10. We note that single-mode laser operation is particularly important for optical-fiber communications, due to the chromatic dispersion of an optical fiber.

In the case of the edge-emitting lasers discussed in the previous section, the output beam is typically elliptical. Due to small beam size in the direction perpendicular to the junction, the output beam is always diffraction limited in the plane perpendicular to the junction. If the stripe width is shorter than ~ 10 μm, the beam is diffraction-limited also in the

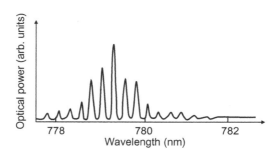

Figure 8.11   Output spectrum of a multilongitudinal mode gain-guided laser.

plane parallel to the junction. If a Gaussian beam profile is assumed in both transverse directions, the beam divergences $\theta_\parallel$ and $\theta_\perp$ in the planes parallel and perpendicular to the junction, respectively, are given by $\theta_\parallel = 2\lambda/\pi d_\parallel$ and $\theta_\perp = 2\lambda/\pi d_\perp$ where $d_\parallel$ and $d_\perp$ are the beam dimensions in the two directions, defined as the full-width between $1/e$ points of the electric field amplitude. Therefore, assuming an elliptical beam on the exit face of a laser with dimensions 1 μm × 5 μm, the divergence is larger in the direction perpendicular to the junction, so that the beam ellipticity rotates by 90° after a few tens of micrometers at the output of the laser cavity. When a circular beam is required, suitable optical systems can be used to compensate for the beam astigmatism.

8.8 Exercises

Exercise 8.1   Consider a double-heterostructure, edge-emitting $Al_xGa_{1-x}As$/GaAs laser with the following characteristics: emission wavelength $\lambda = 840$ nm, injection efficiency $\eta_i = 0.95$, transparency density $N_{tr} = 1.2 \times 10^{18}$ cm^{-3}, length of the active medium $L = 300$ μm, differential gain $\sigma = 3.6 \times 10^{-16}$ cm^2, recombination time constant $\tau = 4$ ns, thickness of the active region $d = 100$ nm, width of the active region $w = 100$ μm, logarithmic losses $\gamma = 1.43$, refractive index of the active material $n_1 = 3.6$, refractive index of cladding regions $n_2 = 3.4$. Calculate the threshold current.

Exercise 8.2   Consider a multiple quantum well laser with separate confinement heterostructure with the following characteristics: GaAs well region, number of wells $N_w = 3$, refractive index $n_w = 3.6$, thickness $t_w = 5$ nm. $Al_{0.3}Ga_{0.7}As$ barrier, $N_b = 4$, $n_b = 3.4$, $t_b = 4$ nm. $Al_{0.7}Ga_{0.3}As$ cladding, $n_c = 3.18$. Emission wavelength $\lambda = 850$ nm.
a) Plot the refractive index of the structure as a function of the propagation coordinate.
b) Calculate the confinement factor.

Exercise 8.3   Consider the multiple quantum well laser of Exercise 8.2, with the following additional characteristics: differential gain $\sigma = 6 \times 10^{-16}$ cm^2, transparency density $N_{tr} = 1.8 \times 10^{18}$ cm^{-3}, injection efficiency $\eta_i = 0.95$, length of the active medium $L = 300$ μm, width of the active region $w = 100$ μm, absorption coefficient $\alpha = 10$ cm^{-1}, intensity reflection coefficient $R = 0.32$, and emission wavelength $\lambda = 840$ nm. Determine the threshold current density and the threshold current.

Exercise 8.4   Consider the multiple quantum well of Exercise 8.3, which is working at $\Delta V = 1.8$ V forward driving voltage. Calculate the slope efficiency.

Exercise 8.5   Consider a double-heterostructure (DH) laser with GaAs active layer and $Al_{0.4}Ga_{0.6}As$ cladding layers. The laser is 200-μm long and 1-μm wide. The active layer is $0.1 - \mu$m thick. Both mirrors are cleaved facets with intensity reflectivity of 30%, the loss coefficient is $\alpha = 20$ cm^{-1}. The carrier lifetime is 1 ns, the differential gain is $\sigma = 10^{-16}$ cm^2, and the transparency density is $N_{tr} = 10^{18}$ cm^{-3}. The relative dielectric coefficient of $Al_xGa_{1-x}As$ ($0 < x < 0.45$) is $13.1 - 3x$. Calculate:

a) the V parameter of the DH waveguide and the number of modes which can be supported by the waveguide;

b) the confinement factor;

c) the threshold gain and threshold current of the laser;

d) the external quantum efficiency.

Exercise 8.6 Estimate the minimum injected electron concentration required to satisfy the Bernard–Duraffourg condition in an $n^+ - p$ GaAs homojunction laser diode at room temperature, assuming that the intrinsic electron population is 10^7 cm^{-3} and that the energy gap is 1.425 eV.

Exercise 8.7 Consider a multiple quantum well InP/InGaAsP laser. The laser is 300-μm long and 30-μm wide, with internal loss $\alpha = 15$ cm^{-1}. The active region is composed by four InGaAs, 3-nm-thick quantum wells. The gain of the active material can be written as $g = g_0 \ln (N/N_{tr})$, with $g_0 = 1{,}200$ cm^{-1} and $N_{tr} = 1.2 \times 10^{18}$ cm^{-3}. The transverse confinement factor is 1% per well and $\eta_i = 0.8$. Calculate:

a) the threshold modal gain;

b) the threshold current;

c) the threshold current when the laser is composed by 8 quantum wells, assuming that the internal modal loss α remains the same.

Exercise 8.8 Consider a semiconductor laser with a cavity length $L = 300$ μm. Assuming that the gain line has a bandwidth $\Delta \nu_L = 380$ GHz and that the group index of the semiconductor is $n_g = 3.6$, calculate the number of longitudinal modes which fall within the laser line. Determine the cavity length for single mode operation.

Exercise 8.9 Consider a semiconductor laser emitting at 850 nm, with a transverse field distribution along the direction parallel and perpendicular to the junction with Gaussian profiles, with spot sizes $w_1 = 2.5$ μm and $w_2 = 0.5$ μm, respectively (assume that, for both field distributions, the beam waists are located at the exit face). Calculate the distance from the exit face at which the beam becomes circular.

Exercise 8.10 Consider a GaAs/AlGaAs quantum well laser, 800-μm long and 3-μm wide. The active medium is composed by 4 quantum wells each of 6-nm thickness, with the following characteristics: internal losses $\alpha_i = 14$ cm^{-1}, injection efficiency $\eta_i = 0.75$, intensity reflection coefficient $R = 0.32$, transparency density $N_{tr} = 2.6 \times 10^{18}$ cm^{-3}, gain $g = g_0 \ln (N/N_{tr})$, with $g_0 = 2{,}400$ cm^{-1}, bimolecular recombination coefficient $B = 10^{-10}$ cm^3/s, and Auger recombination coefficient $C = 4 \times 10^{-30}$ cm^6/s. Calculate the threshold current and the contribution from the Auger nonradiative current.

Exercise 8.11 We want to design an InGaAs/AlGaAs/GaAs double-heterostructure laser with emission wavelength 980 nm, with an output power of 10 mW, threshold current density $J_{th} < 10^3$ A cm^{-2} and single-mode operation. Assume an active layer with a thickness $d = 0.3$ μm and 15 $- \mu$m wide.

a) Determine the composition of the active material using Figure 6.1, which reports the lattice constants as a function of energy gap, and assuming that the lattice constant of In$_{1-x}$Ga$_x$As at 300 K (in Å) can be well approximated by $a = 6.0583 - 0.405x$.

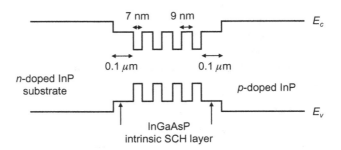

Figure 8.12 InGaAsP/InP quantum well of Exercise 8.12.

b) Determine the composition of $Al_xGa_{1-x}As$ cladding layers in order to have a difference between the energy gaps of the active and cladding layers of the order of at least 0.3 eV and a difference between the refractive indexes of active and cladding layers of the order of 0.01. The refractive index of $Ga_xIn_{1-x}As$ is given by $n = 3.52 + 0.08x$, and the refractive index of $Al_xGa_{1-x}As$ is given by $n = 3.59 - 0.71x + 0.091x^2$. The energy gap of $Al_xGa_{1-x}As$ at 300 K is (in eV) $\mathcal{E}_g = 1.424 + 1.247x$ ($x < 0.45$).

c) Assume internal absorption $\alpha = 15$ cm^{-1}, injection efficiency $\eta_i = 0.95$, recombination time constant $\tau = 4$ ns, differential gain coefficient $\sigma = 3.6 \times 10^{-16}$ cm^2, and transparency density $N_{tr} = 1.2 \times 10^{18}$ cm^{-3}. Calculate the threshold current as a function of the cavity length.

d) Estimate the cavity length to achieve an output power of about 10 mW.

e) Calculate the mode separation $\Delta\lambda$ (assume that $dn/d\lambda = 2.5$ μm^{-1}) and the corresponding frequency separation $\Delta\nu$.

f) Calculate the beam divergence in the junction plane and in the direction perpendicular to the junction plane.

Exercise 8.12   Consider an InGaAsP/InP quantum well laser whose structure is shown in Figure 8.12. Refractive indexes: InP: 3.17, InGaAs SCH layer: 3.386, InGaAsP QW barrier: 3,386, InGaAsP QW: 3.550 at 1,550 nm. Intrinsic material losses: 40 cm^{-1} per 10^{18} cm^{-3} doping in the p-doped regions, 5 cm^{-1} per 10^{18} cm^{-3} doping in the n-doped regions. Length of the laser cavity $L = 500$ μm. End facet intensity reflectivities $R_1 = R_2 = 0.3$, injection efficiency $\eta_i = 0.8$. Gain of each quantum well is given by $g = g_0 \ln(N/N_{tr})$, with $g_0 = 1800$ cm^{-1}, $N_{tr} = 1.6 \times 10^{18}$ cm^{-3}. Recombination constants: $A = 0$, $B = 1.5 \times 10^{-10}$ cm^3/s, $C = 5 \times 10^{-29}$ cm^6/s. Confinement per single well $\Gamma_w = 0.16$. Calculate:

a) the photon lifetime;

b) the threshold gain and the threshold carrier density;

c) the threshold current.

Exercise 8.13   Consider the quantum well laser of Exercise 8.12. Assume that the confinement factor of the active region can be written as $\Gamma = N_w\Gamma_w$, where N_w is the number of wells in the active region.

a) Calculate the threshold current of the laser for N_w ranging from 1 to 10.

b) Plot the results and determine the number of wells giving rise to the lowest threshold current.

Exercise 8.14   Demonstrate that above threshold the ratio $\eta_P = P/(IV)$, where P is the output power and V is the applied bias voltage, is always less than unity.

Exercise 8.15   The gain for a single quantum well laser can be written as:

$$g_1 = \Gamma g_0 \ln\left(\frac{J_1}{J_{tr}}\right).$$

Show that the optimal number of wells is given by:

$$n_{opt} = \mathrm{Int}\left[\frac{L}{\Gamma g_0 \gamma}\right] + 1,$$

where γ represents the logarithmic losses.

Hint: the optimal number of wells can be obtained by setting $dJ_n/dn = 0$, where J_n is the current density for the multiple quantum well structure.

Exercise 8.16   Consider the multiple quantum well laser of Exercise 8.15. Demonstrate that there is a minimum threshold current that is achieved at a cavity length:

$$L = \frac{1}{n\Gamma g_0} \ln\frac{1}{R}.$$

Exercise 8.17   Consider a quantum well laser where the well is formed by a semiconductor with an energy gap of 0.78 eV at room temperature, with effective masses $m_c = 0.06\, m_0$ and $m_v = 0.6\, m_0$. Calculate the well widths required to obtain laser action at 1,550 nm and at 1,300 nm.

Exercise 8.18   Consider a semiconductor laser emitting at 1,550 nm. The cavity length is $L = 300\ \mu\mathrm{m}$, the confinement factor is $\Gamma = 0.2$, the material absorption coefficient is $\alpha = 80\ \mathrm{cm}^{-1}$, and the refractive index inside the laser cavity is $n = 3.43$. Calculate:

a) the spacing, in nanometers, between adjacent laser modes;

b) the threshold gain coefficient;

c) the external efficiency.

Quantum Dot Lasers

9.1 Introduction

In Chapter 6, we have seen that quantum confinement effects in quantum well structures (1D-confinement) may determine significant improvements in optical properties, particularly important for laser applications. The fundamental difference between quantum wells and bulk materials, which fundamentally determines their electronic and optical properties, is the different shape of the density of states. As a natural evolutions, it is possible to consider quantum confinement in two and three dimensions, thus leading to quantum wires and quantum dots (QDs). In this chapter, we will analyze QD optical properties, due to their technological importance for laser applications. Carrier motion in QDs is restricted in all three dimensions. The result is that QDs exhibit discrete atomic-like states, whose energy depends on dot size, increasing upon decreasing the QD size. As a consequence, the corresponding energy gap, and therefore the emission wavelength, is size dependent and can be tuned in a broad energy range by using QDs with different sizes composed by II–VI, III–V and IV–VI semiconductors. This broad spectral tunability is very important for technological applications, in particular for the development of lasers emitting in spectral regions not easily accessible with traditional semiconductor lasers. In addition, the large separation between the atomic-like energy levels in QDs inhibits thermal depopulation of the band edge states, thus leading to a reduction of the optical gain threshold for laser operation and to a reduced temperature dependence of the threshold current.

The use of QDs as active material in a laser was first proposed in 1982 by Arakawa and Sakaki [13]. In this first paper, QDs were named three-dimensional quantum wells or quantum boxes. In 1986, Asada *et al.* [14] theoretically discussed the advantages of quantum boxes in III–V material systems in terms of reduction in threshold current density, reduction in total threshold current, enhanced differential gain, and high spectral purity, as a result of the strongly different density of states in quantum boxes, with respect to bulk and quantum wells.

The demonstration of QD lasers was achieved in the 1990s, thanks to significant progress in nanostructure growth. After the first demonstration of a QD laser, reported in 1994 [15], characterized by a low threshold current density (120 A/cm^2) at 77 K, with a characteristic temperature $T_0 = 350$ K, the growth methods and laser designs were rapidly improved, leading to QD lasers outperforming their quantum well counterpart by the year 2000.

After a very brief introduction on a few QD growing techniques, we will proceed as in the previous chapter by analyzing the optical properties of QDs, in terms of absorption and gain. Finally, we will discuss the main advantages of QD lasers compared to quantum well lasers.

9.2 Fabrication Techniques of QDs

In 1985, Goldstein *et al.* reported on the formation of a three-dimensional structure during the growth of InAs/GaAs-strained quantum wells by molecular beam epitaxy (MBE) [16]. The self-assembling formation of QDs was understood as the Stranski–Krastanov (S–K) growth mode, which is now a widely used technique.

There are three different epitaxial growth modes of thin films, called Frank–van der Merwe, Volmer–Weber, and Stranski–Krastanov, depending on the balance of surface energies during the film growth. If the sum of the surface energy, γ_f, of the free surface of the film and the surface energy, γ_i, of the substrate–film interface is lower than the surface energy, γ_s, of the free surface of the substrate, $\gamma_f + \gamma_i < \gamma_s$, the material being deposited wets the substrate. In this case, we have the Frank–van der Merwe growth mode, also called *layer-by-layer* growth, which requires a very low lattice mismatch between the deposited material and the substrate and a strong enough bond between the film and the substrate, in order to reduce γ_i. The adatoms are very compatible with the substrate and prefer to attach directly on it, as schematically shown in Fig. 9.1(a). This epitaxial growth mode is employed for epitaxial synthesis of two-dimensional materials, as silicene and germanene, or multi-layered structures with quantum wells. In the case of insufficient substrate bonding $\gamma_f + \gamma_i > \gamma_s$, so that the film does not wet the substrate, three-dimensional islands are formed on the bare substrate, as shown in Fig. 9.1(b): This is the Volmer–Weber growth mode, which is the typical growth mode in the case of systems with high lattice mismatch. In the case of the Stranski–Krastanov growth mode, which is typical of heteroepitaxial systems with a nonperfect lattice match, the growth starts with the formation of a wetting layer, which consists of a few monolayer-thick film followed by the spontaneous growth of three-dimensional islands on top of this layer, as shown in Fig. 9.1(c).

For example, in the case of InAs deposited onto GaAs, the lattice mismatch is of the order of 7%: A strain is generated between the substrate and the epilayer, which reacts to this strain by generating three-dimensional islands instead of two-dimensional surfaces. The growth sequence is schematically shown in Fig. 9.2. During the growth, the first few layers of InAs form a pseudomorphic two-dimensional layer (wetting layer). Above a critical thickness of the wetting layer, the two-dimensional growth is no longer favorable energetically, and the growth continues with the generation of three-dimensional islands, usually called self-assembled or self-organized QDs. The islands are then embedded in a barrier material, usually GaAs, InGaAs, or InAlGaAs in the case of InAs QDs. These self-assembled dots, which are typically characterized by a pyramidal shape, can be very small, with volume of the order or smaller than 10 nm^3. Self-assembled QDs characterized by high crystal quality have been grown by implementing various growth techniques, including MBE, metal-organic chemical vapor deposition, chemical-beam epitaxy, and metal

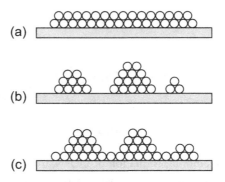

Figure 9.1   Thin film growth modes: (a) Frank–van der Merwe; (b) Volmer–Weber; and (c) Stranski–Krastanov.

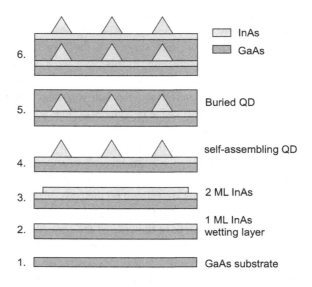

Figure 9.2   Schematic growth sequence in the case of Stranski–Krastanov growth mode.

organic vapor phase epitaxy. Figure 9.3 shows a transmission electron microscope (TEM) image of buried InAs QD layers in a multistack structure used to make lasers.

The main limitations of the S-K method are the following: (i) a two-dimensional wetting layer interconnecting the QDs is unavoidably present, which generates two-dimensional electronic states interconnecting the QDs, thus affecting the optical properties and carrier kinetics of the dots; (ii) limited range of QD morphologies and sizes, due to constrains on elastic, interface, and surface energies; (iii) it is difficult to control separately the QD structure and the surface density; (iv) limited combinations of material compositions for substrate and dots; and (v) limited number of crystal orientations of the substrate.

Another technique used for the creation of QD structures is the *droplet epitaxy* (DE) method, which is based on controlled crystallization of metal nano-droplets into III–V semiconductors. As an important example, we will describe the DE growth of GaAs QDs embedded in AlGaAs, but a similar procedure can be used in a variety of different group

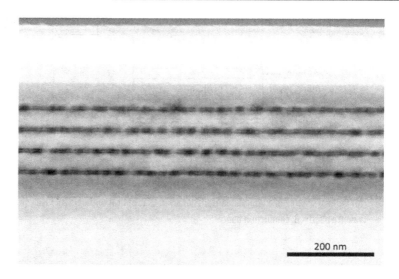

200 nm

Figure 9.3   Transmission electron microscope image of four layers of InAs quantum dots [17].
Source: From J.C. Norman et al., APL Photonics 3, 030901 (2018), CC BY 4.0.' in Figure 9.3.

III–V combinations, possibly changing the temperature ranges due to element-specific binding energies and diffusivity. The first step is the deposition of group III metals (Ga) on the substrate surface (AlGaAs), leading to the spontaneous formation of nanometer-scale droplets, irrespective of the lattice mismatch, due to surface tension, following the Volmer–Weber growth mode. In this case, the adatom–adatom interactions are stronger than the interactions between adatoms and substrate, thus determining the formation of three-dimensional adatom islands. The droplet density, usually in the range between 0.1 and 10^3 droplets per μm^2, can be controlled by adjusting the substrate temperature between 150 and 500 °C. The higher the temperature used during the growth, the lower the surface density and the larger the average size. The droplet size, in the range between 10 and 30 nm, can be controlled by proper choice of metal atom coverage, that is, the total amount of metal deposited per unit surface area. The second step is the droplet crystallization by annealing in the group-V element atmosphere. In contrast with the S-K method, DE QDs maintain their morphology upon capping; moreover, the size and aspect ration can be controlled in a broad range during the island formation adjusting the substrate temperature. Since the formation of the dots is not induced by strain as for the S-K method, it can be applied to the fabrication of both lattice-matched and lattice-mismatched QDs.

9.3 General Scheme of QD Lasers

A general scheme of a typical QD laser is shown in Fig. 9.4(a), where it is present a double heterostructure. The structure consists of an $Al_{0.3}Ga_{0.7}As$ outer cladding, a GaAs inner cladding, and four layers of $In_{0.4}Ga_{0.6}As$ self-organized QDs as the active region. Figure 9.4(b) displays the schematic band diagram of the laser structure. So far the most

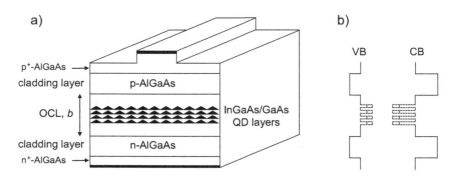

Figure 9.4   (a) General scheme of a typical quantum dot laser. OCL: optical confinement layer. (b) Schematic band diagram. VB: valence band; CB: conduction band.

well-studied QD material system for optoelectronic applications is InAs, often embedded in an InGaAs quantum well (this structure is called *dot in a well*, DWELL). The main advantages of QD lasers compared to quantum well lasers are the following: (i) lower threshold current density; (ii) larger differential gain; (iii) superior temperature stability of the threshold current; (iv) lower linewidth enhancement factor and reduced linewidth; (v) reduced reflection sensitivity; (vi) lower diffusion length of nonequilibrium carriers, thus resulting in reduced leakage from the active region; (vii) reduced sensitivity to defects; (viii) lower sensitivity to nonradiative recombination centers inside or near the active region, due to a reduced interaction with the defects produced by carrier localization in the dots. Many of these advantages have been demonstrated in real devices; on the other hand, the self-assembly techniques, typically used for QD fabrication, inevitably produce significant fluctuations of dot size, determining inhomogeneous broadening, which leads to spreading of the gain over a wider spectral range, thus reducing the maximum peak gain. In real devices, the maximum gain is a function of QD density, the number of dot layers, the dot morphology, and the dot size distribution. Laser action cannot be obtained if the dot density is smaller than a critical density or if inhomogeneous broadening exceeds a critical value.

In the following, we will first calculate the electronic states in cubic and spherical QDs. After a brief session on carrier statistics in QDs, as we have done in the case of bulk and quantum well semiconductors, the optical properties of QDs will be analyzed, in particular, the absorption and gain spectra will be calculated. Finally, the threshold current density in QD lasers and various properties of QD lasers will be discussed.

9.4 Electronic States in QDs

9.4.1 Particle in a Cubic Box

Let us first consider the simplest case: A particle confined in a cubic infinite potential well with edge L, so that the probability of finding an electron outside the cubic dot is zero. We will assume that the potential is zero inside the box and infinite outside. Inside the box the Schrödinger equation is:

$$-\frac{\hbar^2}{2m}\left(\frac{\partial^2}{\partial x^2} + \frac{\partial^2}{\partial y^2} + \frac{\partial^2}{\partial z^2}\right)\psi(x, y, z) = \mathcal{E}\psi(x, y, z), \tag{9.1}$$

with boundary conditions:

$$\begin{aligned}
\psi(0, y, z) &= \psi(L, y, z) = 0 &&\text{for all } y \text{ and } z \\
\psi(x, 0, z) &= \psi(x, L, z) = 0 &&\text{for all } x \text{ and } z \\
\psi(x, y, 0) &= \psi(x, y, L) = 0 &&\text{for all } x \text{ and } y.
\end{aligned} \tag{9.2}$$

Since the Hamiltonian is the sum of three terms with totally separate variables, the wave function can be written as the product of three one-dimensional wave functions:

$$\psi(x, y, z) = X(x)Y(y)Z(z), \tag{9.3}$$

and the corresponding boundary conditions are:

$$X(0) = X(L) = 0, \ Y(0) = Y(L) = 0, \ Z(0) = Z(L) = 0. \tag{9.4}$$

Substituting Eq. 9.3 in Eq. 9.1, we obtain

$$\frac{X''(x)}{X(x)} + \frac{Y''(y)}{Y(y)} + \frac{Z''(z)}{Z(z)} = -\frac{2m}{\hbar^2}\mathcal{E}. \tag{9.5}$$

The only way in which the previous equation can be satisfied is the following:

$$\begin{aligned}
\frac{X''(x)}{X(x)} &= -k_x^2 \\
\frac{Y''(y)}{Y(y)} &= -k_y^2 \\
\frac{Z''(z)}{Z(z)} &= -k_z^2,
\end{aligned} \tag{9.6}$$

where:

$$k_x^2 + k_y^2 + k_z^2 = \frac{2m}{\hbar^2}\mathcal{E}. \tag{9.7}$$

Each of the Eq. 9.6, with the corresponding boundary conditions 9.4, is equivalent to the one-dimensional problem of Chapter 6. Therefore, the normalized functions $X(x)$, $Y(y)$, and $Z(z)$ can be written as:

$$\begin{aligned}
X(x) &= \sqrt{\frac{2}{L}}\sin k_x x, & k_x &= n_x\frac{\pi}{L}, & n_x &= 1, 2, ... \\
Y(y) &= \sqrt{\frac{2}{L}}\sin k_y y, & k_y &= n_y\frac{\pi}{L}, & n_y &= 1, 2, ... \\
Z(z) &= \sqrt{\frac{2}{L}}\sin k_z z, & k_z &= n_z\frac{\pi}{L}, & n_z &= 1, 2, ...
\end{aligned} \tag{9.8}$$

and the allowed energy levels are:

$$\mathcal{E} = \frac{\pi^2\hbar^2}{2mL^2}(n_x^2 + n_y^2 + n_z^2) = \mathcal{E}_1(n_x^2 + n_y^2 + n_z^2), \tag{9.9}$$

Table 9.1 Energy and degeneracy of the first six energy levels of a cubic, infinite potential well.

Energy level	n_x	n_y	n_z	Energy	Degeneracy
1	1	1	1	$3\mathcal{E}_1$	1
2	2	1	1	$6\mathcal{E}_1$	3
	1	2	1		
	1	1	2		
3	2	2	1	$9\mathcal{E}_1$	3
	1	2	2		
	2	1	2		
4	3	1	1	$11\mathcal{E}_1$	3
	1	3	1		
	1	1	3		
5	2	2	2	$12\mathcal{E}_1$	1
6	1	2	3	$14\mathcal{E}_1$	6
	1	3	2		
	2	1	3		
	2	3	1		
	3	1	2		
	3	2	1		

where

$$\mathcal{E}_1 = \frac{\pi^2 \hbar^2}{2mL^2}. \tag{9.10}$$

The lowest energy level corresponds to $n_x = n_y = n_z = 1$, and the corresponding energy is $\mathcal{E} = 3\mathcal{E}_1$. As an example, Table 9.1 displays the energy and the degeneracy of the first six energy levels.

The three-dimensional, normalized wave function is:

$$\psi(x, y, z) = \sqrt{\frac{8}{L^3}} \sin\left(n_x \frac{\pi x}{L}\right) \sin\left(n_y \frac{\pi y}{L}\right) \sin\left(n_z \frac{\pi z}{L}\right). \tag{9.11}$$

These eigenfunctions form an orthonormal set, so that:

$$\int_0^L \int_0^L \int_0^L \psi^*_{n_x' n_y' n_z'}(x, y, z) \psi_{n_x n_y n_z}(x, y, z) \, dxdydz = \delta_{n_x' n_x} \delta_{n_y' n_y} \delta_{n_z' n_z}. \tag{9.12}$$

Therefore, two eigenfunctions are orthogonal unless all three quantum numbers n_x, n_y, and n_z are equal.

9.4.2 Spherical QDs

If the box confining the particle is spherical, the derivation of the eigenfunctions and eigenvalues is mathematically more complex. The stationary Schrödinger equation of a particle in a spherical potential well $V(\mathbf{r})$ is:

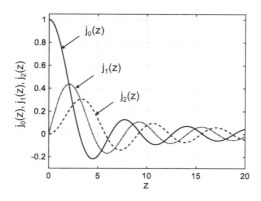

Figure 9.5 First three spherical Bessel functions.

$$\left[-\frac{\hbar^2}{2m} \nabla^2 + V(\mathbf{r}) \right] \psi(\mathbf{r}) = \mathcal{E}\psi(\mathbf{r}), \tag{9.13}$$

where $V(\mathbf{r}) = V(r)$ is centro-symmetric and, in the case of infinite potential well, can be written as:

$$V(r) = \begin{cases} 0, & \text{for } 0 \le r \le a \\ \infty, & \text{for } r > a, \end{cases} \tag{9.14}$$

where a is the radius of the sphere. The wave function is nonzero only inside the sphere. The boundary conditions are the following: $\psi(a) = 0$ and $\psi(\mathbf{r})$ well behaved (i.e., square-integrable) at $r = 0$. The solution of the corresponding time-independent Schrödinger equation can be written as:

$$\psi(r, \theta, \phi) = R_{n\ell}(r) Y_{\ell m}(\theta, \phi), \tag{9.15}$$

where the radial function, $R_{n\ell}(r)$, can be calculated by using the following equation:

$$\frac{d^2 R_{n\ell}}{dr^2} + \frac{2}{r} \frac{dR_{n\ell}}{dr} + \left[k^2 - \frac{\ell(\ell+1)}{r^2} \right] R_{n\ell} = 0, \tag{9.16}$$

where:

$$k^2 = \frac{2m\mathcal{E}}{\hbar^2}. \tag{9.17}$$

If we take $z = kr$, Eq. (9.16) can be written as:

$$\frac{d^2 R_{n\ell}}{dz^2} + \frac{2}{z} \frac{dR_{n\ell}}{dz} + \left[1 - \frac{\ell(\ell+1)}{z^2} \right] R_{n\ell} = 0. \tag{9.18}$$

The physical (i.e., well behaved in $r = 0$) solutions of this differential equation are the spherical Bessel functions $j_\ell(z)$:

$$j_\ell(z) = z^\ell \left(-\frac{1}{z} \frac{d}{dz} \right)^\ell \left(\frac{\sin z}{z} \right). \tag{9.19}$$

The first three spherical Bessel functions, as shown in Fig. 9.5, are:

Table 9.2 Zeros, $z_{n\ell}$, of the spherical Bessel function of order ℓ, for a few values of ℓ and n.

	$n = 1$	$n = 2$	$n = 3$	$n = 4$
$\ell = 0$	π	2π	3π	4π
$\ell = 1$	4,493	7,725	10,904	14,066
$\ell = 2$	5,763	9,095	12,323	15,515
$\ell = 3$	6,988	10,417	13,698	16,924

$$j_0(z) = \frac{\sin(z)}{z} \tag{9.20}$$

$$j_1(z) = \frac{\sin(z)}{(z)^2} - \frac{\cos(z)}{z} \tag{9.21}$$

$$j_2(z) = \left[\frac{3}{(z)^3} - \frac{1}{z}\right]\sin(z) - \frac{3\cos(z)}{(z)^2}. \tag{9.22}$$

Since at $r = a$ the wave function vanishes, $R_{n\ell}(ka) = 0$, k must be chosen so that $z = ka$ corresponds to one of the zeros of $j_\ell(z)$. If $z_{n\ell}$ is the n-th zero of $j_\ell(z)$, we have:

$$ka = z_{n\ell}. \tag{9.23}$$

The first few values of $z_{n\ell}$ are listed in Table 9.2.

The corresponding eigenvalues are:

$$\mathcal{E}_{n\ell} = z_{n\ell}^2 \frac{\hbar^2}{2ma^2}. \tag{9.24}$$

Figure 9.6 shows the six lowest energy levels of a particle in a three-dimensional box with an infinite cubic and spherical potential barrier, assuming that the edge length, L, of the cube is equal to the diameter, $2a$, of the sphere.

The normalized solution in polar coordinates can be written as:

$$\psi_{n\ell m}(r,\theta,\phi) = \sqrt{\frac{2}{a^3}} \frac{1}{j_{\ell+1}(z_{n\ell})} j_\ell\left(z_{n\ell}\frac{r}{a}\right) Y_{\ell m}(\theta,\phi), \tag{9.25}$$

where: $n = 1, 2, 3, ...$, $\ell = 0, 1, 2, ...$ and $-\ell \le m \le \ell$. The $n\ell$ eigenstates are usually referred to as nS, nP, nD, etc. Since the energy is independent of quantum number m, the energy level $n\ell$ is $(2\ell + 1)$-fold degenerate. The interpretation of the three quantum numbers, n, ℓ, and m, is straightforward. The radial quantum number n determines the number of nodes of the wavefunction in the radial direction, in the range between 0 and a, not counting any nodes at $r = 0$ or $r = a$. Thus, $n = 1$ corresponds to no nodes, $n = 2$ to one node and so on. The polar quantum number ℓ determines the number of nodes of the wavefunction when the polar angle θ varies between 0 and π. Finally, the azimuthal quantum number, m, determines the number of nodes of the wavefunction when the azimuthal angle ϕ varies from 0 to 2π.

We note that the spherical Bessel functions are mutually orthogonal:

$$\int_0^a j_\ell^*\left(z_{n\ell}\frac{r}{a}\right) j_\ell\left(z_{n'\ell}\frac{r}{a}\right) r^2 dr = 0, \tag{9.26}$$

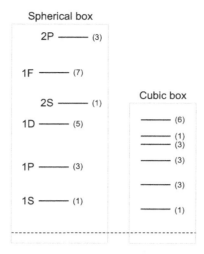

Figure 9.6 Six lowest energy levels of a particle in a three-dimensional box with an infinite cubic and spherical potential barrier, assuming that the edge length, L, of the cube is equal to the diameter, $2a$, of the sphere. The numbers between parenthesis indicate level degeneracy.

when $n \neq n'$. Since the spherical harmonic functions $Y_{\ell m}(\theta, \phi)$ are mutually orthogonal:

$$\int_{\phi=0}^{2\pi} \int_{\theta=0}^{\pi} Y_{\ell m}^*(\theta, \phi) Y_{\ell' m'}(\theta, \phi) \sin\theta \, d\theta \, d\phi = \delta_{\ell\ell'} \delta_{mm'}, \tag{9.27}$$

we conclude that wave functions corresponding to distinct sets of values of the quantum numbers n, ℓ, and m, are mutually orthogonal.

The use of infinite potential barriers leads to an overestimation of the confinement energy. Upon considering a particle in a spherical well with finite potential barrier, V_0, the energy of the lowest energy level can be written as:

$$\mathcal{E}_1 = \eta^2 \frac{\hbar^2 \pi^2}{2ma^2}, \tag{9.28}$$

where η is a function of V_0, m, and a, and it is smaller than 1. In the case of infinite barrier, $\eta \to 1$.

In the case of QDs, the mass to be considered in Eq. 9.24 is the effective mass of holes, m_h, and of electrons, m_e, so that the corresponding band gap, $\mathcal{E}_g$ can be approximated as the sum of a bulk band gap, $\mathcal{E}_{gb}$, and the size-dependent confinement energies of the band edge electron, $1S_e$, and hole, $1S_h$ states, as shown in Fig. 9.7:

$$\mathcal{E}_g = \mathcal{E}_{gb} + \mathcal{E}_{1S_e} + \mathcal{E}_{1S_h} = \mathcal{E}_{gb} + \frac{\pi^2\hbar^2}{2m_e a^2} + \frac{\pi^2\hbar^2}{2m_h a^2} = \mathcal{E}_{gb} + \frac{\pi^2\hbar^2}{2m_r a^2}, \tag{9.29}$$

where m_r is the electron–hole pair reduced mass. Equation 9.29 shows that the band gap of the QD increases as a^{-2} upon decreasing the dot radius. As in the case of quantum wells, the increase of the band gap going from the bulk to the QD is related to the quantum confinement and can reach hundreds of millielectronvolts in sufficiently small QDs.

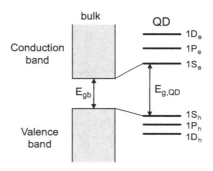

Figure 9.7   Discrete QD energy levels compared to the bulk counterpart.

This represents one of the advantages of QDs, since it is possible to control the emission wavelength by acting on the dot size. In a QD ground state, the level in the valence band contains two electrons, each of which can absorb an incident photon. This excitation generates a single electron–hole pair (exciton).

In the case of QDs with different shapes, for example, in the case of conical or pyramid-shaped QDs, the corresponding energy levels are different but with common features: The energy levels have a quadratic dependence on quantum numbers, a reverse quadratic dependence upon QD dimensions, and a reverse dependence on the effective mass.

9.4.3 Coulomb Interaction

The Coulomb interaction between electron and hole in a QD leads to an additional contribution, $\Delta \mathcal{E}_{e-h}$, to the energy gap, called *Coulomb excitonic contribution*. The corresponding Hamiltonian is:

$$H_{eh} = -\frac{e^2}{4\pi \epsilon_0 \epsilon_r |\mathbf{r}_e - \mathbf{r}_h|} = -\frac{e^2}{4\pi \epsilon |\mathbf{r}_{eh}|}, \tag{9.30}$$

where $\epsilon = \epsilon_0 \epsilon_r$ is the semiconductor dielectric constant (ϵ_r is the relative dielectric constant of the semiconductor). In order to take into account the Coulomb potential, two regimes are generally considered, depending on the ratio of the QD radius to the Bohr radius of the bulk exciton, $R_B = 4\pi \epsilon \hbar^2 / m_r e^2$ (Eq. 6.117). In the case of strong confinement ($a/R_B \leq 2$), the electron–hole Coulomb interaction can be considered as a small perturbation against the single particle terms and the exciton states should be considered as uncorrelated electronic and hole states. In the weak confinement regime, when $a/R_B \geq 4$, the electron–hole pair states are exciton bound states. In this case, the contribution of the Coulomb interaction to the exciton ground state cannot be considered as a perturbation to the confinement energy, but its contribution can still be treated as a perturbation to the infinite confining potential well.

In the strong confinement regime, the ground state electron–hole pair wavefunction can be written as:

$$\phi_{e,h}(\mathbf{r}_e, \mathbf{r}_h) = \psi_{100}(\mathbf{r}_e)\psi_{100}(\mathbf{r}_h)\phi_r(\mathbf{r}_{eh}), \tag{9.31}$$

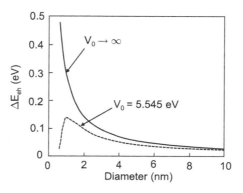

Figure 9.8 Coulomb energy, $\Delta\mathcal{E}_{eh}$, calculated for finite potential barriers, with $V_0 = 5.545$ eV (dashed curve), and for infinite barrier (solid curve), in PbS QDs embedded in SiO$_2$. Adapted with permission from [18].

where the factor $\psi_{100}(\mathbf{r}_e)\psi_{100}(\mathbf{r}_h)$ implies that the electron and hole occupy their respective ground states when the Coulomb interaction is completely neglected. The term $\phi_r(\mathbf{r}_{eh})$, given by:

$$\phi_r(\mathbf{r}_{eh}) = \exp\left(-\sigma\, r_{eh}/2\right), \tag{9.32}$$

is added in such a way that the electron–hole pair is characterized by the behavior of an excitonic bound state, analogous to the ground state of an hydrogen-like atom with mass μ (see Eq. 6.137), therefore $\sigma \propto R_B^{-1}$. The Coulomb excitonic contribution is evaluated by:

$$\Delta\mathcal{E}_{eh} \simeq -\langle\phi_{e,h}(\mathbf{r}_e, \mathbf{r}_h)|H_{eh}|\phi_{e,h}(\mathbf{r}_e, \mathbf{r}_h)\rangle. \tag{9.33}$$

Using the wave function for the infinite potential well, $\Delta\mathcal{E}_{eh}$ can be calculated as:

$$\Delta\mathcal{E}_{eh} \simeq -\frac{1.8e^2}{4\pi\,\epsilon_0\epsilon_r a} - 0.25\text{Ry}^*, \tag{9.34}$$

where Ry* is the binding exciton Rydberg energy (Eq. 6.134). Therefore, for many applications, the energy gap of the QD can be written as:

$$\mathcal{E}_g = \mathcal{E}_{g,b} + \frac{\hbar^2\pi^2}{2a^2}\left(\frac{\eta_e^2}{m_e} + \frac{\eta_h^2}{m_h}\right) - \frac{1.8e^2}{4\pi\,\epsilon_0\epsilon_r a} - 0.25\text{Ry}^*. \tag{9.35}$$

In the case of finite barrier, $\Delta\mathcal{E}_{eh}$ can be evaluated numerically. As an example, Fig. 9.8 shows the Coulomb energy, $\Delta\mathcal{E}_{eh}$ in the case of infinite barrier and finite barrier in the case of PbS nanoparticles embedded in SiO$_2$, from which it is clear that the Coulomb energy decreases as the barrier height decreases. In the case of finite barrier, upon decreasing the particle size, the Coulomb energy first increases and then decreases. The increase of $\Delta\mathcal{E}_{eh}$ is related to the increase of the confinement of both electron and hole as the particle size decreases. Then $\Delta\mathcal{E}_{eh}$ starts to decrease since the probability of finding an electron outside the barrier increases, thus resulting in a decrease of the Coulomb energy. An additional contribution to the Coulomb energy is related to the fact that when a charge is placed near a dielectric interface, it induces a surface polarization charge, so that its potential energy depends on the distance to the interface. The induced surface charge acts also on other charged particles and therefore renormalizes the Coulomb interaction energy. The polarization term depends on the dielectric mismatch, that is, on the ratio between the dielectric constants inside and outside the QD.

9.5 Carrier Statistics in QDs

Carriers can escape from the QDs, since the localization energies $\mathcal{E}_{loc,e,h}$ are finite (see Fig. 9.9). Carrier escape increases with increasing temperature and decreasing localization energies. Steady state is reached when the average recombination rate is equal to the net capture rate, defined as the difference between the capture rate and the escape rate in the dot. In analogy with the Schockley–Read–Hall model (see 5.6.1), the net capture rate for electrons can be written as:

$$C_n n(1 - \langle f_n \rangle) - \frac{\langle f_n \rangle}{\tau_e}, \tag{9.36}$$

where f_n is the occupation probability of the dot ground state for electrons in the dot well, n is the electron density in the barrier, $C_n = v_{th}\sigma_n$, where v_{th} is the thermal velocity and σ_n is the cross-section of electron capture into the QD levels, $\langle ... \rangle$ gives the average over all QDs and τ_e is the characteristic time for the thermally induced escape of electrons from the QD into the barrier region. Equation 9.36 can be rewritten as:

$$C_n[n(1 - \langle f_n \rangle) - n_1 \langle f_n \rangle], \tag{9.37}$$

where n_1 is related to the escape time by the following expression:

$$n_1 = \frac{1}{C_n \tau_e}. \tag{9.38}$$

The same equations can be written for holes, so that the net capture rate for holes, which is equal to that for electrons, can be written as:

$$C_p[p(1 - \langle f_p \rangle) - p_1 \langle f_p \rangle], \tag{9.39}$$

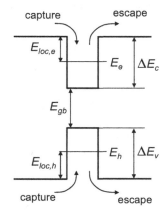

Figure 9.9   Schematic energy diagram for the ground state of a quantum dot.

where:

$$p_1 = \frac{1}{C_p \tau_h}, \qquad C_p = v_{th}\sigma_p, \tag{9.40}$$

and τ_h is the characteristic time for the thermally induced escape of holes from the QD into the barrier region. The average electron–hole recombination rate can be written as the ratio between the product of the occupation probabilities for electrons and holes and the average recombination time τ_D. In steady state, we can write:

$$C_n[n(1 - \langle f_n \rangle) - n_1 \langle f_n \rangle] = C_p[p(1 - \langle f_p \rangle) - p_1 \langle f_p \rangle] = \frac{1}{\tau_D}\langle f_n f_p \rangle. \tag{9.41}$$

There are two limiting cases. If the characteristic times of thermal escape of carriers from the QDs are much shorter than the time of radiative recombination in the dots, that is, if $\tau_e, \tau_h \ll \tau_D$, carriers can redistribute among the dots, thus establishing a quasiequilibrium distribution, common to all dots, with the corresponding quasi-Fermi levels. As a consequence of such a redistribution, the level occupancies and the number of carriers in various QDs will differ. In the second limiting case, $\tau_e, \tau_h \gg \tau_D$, the carriers do not have time to move from one QD to another, so that they recombine in those QDs into which they have been injected. In this case, QDs do not interact with each other.

In the first case, the carrier population is governed by the Fermi–Dirac distribution function, with quasi-Fermi levels $\mathcal{E}_{Fc}$ and $\mathcal{E}_{Fv}$:

$$f_n = f_c = \left[1 + \exp\left(\frac{\mathcal{E}_e - \mathcal{E}_{Fc}}{k_B T}\right)\right]^{-1}, \, f_p = 1 - f_v = \left[1 + \exp\left(\frac{\mathcal{E}_{Fv} - \mathcal{E}_h}{k_B T}\right)\right]^{-1}, \tag{9.42}$$

and the recombination rate in the QDs can be neglected in Eq. 9.41, so that the free carrier densities in the barrier can be written as:

$$n = n_1\frac{\langle f_n \rangle}{1 - \langle f_n \rangle}, \qquad p = p_1\frac{\langle f_p \rangle}{1 - \langle f_p \rangle}. \tag{9.43}$$

If the Boltzmann statistics can be applied (i.e., if the quasi-Fermi level $\mathcal{E}_{Fc}$ is a few $k_B T$ below the barrier), n can be written as:

$$n = N_c \exp\left(-\frac{\Delta\mathcal{E}_c - \mathcal{E}_{Fc}}{k_B T}\right), \tag{9.44}$$

where N_c is the conduction band effective density of states for the optical confinement layer material. Therefore, we have ($f_n \approx \exp[-(\mathcal{E}_e - \mathcal{E}_{Fc})/k_B T]$, $1 - f_n \simeq 1$):

$$n_1 = N_c \exp\left(-\frac{\Delta\mathcal{E}_c - \mathcal{E}_e}{k_B T}\right) = N_c \exp\left(-\frac{\mathcal{E}_{loc,e}}{k_B T}\right). \tag{9.45}$$

In the same way:

$$p_1 = N_v \exp\left(-\frac{\Delta\mathcal{E}_v - \mathcal{E}_h}{k_B T}\right) = N_v \exp\left(-\frac{\mathcal{E}_{loc,h}}{k_B T}\right). \tag{9.46}$$

$\mathcal{E}_e$ and $\mathcal{E}_h$ are the quantized energy levels of an electron and hole in a mean-size QD, measured from the corresponding band edges. The condition for the equilibrium filling of the QDs can be written as $T > T_g$, where:

$$T_g = \max\left(\frac{\Delta\mathcal{E}_c - \mathcal{E}_e}{\ln(C_n N_c \tau_D)}, \frac{\Delta\mathcal{E}_v - \mathcal{E}_h}{\ln(C_p N_v \tau_D)}\right). \tag{9.47}$$

The temperature T_g increases upon increasing the localization energies of the carriers in the QDs, $\Delta\mathcal{E}_{c,v} - \mathcal{E}_{e,h}$, and therefore upon increasing the band gap of the optical confinement layer (OCL) (see Fig. 9.4). T_g depends also on the dot size, in particular because the quantized energies $\mathcal{E}_e$ and $\mathcal{E}_h$ depend on the dot size. Moreover, also the capture cross-sections σ_n and σ_p and the carrier relaxation effects are size dependent.

In the second limiting case ($\tau_e, \tau_h \gg \tau_D$, i.e., $n_1 \ll (C_n\tau_D)^{-1}$), we can neglect n_1 and p_1, and from Eq. 9.41 we get:

$$n = \frac{1}{C_n\tau_D}\frac{\langle f_n f_p\rangle}{1 - \langle f_n\rangle}, \qquad p = \frac{1}{C_p\tau_D}\frac{\langle f_n f_p\rangle}{1 - \langle f_p\rangle}, \tag{9.48}$$

where $f_{n,p}$ represent nonequilibrium population and do not, or only weakly, depend on the dot energy/size. In this case, since redistribution of carriers from one dot to another does not occur, quasi-Fermi levels of the conduction and valence bands are not established.

Real QDs are in between these two limiting cases: Both escape and recombination play a role. Moreover, in real QDs, excited states are generally present, which may act as intermediate levels for escape and capture. A good approximation for the characteristic escape time in thermal equilibrium is given by:

$$\tau_e, \tau_h \approx \tau_0 \exp\left(\frac{\mathcal{E}_{loc}}{k_B T}\right), \tag{9.49}$$

where $\tau_0 \approx 10$ ps. A typical localization energy at room temperature is $\mathcal{E}_{loc} = 120$ meV, which gives $\tau_e \approx 1.2$ ns, comparable to a typical radiative recombination time. Therefore, for typical QDs, $\tau_{e,h}$ and τ_D can be similar, in particular at high temperature.

9.6 Optical Transitions

It is worth pointing out that the characteristic size of QDs is at least one or two orders of magnitude larger than the characteristic length of the unit cells of semiconductors. Therefore, the QD has a macroscopic size in comparison to the unit cell, but it is small on the macroscopic scale: QDs are often called *mesoscopic* structures. For this reason, it is reasonable to use the usual envelope function approximation for the spatial single-particle wavefunctions $\phi(\mathbf{r})$:

$$\phi(\mathbf{r}) = \psi(\mathbf{r})u_{\mathbf{k}}(\mathbf{r}), \tag{9.50}$$

where $u_{\mathbf{k}}(\mathbf{r})$ is the periodic part of the Bloch wavefunction and $\psi(\mathbf{r})$ is the envelope function, which varies on the scale of several unit cells. $\psi(\mathbf{r})$ is given by Eq. 9.25, in the case of spherical QDs with infinite potential barrier. Moreover, we can assume that $u_{\mathbf{k}}(\mathbf{r})$ can be well approximated by its value near the band edge ($\mathbf{k} = 0$), that is, $u_{\mathbf{k}}(\mathbf{r}) \approx u_0(\mathbf{r})$. With this approximation, the wavefunctions $\phi_v(\mathbf{r})$ and $\phi_c(\mathbf{r})$, in the valence and conduction bands, respectively, can be written as:

$$\phi_v(\mathbf{r}) = u_{v0}(\mathbf{r})\psi_v(\mathbf{r}) \tag{9.51}$$

$$\phi_c(\mathbf{r}) = u_{c0}(\mathbf{r})\psi_c(\mathbf{r}). \tag{9.52}$$

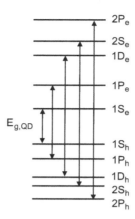

Figure 9.10 Scheme of dipole-allowed optical transitions within the noninteracting particle approximation.

Following the same procedure adopted in the case of quantum wells, it is simple to obtain the transition selection rules in the case of QDs for dipole-allowed optical transitions, within the noninteracting particle approximation. The interaction Hamiltonian is:

$$H'(\mathbf{r}) = \frac{e}{m_0} A_0 \hat{a} \cdot \mathbf{p}. \tag{9.53}$$

Considering the interband transitions we have:

$$H'_{cv} = \frac{eA_0}{2m_0} \langle u_{c0} | \hat{a} \cdot \mathbf{p} | u_{v0} \rangle \langle \psi^c_{n\ell m} | \psi^v_{n'\ell'm'} \rangle = \frac{eA_0}{2m_0} \langle u_{c0} | \hat{a} \cdot \mathbf{p} | u_{v0} \rangle \delta_{nn'} \delta_{\ell\ell'} \delta_{mm'}. \tag{9.54}$$

Due to the orthonormality of the envelope functions $\psi_{n\ell m}$, the term $\langle \psi^c | \psi^v \rangle$ yields delta functions. The transitions between noninteracting electron and hole states, which conserve n and ℓ, are allowed, that is, the transitions $1S_e \rightarrow 1S_h$, $1P_e \rightarrow 1P_h$, $1D_e \rightarrow 1D_h$ are dipole-allowed, as shown in the scheme of Fig. 9.10. The oscillator strength is proportional to $(2\ell + 1)$, since we have to sum over all states with $-\ell \leq m \leq \ell$ contributing to each of these transitions.

In the case of intraband transitions, the dipole transition matrix is proportional to $|\langle \psi^{c,h}_{n\ell m} | \hat{a} \cdot \mathbf{p} | \psi^{c,h}_{n'\ell'm'} \rangle|^2$. In the case of spherical dots, dipole transitions are allowed only between states satisfying the selection rules $\Delta \ell = \pm 1$, $\Delta m = 0$.

9.7 Absorption Spectrum

The density of states in the conduction band can be written as:

$$\rho_c(\mathcal{E}) = 2N_D \sum_{mn\ell} \delta(\mathcal{E} - \mathcal{E}^{n\ell m}_c), \tag{9.55}$$

where N_D is the QD density and the factor 2 takes into account the spin degeneracy. The electron density in the conduction band can be written as:

$$n = \int \rho_c(\mathcal{E})f_c(\mathcal{E})\,d\mathcal{E} = 2N_D \sum_{n\ell m} f_c(\mathcal{E}_c^{n\ell m}). \tag{9.56}$$

Note that if $f_c(\mathcal{E})$ is slowly changing with respect to $\rho_c(\mathcal{E})$, Eq. 9.56 is essentially correct even when $\rho_c(\mathcal{E})$ is not a δ-function.

Let us consider two energy levels in the valence and conduction bands, characterized by the same quantum numbers n, ℓ, and m. The corresponding transition energy, in the case of spherical dots with infinite potential barrier, is:

$$\mathcal{E}_{cv}^{n\ell m} = \mathcal{E}_{gb} + z_{n\ell}^2 \frac{\hbar^2}{2m_r a^2}. \tag{9.57}$$

The joint density of states is:

$$\rho_j(\hbar\omega) = \frac{2}{V_0}\delta(\hbar\omega - \mathcal{E}_{cv}^{n\ell m}), \tag{9.58}$$

where V_0 is the average volume of the QD. By using Eq. 5.52:

$$\alpha(\hbar\omega) = \frac{\pi e^2}{cn\epsilon_0 m_0^2 \omega}|\hat{e}\cdot\mathbf{p}_{cv}|^2 \rho_j(\hbar\omega)[f_v(\hbar\omega) - f_c(\hbar\omega)], \tag{9.59}$$

the interband absorption spectrum, $\alpha(\hbar\omega)$, of a single dot can be written as:

$$\alpha(\hbar\omega) = C_0\frac{2}{V_0}\sum_{n\ell m}|\hat{e}\cdot\mathbf{p}_{cv}|^2\delta(\hbar\omega - \mathcal{E}_{cv}^{n\ell m})[f_v(\hbar\omega) - f_c(\hbar\omega)], \tag{9.60}$$

where:

$$C_0 = \frac{\pi e^2}{cn\epsilon_0 m_0^2 \omega}. \tag{9.61}$$

Equation 9.60 can be written as:

$$\begin{aligned}\alpha(\hbar\omega) &= C_0\frac{2}{V_0}\sum_{n\ell}(2\ell + 1)|\hat{e}\cdot\mathbf{p}_{cv}|^2\delta\left(\hbar\omega - \mathcal{E}_{cv}^{n\ell m}\right)[f_v(\hbar\omega) - f_c(\hbar\omega)] = \\ &\quad \alpha_0(\hbar\omega)[f_v(\hbar\omega) - f_c(\hbar\omega)].\end{aligned} \tag{9.62}$$

If homogeneous broadening caused by carrier scattering processes is considered, the absorption spectrum can be written as:

$$\alpha(\hbar\omega) = C_0\frac{2}{V_0}\sum_{n\ell}(2\ell + 1)|\hat{e}\cdot\mathbf{p}_{cv}|^2 g\left(\hbar\omega - \mathcal{E}_{cv}^{n\ell m}\right)[f_v(\hbar\omega) - f_c(\hbar\omega)], \tag{9.63}$$

where the δ-function has been replaced by a Lorentzian lineshape function, $g(\hbar\omega)$.

In the case of an ensemble of QDs, one has to take into account that each dot is characterized by slightly different properties, caused by fluctuations in size, shape, mechanical strain, etc. Such fluctuations produce a variation of energy levels, thus leading to an inhomogeneous broadening of the lineshape. Typically, the main origin of inhomogeneous broadening is the distribution of dot sizes around a mean value a_0. When inhomogeneous broadening is taken into account, the electron density can be calculated as:

$$n = 2N_D \sum_{n\ell m} \int_0^\infty d\mathcal{E} \ G_c(\mathcal{E}) f_c(\mathcal{E}), \tag{9.64}$$

where $G_c(\mathcal{E})$ is a Gaussian function given by:

$$G_c(\mathcal{E}) = \frac{1}{\sqrt{2\pi}\,\sigma_c} \exp\left[-\frac{(\mathcal{E} - \mathcal{E}_c^{n\ell m})^2}{2\sigma_c^2}\right]. \tag{9.65}$$

Similar expressions can be written for the holes in the valence band. If the effects of inhomogeneous broadening are taken into account, and we assume that $f_v(\hbar\omega) \approx 1$ and $f_c(\hbar\omega) \approx 0$, the absorption spectrum can be calculated as:

$$\begin{aligned}
\alpha_0(\hbar\omega) &= C_0 \frac{2}{V_0} \sum_{n\ell} \int_0^\infty d\mathcal{E} \ (2\ell + 1)|\hat{e} \cdot \mathbf{p}_{cv}|^2 \cdot \\
&\quad \cdot \frac{1}{\sqrt{2\pi}\,\sigma} \exp\left[-\frac{(\hbar\omega - \mathcal{E})^2}{2\sigma^2}\right] \cdot \delta(\mathcal{E} - \mathcal{E}_{cv}^{n\ell m}) = \\
&= C_0 \frac{2}{V_0} \sum_{n\ell} (2\ell + 1)|\hat{e} \cdot \mathbf{p}_{cv}|^2 \frac{1}{\sqrt{2\pi}\,\sigma} \exp\left[-\frac{(\hbar\omega - \mathcal{E}_{cv}^{n\ell m})^2}{2\sigma^2}\right], \quad (9.66)
\end{aligned}$$

where $\sigma^2 = \sigma_c^2 + \sigma_v^2$, in order to account for inhomogeneous broadening effects on the electron and hole energies. The absorption spectrum for the QD ground state transition in an inhomogeneous ensemble can be written as:

$$\begin{aligned}
\alpha(\hbar\omega) &= C_0 \frac{2}{V_0} |\hat{e} \cdot \mathbf{p}_{cv}|^2 \frac{1}{\sqrt{2\pi}\,\sigma} \exp\left[-\frac{(\hbar\omega - \mathcal{E}_g)^2}{2\sigma^2}\right] [f_v(\hbar\omega) - f_c(\hbar\omega)] = \\
&= C_g P(\hbar\omega, \sigma)[f_v(\hbar\omega) - f_c(\hbar\omega)] = \alpha_0(\hbar\omega)[f_v(\hbar\omega) - f_c(\hbar\omega)], \quad (9.67)
\end{aligned}$$

where

$$C_g = C_0 \frac{2}{V_0} |\hat{e} \cdot \mathbf{p}_{cv}|^2. \tag{9.68}$$

$$P(\hbar\omega, \sigma) = \frac{1}{\sqrt{2\pi}\,\sigma} \exp\left[-\frac{(\hbar\omega - \mathcal{E}_g)^2}{2\sigma^2}\right] \tag{9.69}$$

$$\mathcal{E}_g = \mathcal{E}_{gb} + \frac{\pi^2 \hbar^2}{2m_r a^2}. \tag{9.70}$$

In the following, we will calculate the absorption spectrum for an ensemble of dots characterized by a Gaussian distribution, $P(a, \sigma_a)$, of dot radii, with an average radius a_0 and a standard deviation σ_a:

$$P(a, \sigma_a) = \frac{1}{\sqrt{2\pi}\,\sigma_a} \exp\left[-\frac{(a - a_0)^2}{2\sigma_a^2}\right]. \tag{9.71}$$

The absorption spectrum, assuming $f_v(\hbar\omega) \approx 1$ and $f_c(\hbar\omega) \approx 0$, can be calculated as:

$$\begin{aligned}
\alpha_0(\hbar\omega) &= \frac{2C_0}{\sqrt{2\pi}\,\sigma_a} \frac{3}{4\pi} \sum_{n\ell} (2\ell + 1)|\hat{e} \cdot \mathbf{p}_{cv}|^2 \int \frac{da}{a^3} \exp\left[-\frac{(a - a_0)^2}{2\sigma_a^2}\right] \cdot \\
&\quad \cdot \delta\left(\hbar\omega - \mathcal{E}_g - z_{n\ell}^2 \frac{\hbar^2}{2m_r a^2}\right). \quad (9.72)
\end{aligned}$$

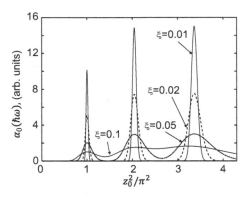

Figure 9.11 Absorption spectrum of ensembles of spherical dots with Gaussian distribution of the radius and different relative standard deviations ξ.

With simple calculations, $\alpha_0(\hbar\omega)$, can be written as:

$$\alpha_0(\hbar\omega) = \frac{2C_0 m_r}{\sqrt{2\pi}\,\hbar^2}|\hat{e}\cdot\mathbf{p}_{cv}|^2 \frac{3}{4\pi a_0}\sum\frac{2\ell+1}{\xi z_{n\ell}^2}\exp\left[-\frac{(z_{n\ell}/z_0-1)^2}{2\xi^2}\right],\tag{9.73}$$

where

$$z_0^2 = \frac{\hbar\omega-\mathcal{E}_g}{\hbar^2/2m_r a_0^2} = \pi^2\frac{\hbar\omega-\mathcal{E}_g}{\mathcal{E}_0}.\tag{9.74}$$

$\mathcal{E}_0 = \pi^2\hbar^2/2m_r a_0^2$ is the average dot ground state quantization energy and $\xi = \sigma_a/a_0$ is the relative standard deviation. Figure 9.11 shows the absorption spectrum of ensembles of spherical dots with Gaussian distribution of the radius and different relative standard deviations ξ. For small ξ, the spectrum is composed by a series of Gaussian peaks centered at $z_{n\ell}/z_0 = 1$, which corresponds to:

$$\hbar\omega_{n\ell} = \mathcal{E}_g + z_{n\ell}^2\frac{\hbar^2}{2m_r a_0^2},\tag{9.75}$$

and linewidth FWHM given by $2\sqrt{2\ln 2}\,\sigma_{n\ell}$ where:

$$\sigma_{n\ell} \approx 2\xi\mathcal{E}_0 z_{n\ell}^2 = 2\xi\frac{\hbar^2}{2m_r}\frac{z_{n\ell}^2}{a_0^2} = 2\xi(\hbar\omega_{n\ell}-\mathcal{E}_g).\tag{9.76}$$

9.8 Gain in QDs

The material gain of a QD ensemble can be obtained using Eq. 9.67:

$$g(\hbar\omega) = C_g P(\hbar\omega,\sigma)[f_c(\hbar\omega)-f_v(\hbar\omega)].\tag{9.77}$$

From the previous expression, it is evident that the peak material gain is inversely proportional to the energetic broadening. The gain of a given dot depends on its population. Considering only the ground state, maximum gain from a single dot can be achieved with

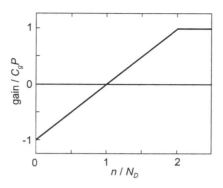

Figure 9.12
Gain of quantum dots as a function of carrier density.

$N_e = 2$ electrons in the conduction band and $N_h = 0$ holes in the valence band. A given gain can be achieved with different electron/hole populations. For example, a gain equal to $g_{max}/2$ can be obtained with different populations of the ground state: $N_e = 2$ electrons and $N_h = 1$ hole or $N_e = 1$ electron and $N_h = 2$ holes, thus with the same carrier density $n_c = (N_e + N_h)/2$. Considering only the ground-state transition, the electron density (Eq. 9.56) can be written as:

$$n = 2f_c N_D,$$ (9.78)

if we assume an overall charge neutrality (i.e., $f_c = 1 - f_v$), Eq. 9.77 can be written as:

$$g(\hbar\omega) = C_g P(\hbar\omega, \sigma)[2f_c(\hbar\omega) - 1] = \frac{n - N_D}{N_D} C_g P(\hbar\omega, \sigma).$$ (9.79)

Therefore, the material gain increases linearly with n (as shown in Fig. 9.12), from $-C_g P$, when $n = 0$, to $C_g P$, when all QD ground states are filled with 2 electrons (and 2 holes) and $n = 2N_D$. The transparency carrier density is $N_{tr} = N_D$: in this case $g = 0$. The corresponding modal gain is:

$$g_{mod}(\hbar\omega) = \Gamma g(\hbar\omega),$$ (9.80)

where Γ is the confinement parameter. The shape of the gain function depends on the shape of the distribution function, $P(\hbar\omega, \sigma)$, so that it strongly depends on the uniformity of the QD size and shape. Since the maximum gain is inversely proportional to the inhomogeneous broadening, it is evident that controlling the uniformity of QD size is of crucial importance to increase the gain.

9.9 Threshold Current Density

Since the same gain can be obtained with quite different electron and hole populations, it is not simple to calculate the injection current density required to generate a given carrier density in the dot ensemble. In the framework of the mean field theory, making use of carrier populations averaged over the dot ensemble, we can write the current density as the sum of two components. The first one is related to the recombination in the QD:

$$j_D = \frac{2eN_D}{A\tau_D}f_c(1-f_v) = \frac{2e}{A_D\tau_D}\zeta f_c(1-f_v),$$ (9.81)

where A is the laser area (cavity length $\times$ cavity width), A_D is the area occupied by a single dot, and ζ is the area coverage given by:

$$\zeta = \frac{N_D A_D}{A}.$$ (9.82)

The second component of the current density is related to spontaneous recombination in the optical confinement layer of thickness b (see Fig. 9.4):

$$j_n = ebBnp,$$ (9.83)

where B is the bimolecular recombination coefficient. At transparency, $f_n + f_p - 1 = f_c - f_v = 0, f_c = f_v = 1/2$, and considering Eq. 9.43, the corresponding current density can be written as:

$$j_{tr} = \frac{e}{2A_D\tau_D}\zeta + ebBn_1p_1.$$ (9.84)

By using a more detailed model, which takes into account the random character of carrier capture and recombination and that does not assume mean field values, the first component of the transparency current density can be written as:

$$j_{D,tr} = \frac{5}{8}\frac{e}{A_D\tau_D}\zeta.$$ (9.85)

The lasing threshold is given by:

$$g^m = \alpha_{tot},$$ (9.86)

where g^m is the peak value of the modal gain spectrum, that is, the effective gain of the active layer with QDs, and α_{tot} is the total loss in the system. By using Eq. 9.79, we can write:

$$\Gamma\frac{N_{th} - N_D}{N_D}\frac{C_g}{\sqrt{2\pi}\sigma} = \alpha_{tot},$$ (9.87)

where N_{th} is the carrier density at threshold. The optical confinement factor, Γ, can be written as the product of an in-plane, Γ_{xy}, and a vertical, Γ_z, confinement factors:

$$\Gamma = \Gamma_{xy}\Gamma_z,$$ (9.88)

where the in-plane confinement factor is given by the area coverage, ζ, given by Eq. 9.82 and the vertical confinement factor is given by the vertical overlap of QDs and optical mode in the laser cavity, averaged over the area A:

$$\Gamma_z = \frac{1}{A}\int_{QD}|E(z)|^2dz/\int_{cavity}|E(z)|^2dz.$$ (9.89)

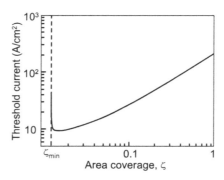

Figure 9.13 Threshold current density as a function of area coverage, in the case of a dot ensemble of 4×10^{10} dots/cm^2, with $\tau_D = 1$ ns, total loss $\alpha_{tot} = 10$ cm^{-1}. Adapted with permission from [19].

Note that the vertical confinement factor is proportional to the thickness of the active layer, therefore it increases upon increasing the number of QD layers in the active region. From Eq. 9.87, we obtain:

$$N_{th} = N_D \left(1 + \frac{1}{\zeta} \frac{\sqrt{2\pi} \sigma \alpha_{tot}}{\Gamma_z C_g} \right).$$ (9.90)

Since the carrier density N_{th} cannot be larger than $2N_D$ (we are considering only the gain from the ground state), the minimum area coverage is:

$$\zeta_{min} = \frac{\sqrt{2\pi} \sigma \alpha_{tot}}{\Gamma_z C_g},$$ (9.91)

and N_{th} can be written as:

$$N_{th} = N_D \left(1 + \frac{\zeta_{min}}{\zeta} \right).$$ (9.92)

Since the total number of carriers can be written as $n = 2f_c N_D$, at threshold we have:

$$f_c = 1 - f_v = \frac{1}{2} \left(1 + \frac{\zeta_{min}}{\zeta} \right).$$ (9.93)

By using the previous equation, the threshold current density as a function of ζ can be written as:

$$j_{th}(\zeta) = \frac{1}{2} \frac{e}{A_D \tau_D} \zeta \left(1 + \frac{\zeta_{min}}{\zeta} \right)^2 + ebBn_1 p_1 \frac{(1 + \zeta_{min}/\zeta)^2}{(1 - \zeta_{min}/\zeta)^2}.$$ (9.94)

Figure 9.13 shows the threshold current density as a function of ζ in the case of a dot ensemble of 4×10^{10} dots/cm^2, with $\tau_D = 1$ ns, total loss $\alpha_{tot} = 10$ cm^{-1}. The dependence of j_{th} on ζ is nonmonotonic. At $\zeta \to \zeta_{min}$, the recombination current in the optical confinement region goes to infinity. Lasing does not occur when $\zeta < \zeta_{min}$ since the maximum saturated gain is always smaller than the total absorption. When $\zeta \gg \zeta_{min}$, the threshold current density tends to the transparency current density given by Eq. 9.84.

It is worth to note that in Eqs. 9.93 and 9.94, ζ_{min}/ζ can be equivalently replaced by L_{min}/L, where L is the cavity length, or by δ/δ_{max}, where δ is the rms of relative fluctuation

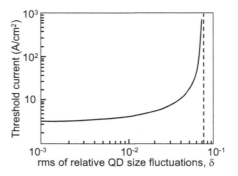

Figure 9.14   Threshold current density as a function of the rms of relative fluctuations of the QD size. Adapted with permission from [20].

of the QD size. In this last case, as $\delta \to 0$, the threshold current density decreases toward the transparency current density, while, when δ approaches its maximum value, δ_{max}, j_{th} rapidly increases toward infinity, as shown in Fig. 9.14. In the case of nonequilibrium filling of the QDs ($\tau_e, \tau_h \gg \tau_D$), the average filling of the ground state with electrons and holes are still given by Eq. 9.93 as a function of the area coverage. The threshold current density can be obtained by using Eq. 9.93 into 9.81 and 9.83, thus obtaining:

$$j_{th}(\zeta) = \frac{1}{2}\frac{e}{A_D\tau_D}\zeta\left(1 + \frac{\zeta_{min}}{\zeta}\right)^2 + \frac{1}{4}\frac{ebB}{C_nC_p\tau_D^2}\frac{(1+\zeta_{min}/\zeta)^4}{(1-\zeta_{min}/\zeta)^2}. \tag{9.95}$$

Upon using k layers of identical, uncoupled QDs in the active region, the formulas discussed above remain valid if the variables are properly scaled. In particular, the minimum area coverage is decreased by a factor k, while the maximum gain is increased by a factor k, thus attenuating the problem of gain saturation.

9.9.1 Temperature Dependence of Threshold Current

In the ideal case, initially investigated by Arakawa and Sakaki [13], the threshold current density of a QD laser should be temperature independent and, therefore, the characteristic temperature, T_0, should be infinitely high. Indeed, in a conventional laser (not based on QDs), the dependence of j_{th} on temperature is related to the thermal spreading of the injected carriers over a broad range of energy states, thus leading to a decrease of the maximum gain at a given level of carrier injection. In the case of QDs, thermal spreading of carriers vanishes, since the density of states is a δ-function. Therefore, the temperature dependence of j_{th} completely disappears, as long as the electron population in higher energy levels can be assumed as negligible. In real QD lasers, j_{th} depends on temperature, although with superior temperature stability compared to quantum well lasers and the characteristic temperature is very high but not infinite.

Two conditions should be verified in a real device in order to have temperature-independent j_{th}: The whole injection current should go into recombination in the QDs and charge neutrality should be verified. In a real laser, QDs are always embedded in a waveguide material (the optical confinement region), into which carriers are injected before reaching the QDs. As already observed, the carrier recombination in the waveguide (barrier) region gives rise to a component, $j_{th,w}$, of the threshold current density: This component is related to thermal escape of carriers from QDs, which depends exponentially on temperature. This is the origin of the temperature dependence of the threshold current density in the case of equilibrium filling of QDs ($T > T_g$). In the case of nonequilibrium filling ($T < T_g$), j_{th} is almost temperature independent: The weak dependence on T arises from the temperature dependence of the cross-sections of carrier capture into the QD, $\sigma_{n,p}$, of the thermal velocities of electrons and holes, and of the bimolecular recombination coefficient, B. If charge neutrality is assumed, the occupation probabilities $f_n = f_p$ ($f_c = 1 - f_v$) are completely temperature independent, as shown by Eq. 9.93, so that the component $j_{th,D}$ of j_{th}, related to the recombination in the QDs, does not depend on temperature. If violation of the charge neutrality in the QDs is taken into account, f_n and f_p, and therefore $j_{th,D}$ become temperature dependent. This is the origin of temperature dependence of j_{th} and of finite values of T_0 also at temperatures at which thermal escape of carriers from the QDs and the recombination outside the dots are almost completely suppressed.

We observe that the characteristic temperature T_0 can be written as:

$$\frac{1}{T_0} = \frac{j_D}{j_D + j_w}\frac{1}{T_0^D} + \frac{j_w}{j_D + j_w}\frac{1}{T_0^w},\tag{9.96}$$

where:

$$T_0^D = \left(\frac{\partial \ln j_{th,D}}{\partial T}\right)^{-1}, \qquad T_0^w = \left(\frac{\partial \ln j_{th,w}}{\partial T}\right)^{-1}.\tag{9.97}$$

Since $j_{th,D}$ increases with temperature much more slowly than $j_{th,w}$, T_0^D turns out to be much greater than T_0^w.

An important factor, which limits the temperature characteristics of QD lasers, is related to the highly asymmetric conduction and valence band offsets in typically used material systems, InAs/GaAs and InAs/InP, which leads to an unfavorable offset in the valence band so that the hole states are far less confined than the electron states. The problem can be minimized by p-type modulation doping in the QD active region to offset the effects of thermalization. In general, QD lasers present superior temperature characteristics compared to their quantum well counterparts, in terms of characteristic temperature, T_0, and high temperature CW operation, with record high temperature CW operation up to 220 °C.

9.10 Additional Advantages of QD Lasers

For various applications, where a narrow emission linewidth and an improved stability against undesired feedback are important issues, the linewidth enhancement factor, usually denoted by α or α_H, is a particularly important parameter. The linewidth enhancement factor is defined as the ratio of the change in the real part of the refractive index, n_r, of the laser medium with carrier density to that of the imaginary part, n_i , with respect to carrier density, N, which can be also written in terms of the wavelength, λ, and differential gain, dg/dN as follows:

$$\alpha_H = -\left(\frac{dn_r}{dN}\right)\left(\frac{dn_i}{dN}\right)^{-1} = -\frac{4\pi}{\lambda}\left(\frac{dn_r}{dN}\right)\left(\frac{dg}{dN}\right)^{-1}. \qquad (9.98)$$

The sign in the definition of α_H is chosen to have $\alpha_H > 0$ at semiconductor laser wavelengths. In a laser, the fluctuations of electron–hole density produced by spontaneous emission cause fluctuations of the refractive index of the laser medium. The resulting fluctuations in the optical cavity length produce fluctuations of the cavity frequency and therefore of the oscillation frequency. For this reason, in a semiconductor laser, the spectral width given by the Schawlow–Townes formula, which gives the quantum limit to laser linewidth, must be multiplied by α_H^2.

On the basis of the Kramers–Kronig relations, the symmetric density of states of QDs should result in a vanishing α_H factor. Moreover, since the gain is decoupled from the carrier reservoir, changes in the real and imaginary parts of the refractive index are decoupled at higher injection levels leading to a low ratio of $\Delta n_r/\Delta g$. While typical values of α_H in quantum well lasers are in the range between 3 and 5 near threshold, very small values of α_H, down to 0.1, have been measured in QD lasers.

A low α_H not only leads to a reduced linewidth, but it also determines a reduction in the laser sensitivity to undesired optical feedback, caused by unintentional reflections from various interfaces, waveguide crossing, process imperfections, etc. This unwanted feedback may originate deleterious effects such as linewidth broadening, mode hopping and increased amplitude noise. It was observed that InAs/GaAs QD lasers on native GaAs substrates can be much less sensitive to feedback than quantum well lasers. Theoretical analysis have correlated this behavior to the lower linewidth enhancement factor and to the highly damped relaxation oscillations in QD lasers compared to the quantum well counterpart. High levels of feedback can induce multimode operation or even total coherence collapse. The critical feedback level, f_{crit}, to induce optical instability scales as:

$$f_{crit} \propto \frac{1 + \alpha_H^2}{\alpha_H^4}. \qquad (9.99)$$

Another important advantage of QD lasers compared to quantum well lasers is the very low sensitivity to crystalline defects as a result of the in-plane quantum confinement, of the reduction of the diffusion length of carriers outside the QD, and of the mechanical hardening of the crystal lattice in the vicinity of the dots. The strong carrier confinement in three directions leads to a very effective trapping of carriers that could otherwise diffuse toward a dislocation to recombine. Moreover. assuming a QD density as high as 6×10^{10} cm^{-2} and typical dislocation densities in optimized buffers of $\sim 10^6$ cm^{-2}, the likelihood of charge carriers finding a dot before a dislocation is extremely high.

A source of semiconductor laser failure is often related to the dark-line defects (DLDs), which are typically dense three-dimensional networks of dislocation loops developing around a threading or a misfit dislocation crossing the active region. The growth of DLDs is a process of dislocation elongation, produced by the interaction between point defects and the minority carriers introduced in the active region of the laser by electric injection and by optical generation due to self-absorption of the laser light inside the laser cavity. The degradation of laser performances is related to the formation and fast growth of large

DLDs. Recent advances in fabrication of QD lasers have produced almost degradation-free operation near room temperature at a dislocation density up to 7×10^6 cm^{-2}.

9.11 Exercises

Exercise 9.1 Consider a semiconductor with band gap $\mathcal{E}_g = 1$ eV, effective mass in the conduction band $m_c = 0.05\ m_0$, and effective mass in the valence band $m_v = 0.6\ m_0$. Consider a cubic quantum dot of this semiconductor with edge $L = 2$ nm. Determine the ground state transition energy.

Exercise 9.2 Consider a spherical quantum dot and demonstrate that the radial function, $R_{nl}(r)$, can be calculated by using Eq. 9.16.

Exercise 9.3 Consider a quantum dot ensemble characterized by a Gaussian distribution $P(a, \sigma_a)$ of dot radii (Eq. 9.71). Demonstrate that the absorption spectrum can be calculated by using Eq. 9.73.

Exercise 9.4 The normalized inhomogeneous Gaussian distribution in energy of a quantum dot ensemble is given by Eq. 9.65. Calculate its FWHM and the peak value in terms of the standard deviation σ.

Exercise 9.5 Consider an InAs quantum dot with edge L. Calculate L in order to have emission at $\lambda = 1.28\ \mu$m. Effective mass of electrons in InAs is $m^* = 0.023\ m_0$.

Exercise 9.6 Demonstrate that the normalized wavefunction of a particle in an infinite spherical potential well can be written as in Eq. 9.25.

Distributed Feedback Lasers

10.1 Basic Concepts

The structure of a distributed feedback (DFB) laser is schematically shown in Fig. 10.1, and it is rather different in comparison to the typical laser schemes that we have seen in Chapter 9. The laser consists of an active medium in which a periodic thickness variation of the cladding layers forms a grating, which provides the feedback required for laser operation. The laser cavity mirrors are distributed along the length of the cavity. The example of Fig. 10.1 refers to a DFB laser oscillating at 1.55 μm. The thickness of the upper InGaAsP confinement layer is periodically modulated, with a period Λ. Due to the thickness modulation, the refractive index experienced by the oscillating mode is periodically modulated, since the refractive index of the InGaAsP confinement layer is larger than that of the InP p- and n-type layers. The effective refractive index can be calculated as $n(z) = \langle n(x,z) \rangle_x$, where the average is calculated over the x-coordinate, perpendicular to the longitudinal propagation z-axis, with a weight determined by the transverse distribution of the beam intensity inside the cavity, $|U(x)|^2$. In the following, we will assume a sinusoidal modulation:

$$n(z) = n_0 + \delta n \cos\left(\frac{2\pi}{\Lambda}z + \phi\right), \tag{10.1}$$

where $\delta n \ll n_0$. The electric field in the forward (positive z) and backward (negative z) directions are effectively coupled when Bragg condition is satisfied:

$$\frac{\lambda_0}{4\bar{n}} = \frac{\Lambda}{2}, \tag{10.2}$$

where λ_0 is the free-space wavelength of the radiation and $\bar{n}$ is the effective refractive index: $\bar{n} = \langle n(z) \rangle_z$, where the average is calculated along the z coordinate and the weight is determined by the standing wave pattern inside the cavity. Note that when the modulation of the refractive index is given by a periodic square-wave function of period Λ, the structure shown in Fig. 10.1 is equivalent to a sequence of alternating layers with high and low refractive index. The Bragg condition for constructive interference of all the waves reflected at each layer interface occurs when the layer thickness is equal to a quarter of the wavelength: $d_1 n_1 = d_2 n_2 = \lambda_0/4$, with $d_1 + d_2 = \Lambda$. Equation 10.2 shows that the grating period Λ determines the wavelength of the mode which can oscillate in the cavity. Figure 10.2 shows a scanning electron microscopy (SEM) image of a DFB grating made by electron beam lithography and subsequent reactive ion etching. The grating period (first order) is about 220 nm, for operation at 1550 nm.

Before proceeding with an analytical treatment of a DFB laser by using the method of the coupled-mode equations, we will anticipate a few general results. We will first assume that

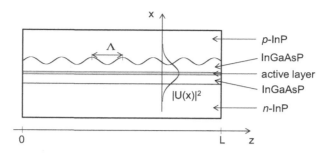

Figure 10.1 Schematic structure of a DFB laser with uniform grating. $|U(x)|^2$ is the transverse distribution of the beam intensity inside the cavity.

Figure 10.2 SEM image of a $\lambda/4$-shifted DFB grating with a period of about 220 nm. Adapted with permission from [8].

the field reflectivities at the two laser end faces are zero ($r_1 = r_2 = 0$). The electric field travelling in the forward direction will be indicated as $E_f(z)$. Moreover, the propagation constant at the Bragg condition is given by:

$$\beta_0 = k_0 \bar{n} = \frac{2\pi}{\lambda_0} \bar{n} = \frac{\pi}{\Lambda}. \tag{10.3}$$

The analytical treatment will show that the intensity transmittance, T, for the forward beam, defined as:

$$T = \frac{|E_f(L)|^2}{|E_f(0)|^2}, \tag{10.4}$$

where L is the cavity length, as a function of the detuning from the Bragg condition, $(\beta - \beta_0)L$, shows maxima, whose values increase upon increasing the gain. The threshold gain corresponds to the gain value required to obtain a strong increase of one (or two) of these maxima. The situation is illustrated in Fig. 10.3, which shows the transmittance T as a function of detuning for two values of the effective gain gL. This figure shows that in the case of a uniform grating (i.e., a grating with constant periodicity along the cavity length), the transmittance is minimum at resonance ($\beta = \beta_0$) and presents various maxima symmetrically located with respect to the Bragg condition. In this case, just above threshold, two modes oscillate at two different frequencies, which correspond to the two maxima

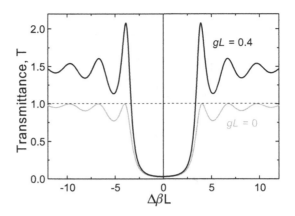

Figure 10.3 Intensity transmittance T of a uniform grating as a function of the detuning $\Delta\beta L$ calculated for two different values of the effective gain gL.

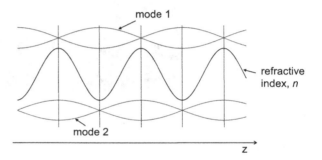

Figure 10.4 Schematic representation of the refractive index n and of the standing-wave patterns of the two lowest-order modes along the cavity of a DFB laser with a uniform grating.

closest to $\beta = \beta_0$. The region between these two peaks defines the so-called *stop-band* of the device. Upon increasing the gain, higher-order modes will start to grow, corresponding to the other transmittance maxima.

The existence of two oscillating modes can be intuitively understood with the help of Fig. 10.4, which shows the modulation of the refractive index along the cavity together with the standing-wave patterns of the two lowest-order oscillating modes. The effective refractive index experienced by mode 1, $\bar{n}_1$, is smaller than that of mode 2, $\bar{n}_2$, since the anti-nodes of mode 1 are located in correspondence of the minima of the refractive index variation along the cavity, while the anti-nodes of mode 2 are placed in correspondence of the maxima of n. The effective refractive index is calculated as a weighted average along the z-coordinate, where the weight function is the standing wave pattern of the mode. Since $\bar{n}_1 < n_0 < \bar{n}_2$, the free-space wavelengths of modes 1 and 2, given by $\lambda_{01,2} = 2\bar{n}_{1,2}\Lambda$, are $\lambda_{01} < \lambda_0 < \lambda_{02}$ (inside the cavity the two modes have the same wavelength $\lambda = 2\Lambda$).

In general, this double-mode operation is not advantageous for applications; various techniques have been introduced to achieve single-mode operation. For example, it is

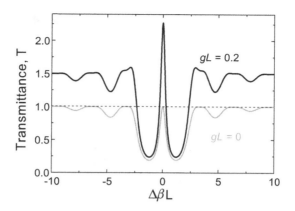

Figure 10.5 Intensity transmittance T of a $\lambda/4$-shifted grating as a function of the detuning, $\Delta\beta L$, calculated for two different values of the effective gain gL.

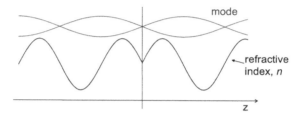

Figure 10.6 Schematic representation of the refractive index, n, and of the standing-wave pattern of the lowest-order mode along the cavity of a DFB laser with a $\lambda/4$-shifted grating.

possible to use an asymmetric structure, characterized by two different mirror reflectivities r_1 and r_2. The best solution is represented by the use of a periodic modulation of the thickness of the cladding layer, which undergoes a phase shift of $\Lambda/2$ at the center of the structure. Since at the Bragg condition (Eq. 10.2) $\Lambda/2 = \lambda_0/4\bar{n}$, this structure is called $\lambda/4$-*shifted DFB laser* (Fig. 10.2 shows a $\lambda/4$-shifted DFB grating). Figure 10.5 shows the transmittance of a $\lambda/4$-shifted DFB device as a function of the detuning, for two values of the effective gain. A maximum is present at the Bragg resonance: Just above the threshold gain, a single mode will oscillate with $\beta = \beta_0$. Upon increasing the gain above the threshold value for single-mode operation, higher-order modes, symmetrically located with respect to the Bragg frequency, will start to grow. In this case, the mode selectivity is higher than in the case of DFB lasers with uniform grating, that is, the relative gain increase with respect to the single-mode condition in order to have oscillation of the higher-order modes is higher than in the case of uniform grating. The existence of a single maximum at the Bragg frequency can be intuitively understood following the same arguments used in the previous case. The modulation of the refractive index along the cavity and the standing-wave pattern of the lowest-order mode is shown in Fig. 10.6. The effective refractive index experienced by the mode is n_0, and the free-space wavelength is $\lambda_0 = \lambda_B = 2\Lambda n_0$.

10.2 Coupled-Mode Theory

We will first consider a passive structure ($g = 0$). The refractive index modulation is given by:

$$n(z) = n_0 + \delta n \, h(z) \cos\left(\frac{2\pi}{\Lambda}z + \phi(z)\right), \tag{10.5}$$

where $\delta n \ll n_0$ and the functions $h(z)$ (with $0 < h(z) < 1$) and $\phi(z)$ have been introduced to analyze a general situation, where both the amplitude and phase of the modulation are functions of the propagation coordinate z. The general wave equation is:

$$\nabla^2 \widetilde{\mathbf{E}} - \frac{n^2}{c^2}\frac{\partial^2 \widetilde{\mathbf{E}}}{\partial t^2} = 0. \tag{10.6}$$

We assume that the electric field distribution in the xy plane, perpendicular to the longitudinal coordinate z, is given by the fundamental mode of the waveguide:

$$\widetilde{\mathbf{E}}(x, y, z, t) = F(x, y)E(z, t)\mathbf{e}, \tag{10.7}$$

where $\mathbf{e}$ is the unit vector in the direction of the field polarization. From Eqs. 10.6 and 10.7, we obtain the following wave equation:

$$\frac{\partial^2 E}{\partial z^2} - \frac{n^2(z)}{c^2}\frac{\partial^2 E}{\partial t^2} = 0. \tag{10.8}$$

We can write the general solution as the sum of two counter-propagating modes:

$$E = A(z, t)\exp\left[i(\beta_0 z - \omega_0 t)\right] + B(z, t)\exp\left[-i(\beta_0 z + \omega_0 t)\right] + \text{c.c.} \tag{10.9}$$

where $A(z, t)$ and $B(z, t)$ are the amplitudes of the forward and backward fields, respectively, and $\beta_0 = \pi/\Lambda = \omega_0 n_0/c$. Since the coupling is weak, we will assume that the field amplitudes vary slowly, both in space and in time. Therefore, we have:

$$\left|\frac{\partial A}{\partial z}\right| \Lambda \ll |A| \qquad \left|\frac{\partial^2 A}{\partial z^2}\right| \ll \frac{1}{\Lambda}\left|\frac{\partial A}{\partial z}\right| = \frac{\beta_0}{\pi}\left|\frac{\partial A}{\partial z}\right| \tag{10.10}$$

$$\left|\frac{\partial A}{\partial t}\right|\frac{2\pi}{\omega_0} \ll |A| \qquad \left|\frac{\partial^2 A}{\partial t^2}\right| \ll \omega_0\left|\frac{\partial A}{\partial t}\right|. \tag{10.11}$$

The same expressions hold for the amplitude $B(z, t)$ of the backward wave. By using the slowly varying envelope approximation, the spatial and temporal derivatives of $E(z, t)$ in Eq. 10.8 can be written as:

$$\frac{\partial^2 E}{\partial z^2} = \left(2i\beta_0\frac{\partial A}{\partial z} - \beta_0^2 A\right)e^{i(\beta_0 z - \omega_0 t)} +$$
$$- \left(2i\beta_0\frac{\partial B}{\partial z} + \beta_0^2 B\right)e^{-i(\beta_0 z + \omega_0 t)} + \text{c.c.} \tag{10.12}$$

$$\frac{\partial^2 E}{\partial t^2} = -\left(2i\omega_0\frac{\partial A}{\partial t} + \omega_0^2 A\right)e^{i(\beta_0 z - \omega_0 t)} +$$
$$- \left(2i\omega_0\frac{\partial B}{\partial t} + \omega_0^2 B\right)e^{-i(\beta_0 z + \omega_0 t)} + \text{c.c.} \tag{10.13}$$

Moreover, $n^2(z)$ can be written as follows:

$$n^2(z) \simeq n_0^2 + 2n_0\delta n\, h(z)\cos\left(\frac{2\pi}{\Lambda}z + \phi(z)\right) = n_0^2 + 2n_0\delta n\, h(z)\cos(2\beta_0 z + \phi), \quad (10.14)$$

or, in a similar way:

$$n^2(z) \simeq n_0^2 + n_0\delta n\, h(z)[\exp(2i\beta_0 z)e^{i\phi} + \exp(-2i\beta_0 z)e^{-i\phi}]. \quad (10.15)$$

By substituting Eqs. 10.12, 10.13, and 10.15 in the wave equation 10.8 and equating the terms containing $\exp[i(\beta_0 z - \omega_0 t)]$, we obtain:

$$2i\beta_0\frac{\partial A}{\partial z} - \beta_0^2 A + \frac{n_0^2}{c^2}\left(2i\omega_0\frac{\partial A}{\partial t} + \omega_0^2 A\right) + $$
$$+ \frac{n_0\delta n\, h(z)}{c^2}e^{i\phi}\left(2i\omega_0\frac{\partial B}{\partial t} + \omega_0^2 B\right) = 0. \quad (10.16)$$

Since $\beta_0 = \omega_0 n_0/c$

$$-\beta_0^2 A + \frac{n_0^2\omega_0^2}{c^2}A = 0. \quad (10.17)$$

If we neglect the term $\partial B/\partial t$ with respect to $\omega_0 B$ from Eq. 10.16, we get:

$$\frac{\partial A}{\partial z} + \frac{n_0}{c}\frac{\partial A}{\partial t} = i\frac{\delta n}{2n_0}\beta_0 h(z)e^{i\phi(z)}B. \quad (10.18)$$

In the same way, by substituting Eqs. 10.12, 10.13, and 10.15 in the wave equation 10.8 and equating the terms containing $\exp[-i(\beta_0 z + \omega_0 t)]$, we can write:

$$\frac{\partial B}{\partial z} - \frac{n_0}{c}\frac{\partial B}{\partial t} = -i\frac{\delta n}{2n_0}\beta_0 h(z)e^{-i\phi(z)}A. \quad (10.19)$$

The last two equations show that the forward and backward waves are coupled by the modulation of the refractive index. A coupling parameter, $q(z)$, is usually introduced:

$$q(z) = \frac{\delta n}{2n_0}\beta_0 h(z)e^{i\phi(z)}, \quad (10.20)$$

so that the coupled-mode equations can be rewritten as follows:

$$\begin{cases} \dfrac{\partial A}{\partial z} + \dfrac{n_0}{c}\dfrac{\partial A}{\partial t} = iq(z)B \\[3mm] \dfrac{\partial B}{\partial z} - \dfrac{n_0}{c}\dfrac{\partial B}{\partial t} = -iq^*(z)A. \end{cases} \quad (10.21)$$

So far we have considered a passive structure; the effect of gain can be simply introduced in Eqs. 10.21:

$$\begin{cases} \dfrac{\partial A}{\partial z} + \dfrac{n_0}{c}\dfrac{\partial A}{\partial t} = iq(z)B + \dfrac{g}{2}A \\[3mm] \dfrac{\partial B}{\partial z} - \dfrac{n_0}{c}\dfrac{\partial B}{\partial t} = -iq^*(z)A - \dfrac{g}{2}B. \end{cases} \quad (10.22)$$

10.2.1 Threshold Conditions

We will derive the threshold conditions by performing a perturbative analysis of the coupled-mode equations 10.22. We will assume that the structure is not excited, that is, there are no injected signals:

$$A(0, t) = 0 \qquad B(L, t) = 0. \tag{10.23}$$

If we now perturb the solution $A = B = 0$ of the coupled-mode equations, the threshold condition is reached when the perturbation grows with time. In the case of a real device, the perturbation is given by the spontaneous emission. We look for a solution as:

$$\begin{aligned}
A(z, t) &= \widetilde{A}(z)e^{at} \\
B(z, t) &= \widetilde{B}(z)e^{at}.
\end{aligned} \tag{10.24}$$

Upon using Eqs. 10.24 in Eqs. 10.22, we obtain:

$$\begin{cases}
\dfrac{d\widetilde{A}}{dz} = \left(\dfrac{g}{2} - \dfrac{n_0}{c} a\right)\widetilde{A} + iq(z)\widetilde{B} \\[2mm]
\dfrac{d\widetilde{B}}{dz} = \left(-\dfrac{g}{2} + \dfrac{n_0}{c} a\right)\widetilde{B} - iq^*(z)\widetilde{A}.
\end{cases} \tag{10.25}$$

The general solution of the previous set of linear equations can be written as:

$$\begin{aligned}
\widetilde{A}(z) &= M_{11}(z)\widetilde{A}(0) + M_{12}(z)\widetilde{B}(0) \\
\widetilde{B}(z) &= M_{21}(z)\widetilde{A}(0) + M_{22}(z)\widetilde{B}(0),
\end{aligned} \tag{10.26}$$

as will be shown in the next section in the case of a uniform grating and in the case of a $\lambda/4$-shifted grating. $M_{ij}(z)$ can be calculated when the coupling parameter $q(z)$ is known. Upon using the boundary conditions 10.23 in Eqs. 10.26, the following equations can be written:

$$\begin{cases}
\widetilde{A}(L) = M_{12}(L)\widetilde{B}(0) \\
0 = M_{22}(L)\widetilde{B}(0).
\end{cases} \tag{10.27}$$

From the second equation, we obtain:

$$M_{22}(L) = 0, \tag{10.28}$$

where, in general, $M_{22}(L)$ depends on the gain coefficient, g, and on the parameter a introduced in Eq. 10.24. Therefore, if the gain is fixed, a can be calculated: $a = a(g)$. It is clear that the applied perturbation increases with time only if $\mathrm{Re}(a) \geq 0$, indeed

$$A(z, t) = \widetilde{A}(z)e^{\mathrm{Re}(a)t}e^{i\mathrm{Im}(a)t}. \tag{10.29}$$

Threshold is reached when $\mathrm{Re}(a) = 0$. The corresponding value of g, which we will indicate as g_{th}, is the threshold gain. At threshold, the parameter a can be written as:

$$a = -i\Omega = -i(\omega - \omega_0). \tag{10.30}$$

By using this result, the set of equations 10.25 can be rewritten as:

$$\begin{cases} \dfrac{d\widetilde{A}}{dz} = i\Delta\kappa\widetilde{A} + iq(z)\widetilde{B} \\ \dfrac{d\widetilde{B}}{dz} = -i\Delta\kappa\widetilde{B} - iq^*(z)\widetilde{A}, \end{cases} \tag{10.31}$$

where:

$$\Delta\kappa = \frac{n_0}{c}\Omega - i\frac{g}{2} = \frac{n_0}{c}(\omega - \omega_0) - i\frac{g}{2} = \Delta\beta - i\frac{g}{2} = -i\left(\frac{g}{2} + i\Delta\beta\right). \tag{10.32}$$

In order to take into account the confinement factor in the active region, Eq. 10.32 can be rewritten as follows:

$$\Delta\kappa = \frac{n_0}{c}(\omega - \omega_0) - i\Gamma\frac{g}{2}. \tag{10.33}$$

The general method illustrated in this section will be applied in the next section to derive the threshold gain and the frequencies of the oscillating modes of a uniform grating DFB laser and of a $\lambda/4$-shifted DFB laser.

Before proceeding, it is interesting to find the link between the perturbative analysis presented in this section and the discussion presented in Section 10.1, where the threshold conditions for laser action have been discussed in terms of the transmittance of the DFB device. Let us use the DFB structure as an amplifier: A signal is injected at $z = 0$ (at the frequency ω), while no signals are injected at $z = L$. From the set of Eqs. 10.26, we have:

$$\begin{cases} M_{11}(L)\widetilde{A}(0) + M_{12}(L)\widetilde{B}(0) = \widetilde{A}(L) \\ M_{21}(L)\widetilde{A}(0) + M_{22}(L)\widetilde{B}(0) = 0, \end{cases} \tag{10.34}$$

If we now divide both equations by $\widetilde{A}(0)$, we obtain:

$$\begin{cases} M_{11}(L) + rM_{12}(L) = t \\ M_{21}(L) + rM_{22}(L) = 0, \end{cases} \tag{10.35}$$

where we have introduced the field reflection, r, and transmission, t, coefficients:

$$r = \frac{\widetilde{B}(0)}{\widetilde{A}(0)}\bigg|_{\widetilde{B}(L)=0}$$
$$t = \frac{\widetilde{A}(L)}{\widetilde{A}(0)}\bigg|_{\widetilde{B}(L)=0,} \tag{10.36}$$

which can be easily calculated from Eqs. 10.35:

$$r = -\frac{M_{21}(L)}{M_{22}(L)}$$
$$t = \frac{\det(M)}{M_{22}(L)}, \tag{10.37}$$

where $\det(M) = M_{11}M_{22} - M_{12}M_{21}$ is the determinant of the matrix $\{M_{ij}\}$. Since the structure is reciprocal (i.e., the transmission through the grating is equivalent for light incident from either side of the structure) $\det(M) = 1$, so that $t = 1/M_{22}$ and the grating transmittance can be written as:

$$T = \frac{1}{|M_{22}(L)|^2}. \tag{10.38}$$

At threshold, with $\omega - \omega_0 = \Omega$, $M_{22}(L) = 0$ and the transmittance increases to infinity as we have seen, on a qualitative ground, in Section 10.1.

10.3   DFB Laser with Uniform Grating

In the case of a uniform grating $\phi(z) = 0$ and $h(z) = 1$. The coupling parameter can be written as:

$$q(z) = q_0 = \frac{\delta n}{2n_0}\beta_0 = \frac{\delta n}{2n_0}\frac{\pi}{\Lambda}. \tag{10.39}$$

The coupled-mode equations 10.31 can be solved by writing:

$$\widetilde{A}(z) = Ae^{\gamma z}, \qquad \widetilde{B}(z) = Be^{\gamma z}. \tag{10.40}$$

The set of Eqs. 10.31 can be written as:

$$\begin{cases} (\gamma - i\Delta\kappa)A - iq_0B = 0 \\ iq_0^*A + (\gamma + i\Delta\kappa)B = 0, \end{cases} \tag{10.41}$$

which can be easily solved. In order to have solutions different from the trivial one, $A = B = 0$, we must have:

$$\begin{pmatrix} \gamma - i\Delta\kappa & -iq_0 \\ iq_0^* & \gamma + i\Delta\kappa \end{pmatrix} = \gamma^2 + \Delta\kappa^2 - |q_0|^2 = 0, \tag{10.42}$$

which gives two eigenvalues:

$$\gamma_1 = \gamma = \sqrt{|q_0|^2 - \Delta\kappa^2}, \qquad \gamma_2 = -\gamma = -\sqrt{|q_0|^2 - \Delta\kappa^2}. \tag{10.43}$$

The corresponding eigenvectors are:

$$\begin{pmatrix} iq_0 \\ \gamma - i\Delta\kappa \end{pmatrix}, \qquad \begin{pmatrix} iq_0 \\ -\gamma - i\Delta\kappa \end{pmatrix}, \tag{10.44}$$

and the general solution can be written as:

$$\begin{pmatrix} \widetilde{A}(z) \\ \widetilde{B}(z) \end{pmatrix} = C_1\begin{pmatrix} iq_0 \\ \gamma - i\Delta\kappa \end{pmatrix}e^{\gamma z} + C_2\begin{pmatrix} iq_0 \\ -\gamma - i\Delta\kappa \end{pmatrix}e^{-\gamma z}. \tag{10.45}$$

When $z = 0$ we have:

$$\begin{cases} \widetilde{A}(0) = iq_0(C_1 + C_2) \\ \widetilde{B}(0) = \gamma(C_1 - C_2) - i\Delta\kappa(C_1 + C_2), \end{cases} \tag{10.46}$$

from which it is simple to calculate C_1 and C_2:

$$\begin{cases} C_1 = \dfrac{1}{2}\left(\dfrac{\widetilde{A}(0)}{iq_0} + \dfrac{\Delta\kappa}{\gamma}\dfrac{\widetilde{A}(0)}{q_0} + \dfrac{\widetilde{B}(0)}{\gamma}\right) \\[2mm] C_2 = \dfrac{1}{2}\left(\dfrac{\widetilde{A}(0)}{iq_0} - \dfrac{\Delta\kappa}{\gamma}\dfrac{\widetilde{A}(0)}{q_0} - \dfrac{\widetilde{B}(0)}{\gamma}\right). \end{cases} \tag{10.47}$$

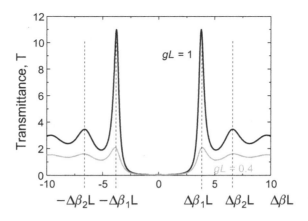

Figure 10.7 Intensity transmittance T of a uniform grating as a function of the detuning $\Delta\beta L$ calculated for two different values of the effective gain gL.

So that we obtain:

$$\begin{cases} \widetilde{A}(z) = \left[\cosh(\gamma z) + i\dfrac{\Delta\kappa}{\gamma}\sinh(\gamma z)\right]\widetilde{A}(0) + \dfrac{iq_0}{\gamma}\sinh(\gamma z)\widetilde{B}(0) \\[2mm] \widetilde{B}(z) = -i\dfrac{q_0^*}{\gamma}\sinh(\gamma z)\widetilde{A}(0) + \left[\cosh(\gamma z) - i\dfrac{\Delta\kappa}{\gamma}\sinh(\gamma z)\right]\widetilde{B}(0), \end{cases} \tag{10.48}$$

which represent Eqs. 10.26 in the particular case of a uniform grating. The threshold condition is $M_{22}(L) = 0$:

$$\cosh(\gamma L) - i\frac{\Delta\kappa}{\gamma}\sinh(\gamma L) = 0, \tag{10.49}$$

which can be also written as:

$$\Delta\kappa\sinh(\gamma L) + i\gamma\cosh(\gamma L) = 0. \tag{10.50}$$

In the previous equation, the real and imaginary components on the left-hand side must be equal to zero, thus giving two equations. The unknown are the threshold gain, g_{th}, and the frequency, ω, of the oscillating mode. Therefore, the solution of Eq. 10.50 gives a set of (g_{th}, ω) values. From a practical point of view, to find the set of solutions (g_{th}, ω) corresponding to the various modes, a particular gain value g is first fixed and the transmission $T = 1/|M_{22}(L)|^2$ is plotted as a function of the detuning $n_0\Omega/c$, as shown in Fig. 10.7. Below threshold, the transmission presents various peaks, symmetrically placed with respect to $\Omega = 0$. Upon increasing the gain the first two maxima (at $\pm\Delta\beta_1$) increase and, at threshold, the transmission at these frequencies goes to infinity, thus giving the value of the threshold gain for the first two fundamental oscillating modes. In the same way, the gain threshold for the modes at $\pm\Delta\beta_2$ can be determined by finding the gain value for which the two maxima at $\pm\Delta\beta_2$ increase to infinity.

We note that the threshold gain decreases upon increasing the value of the coupling parameter q_0. The stop-band of the structure is the frequency region between the two frequencies of the two lowest-order modes. In a real DFB laser with uniform grating, only

one of the two symmetric lowest-order modes can effectively oscillate, since structural imperfections, spatial hole burning and not equal anti-reflection coatings of the end faces can
decrease the threshold gain of one of the two modes with respect to that of the other one.
This is not the ideal situation in order to have a single mode oscillation, since it is difficult
to determine which of the two lowest-order modes will eventually oscillate.

10.3.1 Approximated Evaluation of Oscillating Frequencies and Threshold Gain

In this section, we will calculate the oscillating frequencies and the threshold gain of a
DFB with uniform grating in the case of strong gain, that is, we will assume that $g \gg q_0$.
Equation 10.50 can be rewritten as follows:

$$e^{2\gamma L} = \frac{1 + \gamma/(g/2 + i\Delta\beta)}{1 - \gamma/(g/2 + i\Delta\beta)}. \tag{10.51}$$

The term $\gamma/(g/2 + i\Delta\beta)$ can be written in a different way, under the condition of strong
gain:

$$\frac{\gamma}{g/2 + i\Delta\beta} = \frac{\sqrt{q_0^2 + (g/2 + i\Delta\beta)^2}}{g/2 + i\Delta\beta} = \sqrt{1 + \left(\frac{q_0}{g/2 + i\Delta\beta}\right)^2}$$

$$\approx 1 + \frac{1}{2}\left(\frac{q_0}{g/2 + i\Delta\beta}\right)^2. \tag{10.52}$$

By using this equation, γ can be written as:

$$\gamma = \frac{g}{2} + i\Delta\beta + \frac{q_0^2(g/2 - i\Delta\beta)}{2(g^2/4 + \Delta\beta^2)}. \tag{10.53}$$

By using Eq. 10.52 in Eq. 10.51 and applying the strong-gain condition we have:

$$e^{2\gamma L} = -4\left(\frac{g/2 + i\Delta\beta}{q_0}\right)^2, \tag{10.54}$$

which can be rewritten as follows:

$$\frac{1}{4}\left(\frac{q_0}{g/2 + i\Delta\beta}\right)^2 e^{2\gamma L} = -1. \tag{10.55}$$

Upon taking the square root of both members of the previous equation, we can write:

$$\frac{1}{2}\frac{q_0}{g/2 + i\Delta\beta}e^{\gamma L} = \pm i. \tag{10.56}$$

By equating the phases in Eq. 10.56, we obtain:

$$2\arctan\left(\frac{2\Delta\beta}{g}\right) - 2\Delta\beta L + \frac{q_0^2\Delta\beta L}{(g/2)^2 + \Delta\beta^2} = (2m + 1)\pi, \tag{10.57}$$

where $m = 0, \pm 1, \pm 2, \ldots$. In the limit $g \gg \Delta\beta, q_0$, the oscillating mode frequencies can be
easily obtained from Eq. 10.57:

$$L\Delta\beta \approx -\left(m + \frac{1}{2}\right)\pi. \tag{10.58}$$

The two lowest-order modes correspond to $m = 0$ and $m = -1$. When $m = 0$ the detuning is given by:

$$\Delta\beta = -\frac{\pi}{2L}, \tag{10.59}$$

so that the oscillating frequency is:

$$\omega_{\omega=0} = \omega_0 - \frac{\pi c}{2n_0 L}. \tag{10.60}$$

When $m = -1$, the detuning is:

$$\Delta\beta = \frac{\pi}{2L}, \tag{10.61}$$

and the corresponding oscillating frequency is:

$$\omega_{\omega=-1} = \omega_0 + \frac{\pi c}{2n_0 L}. \tag{10.62}$$

In the previous equations, n_0 can be replaced by n_{eff}, the effective waveguide refractive index, so that:

$$\omega_m = \omega_0 \pm \frac{\pi c}{2n_{eff}L}, \tag{10.63}$$

and the mode frequency spacing is given by:

$$\omega_m - \omega_{m-1} = \frac{\pi c}{n_{eff}L}. \tag{10.64}$$

Therefore, the two lowest-order modes are symmetrically located with respect to the Bragg frequency ω_0.

By equating the amplitudes in Eq. 10.55, one can write:

$$\frac{e^{gL}}{(g/2)^2 + \Delta\beta^2} = \frac{4}{|q_0|^2}, \qquad (g = g_{th}), \tag{10.65}$$

which, by using Eq. 10.58 can be rewritten as:

$$e^{gL} = \frac{4}{(q_0 L)^2}\left[\left(\frac{gL}{2}\right)^2 + \left(m + \frac{1}{2}\right)^2 \pi^2\right], \qquad (g = g_{th}) \tag{10.66}$$

This equation can be solved graphically to obtain the threshold gain, as shown in Fig. 10.8. From Eq. 10.65, we see that, when q_0 is fixed, the threshold gain depends only on $|\Delta\beta|$; therefore, the two lowest-order modes ($m = 0, -1$) have the same threshold. Moreover, upon increasing m (therefore, upon increasing the detuning $\Delta\beta$), the threshold gain increases; indeed, from Eq. 10.65, we have:

$$|q_0|^2 \frac{e^{gL}}{4} - \left(\frac{g}{2}\right)^2 = \Delta\beta^2, \tag{10.67}$$

so that:

$$\frac{|q_0|^2}{4}e^{gL} \approx \Delta\beta^2, \tag{10.68}$$

this is the reason for the spectral selectivity of DFB lasers.

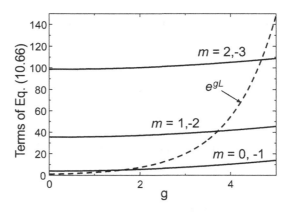

Figure 10.8   Graphical solution of Eq. 10.66. The intersections between the dashed curve (which represents exp (gL)) and the solid curves (which represent the function on the right-hand side of Eq. 10.66) give the threshold gain for the modes m. Note that the lowest-order modes ($m = 0$ and $m = -1$) have the same threshold gain.

10.4   DFB Laser with $\lambda/4$-Shifted Grating

In this section, we will consider DFB laser with a phase shift ϕ_0 at the center of the grating. In the case of a $\lambda/4$-shifted structure $\phi_0 = \pi$. We will consider a grating with a total length of $2L$, composed by two uniform gratings of length L, with a phase shift ϕ_0 between them. The first grating is characterized by a coupling parameter q_0 (given by Eq. 10.39), while the second grating is characterized by a coupling parameter $q_0 e^{i\phi_0}$. We have seen that in the case of a single uniform grating $\widetilde{A}(L)$ and $\widetilde{B}(L)$ can be calculated as follows:

$$
\begin{pmatrix} \widetilde{A}(L) \\ \widetilde{B}(L) \end{pmatrix} = \begin{pmatrix} \cosh(\gamma L) + i\frac{\Delta\kappa}{\gamma}\sinh(\gamma L) & i\frac{q_0}{\gamma}\sinh(\gamma L) \\ -i\frac{q_0^*}{\gamma}\sinh(\gamma L) & \cosh(\gamma L) - i\frac{\Delta\kappa}{\gamma}\sinh(\gamma L) \end{pmatrix} \begin{pmatrix} \widetilde{A}(0) \\ \widetilde{B}(0) \end{pmatrix}.
$$

In the case of two consecutive gratings, the total matrix $\{M_{ij}\}$ is given by the product of the matrices of the two gratings:

$$
M = \begin{pmatrix} \cosh(\gamma L) + i\frac{\Delta\kappa}{\gamma}\sinh(\gamma L) & i\frac{q_0}{\gamma}\,e^{i\phi_0}\sinh(\gamma L) \\ -i\frac{q_0^*}{\gamma}\,e^{-i\phi_0}\sinh(\gamma L) & \cosh(\gamma L) - i\frac{\Delta\kappa}{\gamma}\sinh(\gamma L) \end{pmatrix}.
$$

$$
\cdot \begin{pmatrix} \cosh(\gamma L) + i\frac{\Delta\kappa}{\gamma}\sinh(\gamma L) & i\frac{q_0}{\gamma}\sinh(\gamma L) \\ -i\frac{q_0^*}{\gamma}\sinh(\gamma L) & \cosh(\gamma L) - i\frac{\Delta\kappa}{\gamma}\sinh(\gamma L) \end{pmatrix}. \tag{10.69}
$$

The matrix element $M_{22}(2L)$ is given by:

$$
M_{22}(2L) = \cosh^2(\gamma L) + \sinh^2(\gamma L)\left(\frac{|q_0|^2}{\gamma^2}\,e^{-i\phi_0} - \frac{\Delta\kappa^2}{\gamma^2}\right) - 2i\frac{\Delta\kappa}{\gamma}\cosh(\gamma L)\sinh(\gamma L). \tag{10.70}
$$

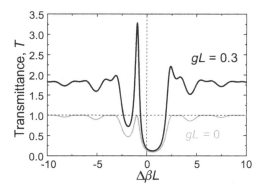

Figure 10.9 Intensity transmittance T of a shifted grating (with $\phi_0 = \pi/2$) as a function of the detuning $\Delta\beta L$ calculated for two different values of the effective gain gL.

The threshold condition can be obtained by equating $M_{22}(2L)$ to zero. In the case of a $\lambda/4$-shifted structure, the threshold condition can be written as:

$$\left[\frac{q_0 \sinh(\gamma L)}{\gamma \cosh(\gamma L) - i\Delta\kappa \sinh(\gamma L)}\right]^2 = 1. \tag{10.71}$$

It is now possible to proceed as described in the case of uniform grating by plotting the transmittance $T = 1/|M_{22}^2(L)|^2$ of the structure for increasing values of the gain coefficient. Laser threshold is reached when the transmittance increases to infinity. In this case, as already pointed out in Section 10.1, the lowest-order mode oscillates at the Bragg frequency: $\Delta\beta = 0$, $\beta = \beta_0 = \pi/\Lambda$. The higher-order modes are symmetrically located with respect to $\Delta\beta = 0$.

The general expression for the matrix element $M_{22}(2L)$ given by Eq. 10.70 allows one to calculate the threshold gain and the mode frequencies when $\phi_0 \neq \pi$. As an example, Fig. 10.9 shows the transmittance of the structure calculated assuming $\phi_0 = \pi/2$. In this case, the frequency of the lowest-order mode is $\beta < \beta_0$.

10.5 Distributed Bragg Reflector (DBR) Laser

In the case of a DFB laser, the grating structure, which provides the coupling between forward and backward waves, is produced along the cavity length of the laser. In the case of a Distributed Bragg reflector (DBR) laser, two Bragg reflectors are placed at the ends of the laser cavity. The Bragg reflectors are formed by two passive (i.e., not pumped) gratings, where the refractive index is periodically modulated. A schematic illustration of a DBR laser is shown in Fig. 10.10(a). An important advantage is that in this case the grating is achieved in a portion of the device which is physically separated from the active region: This characteristic simplifies the production procedures. Nevertheless, the frequency selectivity of a DBR laser is typically lower than that of a DFB laser, due to the presence of various longitudinal modes of the Fabry–Perot cavity. As a matter of fact,

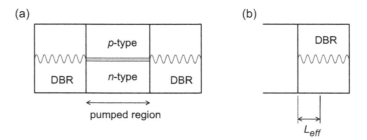

Figure 10.10 (a) Schematic structure of a DBR laser. (b) A DBR around the Bragg frequency can be modeled by a discrete mirror placed at a distance L_{eff} from the input of the DBR.

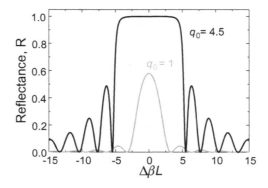

Figure 10.11 Intensity reflectance R of a DBR as a function of the detuning $\Delta\beta L$ calculated for two different values of the coupling parameter q_0.

due to the very short cavity length, typically a single mode is contained within the high-reflectivity band of the structure, but temperature variations may induce mode hopping: For this reason, DBR lasers are less frequently used then DFB lasers.

We can investigate the performances of a DBR laser by using the coupled-mode equation approach used in the case of DFB lasers. A DBR consists of a uniform grating of length L with:

$$\Delta\kappa = \Delta\beta + i\frac{\alpha}{2}, \qquad \gamma = \sqrt{|q_0|^2 - \Delta\beta^2}, \tag{10.72}$$

where α is the absorption coefficient. Its reflectivity can be calculated by using Eq. 10.48:

$$r_g = \frac{\widetilde{B}(0)}{\widetilde{A}(0)}\bigg|_{\widetilde{B}(L)=0} = -\frac{M_{21}(L)}{M_{22}(L)} = \frac{iq_0^* \sinh(\gamma L)}{\gamma \cosh(\gamma L) - i\Delta\kappa \sinh(\gamma L)}. \tag{10.73}$$

The intensity reflectance of the DBR is given by $R = |r_g|^2$ and it is plotted in Fig. 10.11 as a function of the detuning $\Delta\beta L$ for two values of the coupling parameter q_0. The reflectivity is maximum when $\Delta\beta = 0$, that is, when the propagation constant β of the input wave is equal to the Bragg propagation constant $\beta_0 = \pi/\Lambda$ of the grating. The reflectivity at the Bragg frequency can be calculated from Eq. 10.73 (we assume that the absorption in the grating can be neglected, so that $\Delta\kappa = \Delta\beta = 0$ and $\gamma = q_0$):

$$r_g = i \tanh(q_0 L). \qquad (10.74)$$

The intensity reflectance $R = \tanh^2(q_0 L)$ increases with $|q_0|L$. It is also simple to calculate an approximated expression for the effective length L_{eff} of the Bragg reflector. The reflection of a DBR around the Bragg frequency can be modeled by the reflection from a discrete mirror placed at a distance L_{eff} from the input of the DBR, as shown in Fig. 10.10(b). The reflectivity of the DBR given by Eq. 10.73 can be approximated by:

$$r_g \approx r_0 \exp(2i\beta_0 L_{eff}), \qquad (10.75)$$

where:

$$L_{eff} = \frac{1}{2} \frac{d \arg(r_g)}{d\beta}\bigg|_{\beta_0}. \qquad (10.76)$$

The phase $\phi(\beta)$ of the DBR can be expanded with a Taylor series around β_0:

$$\phi(\beta) \approx \phi(\beta_0) + \frac{\partial \phi}{\partial \beta}\bigg|_{\beta_0} (\beta - \beta_0). \qquad (10.77)$$

From Eq. 10.73, assuming a real coupling parameter q_0, the phase of the DBR can be calculated as:

$$\phi_{DBR}(\beta) = \pi - \arctan\left[\frac{\gamma}{\Delta\beta \tanh(\gamma L)}\right]. \qquad (10.78)$$

Around $\beta = \beta_0$, assuming $\alpha = 0$, $\gamma = q_0$ and Eq. 10.78 can be rewritten as:

$$\phi_{DBR}(\beta) = \pi - \arctan\left[\frac{q_0}{\Delta\beta \tanh(q_0 L)}\right]. \qquad (10.79)$$

The effective length is therefore given by:

$$L_{eff} = \frac{1}{2} \frac{\partial \phi_{DBR}}{\partial \beta} = \frac{1}{2q_0} \tanh(q_0 L), \qquad (10.80)$$

where we have derived Eq. 10.79 with respect to β.

10.6 MATLAB Program: Characteristics of a DFB Laser

In this last section, we give a MATLAB program, which allows calculation of the transmission of a DFB laser and of the threshold gain for the fundamental and high-order modes.[1]

function DFB_transmission(gmaxL,delta_n,Lambda,L,phi,n_g)

% DFB_transmission(gmaxL,delta_n,lambda,L,phi)
% inputs in SI:
% gmaxL = max gain to plot
% delta_n = amplitude of the refractive index modulation

---

[1]Courtesy of Dr. Matteo Lucchini, Politecnico di Milano

```
% Lambda = periodicity of the refractive index modulation [μm]
% L = total length of the laser cavity [μm]
% phi = phase shift between the two DFBs of length L/2
% n_g = number of gain points
% outputs:
% figure 1: T vs detuning for different gains
% figure 2: log(T) vs gain and detuning
% figure 3: value of max T vs gain and M vs gain
% figure 4: value of max T vs gain and detuning vs gain
% figure 5: numerical vs graphical solution of the threshold gain for different modes
% figure 6: relative change in threshold gain between different modes
% figures 2 - 6 are plotted only if gmaxL > 0
% figures 5 - 6 are plotted only if at least one threshold is reached

% initialization
  Lambda = Lambda * 1e-6;              % grating period
  L = L/2 * 1e-6;                      % length of the two adjacent gratings
  n = 3;                              % average refractive index
  n_k = 2^14;                         % number of points in detuning
  delta_beta_L = linspace(-10,10,n_k);  % creates the delta_beta*L vector
  delta_beta = delta_beta_L/L;        % given the length, extracts delta_beta
  flag = 1;                           % makes additional plots if gain > 0
  if ~gmaxL
      n_g = 1;
      flag = 0;
  end
  gL_vect = linspace(0,gmaxL,n_g);    % gain vector
  q0 = (delta_n*pi)/(2*n*Lambda);     % coupling parameter
% calculating the transmission coefficient for the different gains
  gLm = bsxfun(@times,gL_vect.',ones(n_g,n_k));      % matrix with changing gain
  D_betam = bsxfun(@times,delta_beta,ones(n_g,n_k)); % matrix with
                                                     delta_beta = n0/c*Omega
  delta_km = D_betam - 1i*(gLm/(2*L));    % associated delta_k
  gamma = sqrt(abs(q0)^2 - delta_km.^2);  % associated gamma
  Mm = cosh(gamma*L).^2 + ...
    sinh(gamma*L).^2.*(abs(q0)^2./gamma.^2*exp(-1i*phi) - delta_km.^2./gamma.^2) -
    ...
    *delta_km./gamma.*sinh(2*L*gamma);   % value of M22
  T=1./abs(Mm).^2;                        % transmission (n_g,n_k)

% plotting T
figure
plot(delta_beta_L,T)
phi title(['\phi_0 =', num2str(phi/pi),'\pi , g_maxL=',num2str(gmaxL)])
set(gca,'FontSize',12)
```

```
ylabel('Transmission','FontSize',16)
xlabel('\Delta\betaL','FontSize',16)
set(gcf,'Color',[1 1 1])
set(gca,'Xgrid','on')
set(gca,'Ygrid','on')
legend(cellstr(string(gL_vect)))

% additional plots
if flag
% T vs gL and Delta_beta_L
figure
% surface plot
surf(delta_beta_L,gL_vect,log(T))
shading flat
title(['\phi_0 =', num2str(phi/pi),'\pi , g_maxL=',num2str(gmaxL)])
set(gca,'FontSize',12)
zlabel('ln(T)','FontSize',16)
xlabel('\Delta\betaL','FontSize',16)
ylabel('gL','FontSize',16)
set(gcf,'Color',[1 1 1])
colormap(gca,'jet')
colorbar

% T and M
% initialization
figure;
T_max = zeros(1,n_g);
M_max = T_max;
% searching maxima and storing
[ ~,idx] = max(T,[],2);
for ij = 1:n_g
   T_max(ij) = T(ij,idx(ij));
   M_max(ij) = Mm(ij,idx(ij));
end
  d_beta_L_max = delta_beta_L(idx(2:end));    % discard the first point for g = 0
  if(phi==pi||phi==0)                          % with pi or 0 we have a symmetric curve
d_beta_L_max = abs(d_beta_L_max); % focus on positive detunings
end
% double y plot
[AX,H1,H2]=plotyy(gL_vect,T_max,gL_vect,abs(M_max));
set(AX(1),'FontSize',12,'YColor','k')
set(AX(2),'FontSize',12,'YColor','r')
xlabel('gL','FontSize',16)
set(get(AX(1),'Ylabel'),'String','T_max','FontSize',16)
```

```
set(get(AX(2),'Ylabel'),'String','M_22(L)','FontSize',16)
set(get(AX(1),'Ylabel'),'Color','k')
set(get(AX(2),'Ylabel'),'Color','r')
set(H1,'Color','k','LineStyle','-','LineWidth',1,'Marker','o')
set(H2,'Color','r','LineStyle','-','LineWidth',1,'Marker','s'
set(gcf,'Color',[1 1 1])
set(gca,'Xgrid','on')
legend('T_max','|M_22(L)|')


% T and Delta_beta_L
figure;
% double y plot
[AX,H1,H2]=plotyy(gL_vect,T_max,gL_vect(2:end),d_beta_L_max);
set(AX(1),'FontSize',12,'YColor','k')
set(AX(2),'FontSize',12,'YColor','b')
xlabel('gL','FontSize',16)
set(get(AX(1),'Ylabel'),'String','T_max','FontSize',16)
set(get(AX(2),'Ylabel'),'String','\Delta\betaL_max','FontSize',16)
set(get(AX(1),'Ylabel'),'Color','k')
set(get(AX(2),'Ylabel'),'Color','b')
set(H1,'Color','k','LineStyle','-','LineWidth',1,'Marker','o')
set(H2,'Color','b','LineStyle','-','LineWidth',1,'Marker','s')
set(gcf,'Color',[1 1 1])
set(gca,'Xgrid','on')
legend('T_max','\Delta\betaL_max')


% finding local maxima
[T_peaks, idx_peaks] = findpeaks(T_max);
if(~isempty(idx_peaks)) % checks that we reached a threshold

% graphical solutions = initial conditions for numerical algorithm
gL_peaks = gL_vect(idx_peaks); % gain
d_beta_L_peaks = d_beta_L_max(idx_peaks-1); % detuning

% plotting the peaks on
hold(AX(1),'on')
plot(AX(1),gL_peaks,T_peaks,'*r')
hold(AX(2),'on')
plot(AX(2),gL_peaks,d_beta_L_peaks,'*g',[gL_vect(1) gL_vect(end)],[0 0],'-k')
legend('T_max','T_peaks','\Delta\betaL_max','\Delta\betaL_peaks')

% searching for the threshold solution numerically
% initialization
gL_th_v = zeros(size(length(T_peaks)));
```

```
DbL_th_v = zeros(size(length(T_peaks)));

% finding solutions numerically with fsolve
for ij = 1:length(T_peaks)
x0 = [d_beta_L_peaks(ij) gL_peaks(ij)]; % initial guess exitflag = -1; % flag that tells if
the solution has been found
while exitflag <0 % loop to find local solution of M_22 = 0
options = optimset('Display','off'); % Option to display output
set(0,'RecursionLimit',1500); % options to set resolution
[x,~,exitflag]=fsolve(@M_22_calculator, x0, options, q0, L, phi); % solver
x0(1) = x0(1) +0.05; % changes initial guess if it did not converge
end
gL_th_v(ij) = x(2); % storing solution
DbL_th_v(ij) = x(1); % storing solution
clear exitflag x
end
% plotting the numerical solution vs graphical
mode_order = 1:length(T_peaks); % mode orders figure
% numerical solution
plot(mode_order,gL_th_v,'b--o','LineWidth',1,'MarkerSize',8,... 'MarkerFaceColor',[1 1
1])
% graphical solution = starting points
hold on
plot(mode_order,gL_peaks,'*r')
set(gca,'FontSize',12)
ylabel('gL_th','FontSize',16)
xlabel('Mode order','FontSize',16
) set(gcf,'Color',[1 1 1])
set(gca,'Xgrid','on')
set(gca,'Ygrid','on','XTick',mode_order)
legend('Numerical','Graphical')

figure
% numerical solution
plot(mode_order,DbL_th_v,'k--o','LineWidth',1,'MarkerSize',8,... 'MarkerFaceColor',[1
1 1])
% graphical solution = starting points
hold on
plot(mode_order,d_beta_L_peaks,'*g')
set(gca,'FontSize',12)
ylabel('\Delta\betaL_{th}','FontSize',16)
xlabel('Mode order','FontSize',16)
set(gcf,'Color',[1 1 1])
set(gca,'Xgrid','on')
```

```
set(gca,'Ygrid','on','XTick',mode_order)
legend('Numerical','Graphical')

% relative change in gain
figure
plot(mode_order,(gL_th_v-gL_th_v(1))/gL_th_v(1),'.--','MarkerSize',16)
set(gca,'FontSize',12)
ylabel('\DeltagL_{th}/gL^{1st}_{th}','FontSize',16)
xlabel('Mode order','FontSize',16)
set(gcf,'Color',[1 1 1])
set(gca,'Xgrid','on')
set(gca,'Ygrid','on','XTick',mode_order)
end
end
end
```

10.7 Exercises

Exercise 10.1 By using a numerical code, plot the intensity transmittance of a DFB device versus $\Delta\beta L$ for different values of the gain (below threshold) for the following structures:
a) uniform grating DFB laser;
b) quarter-wave-shifted DFB laser;
c) uniform grating with one end mirror with antireflection coating and the second mirror with $R = 0.32$;
d) uniform grating with one end mirror with antireflection coating and the second mirror with $R = 0.95$.

Exercise 10.2 By using the numerical code of Exercise 10.1, determine the threshold gain $(\Gamma g_{th} - \alpha_i)L$ for the first two lowest-order modes, considering various coupling parameters, $q_0 L$, and plot the results versus $\Delta\beta L$. Consider the four structures of Exercise 10.1.

Exercise 10.3 Consider an InGaAsP/InP quarter-wave-shifted DFB laser emitting at 1,550 nm, with the following characteristics: confinement factor $\Gamma = 0.08$, injection efficiency $\eta_i = 80\%$, internal losses $\alpha_i = 15$ cm^{-1}, coupling parameter $q_0 = 20$ cm^{-1}, average effective refractive index 3.21, and length of the laser cavity $L = 500$ μm. Both laser facets are antireflection coated. By using the results of the previous exercises, determine:
a) the threshold modal gain;
b) the external quantum efficiency.

Exercise 10.4 Consider an InGaAsP/InP DFB laser with the same characteristics of the laser of Exercise 10.3 but with uniform grating. Calculate:
a) the threshold modal gain;
b) the external quantum efficiency;
c) the emission wavelength.

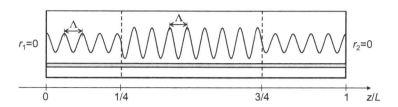

Figure 10.12 Distributed coupling coefficient (DCC) DFB laser.

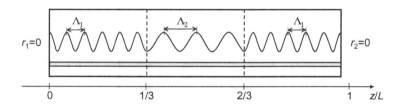

Figure 10.13 Corrugation-Pitch-Modulated (CPM) DFB laser.

Exercise 10.5 Consider a DFB laser with total length $4L$, with three phase shifts of $\Lambda/2$ at $z = L$, $z = 2L$ and $z = 3L$, and zero reflectivity of the facets.
a) By using the coupled-mode equations, determine the threshold condition.
b) Plot the intensity transmittance, T, of the structure for different values of the effective gain.

Exercise 10.6 A DFB laser based on InGaAsP emits at 1,550 nm. The index of refraction of the alloy is $n = 3.4$. Determine the period Λ of the grating.

Exercise 10.7 Derive the threshold condition of a phase-shifted DFB laser with a phase shift ϕ_0 at the cavity center, assuming nonreflecting facets.

Exercise 10.8 Consider a DFB laser with uniform grating. By using a numerical code, plot the contour lines of the power transmittance of the DFB structure in a two-dimensional map, where the horizontal axis is the frequency detuning, $\Delta\beta L$, and the vertical axis is the gain, gL.

Exercise 10.9 Repeat Exercise 10.8 in the case of a DFB $\lambda/4$-shifted structure.

Exercise 10.10 A distributed coupling coefficient (DCC) DFB laser is a DFB laser structure where the amplitude of the refractive index modulation, and therefore the coupling coefficient, change along the laser cavity. By using the coupled-mode equations, investigate the main characteristics of the DCC-DFB laser structure shown in Fig. 10.12.
a) Plot the power transmittance as a function of the frequency detuning, $\Delta\beta L$.
b) Determine the threshold gain upon fixing a given coupling coefficient and cavity length.

Exercise 10.11 Repeat Exercise 10.10 in the case of the DFB structure shown in Fig. 10.13.

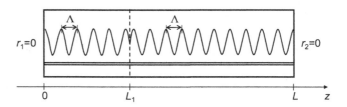

Figure 10.14 $\lambda/4$-shifted DFB laser.

Exercise 10.12 Repeat Exercise 10.8 for the DFB structures analyzed in Exercises 10.10 and 10.11.

Exercise 10.13 Consider a single-phase-shifted DFB laser, whose structure is shown in Fig. 10.14. Assume that the phase correspond to a modulation shift of $\Lambda/2$. Assume a total cavity length $L = 500 \ \mu$m, emission wavelength $\lambda = 1.33 \ \mu$m, coupling parameter $q = q_0 \exp i\phi_0$, where $\phi_0 = \pi$ and $q_0 = 20$ cm^{-1}. By using the numerical code employed in the previous exercises, evaluate the threshold gain of the fundamental mode as a function of the phase shift position, $\xi = L_1/L$.

Exercise 10.14 Consider an InGaAsP DFB laser emitting at $\lambda = 1.55 \ \mu$m. The effective refractive index is $n = 3.4$, and the cavity length is $L = 60 \ \mu$m. Determine the period Λ of the refractive index modulation for a first-order and for a second-order grating. Determine the number of corrugations in the two cases.

11   Vertical Cavity Surface-Emitting Lasers

11.1   Basic Structure

In Chapter 8, we have evidenced that edge-emitting lasers typically oscillated on several modes and the output beam is characterized by an elliptical shape, with high divergence, unless special optical systems are employed. For a number of important applications (in particular when the laser beam must be coupled into a single-mode optical fiber), a different laser geometry can be employed: A vertical cavity is used, where the laser beam propagates in the direction perpendicular to the junction. These lasers are called *Vertical Cavity Surface-Emitting Lasers* (VCSELs). A remarkable characteristic of these lasers is the very short length of the active medium, of the order of the emission wavelength, which determines a very small gain. The laser cavity develops along the epitaxial growth axis of the semiconductor materials. The small gain requires end mirrors with very high reflectivity, in order to reduce internal losses. In this case, the Fresnel reflection at the semiconductor–air interface is by far not enough to guarantee sufficiently small losses in the cavity and Bragg reflectors are used, achieved by epitaxial deposition of alternating layers with low and high index of refraction with controlled thickness. A schematic structure of a VCSEL is shown in Fig. 11.1. Electrons and holes are injected into the device through the *n*-type (bottom DBR in Fig. 11.1) and *p*-type (upper DBR in Fig. 11.1) mirrors, respectively.

11.2   Threshold Conditions

Let us first examine the threshold conditions for a VCEL. The threshold gain is given by Eq. 8.32, $\Gamma g_{th} = \gamma/L$, where L is the cavity length and Γ the confinement factor, which can be decomposed in a confinement factor in the transverse directions, x and y, and in an axial confinement factor in the direction z of beam propagation:

$$\Gamma = \Gamma_{xy}\Gamma_z. \tag{11.1}$$

$\gamma = \alpha L - \log R$ represents the cavity losses, where α is the absorption coefficient of the cavity and R is the average mirror reflectivity $R = \sqrt{R_1 R_2}$, where R_1 and R_2 are the reflectivities of the bottom and upper mirrors in Fig. 11.1.

The schematic geometry of the laser cavity is shown in Fig. 11.2. The standing-wave pattern of the electric field in the cavity can be written as:

$$E(x, y, z) = U(x, y)\sqrt{2}\cos \beta z, \tag{11.2}$$

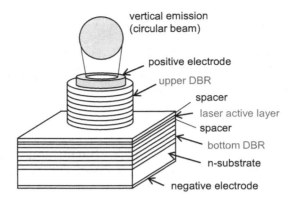

Figure 11.1 Schematic setup of a VCSEL.

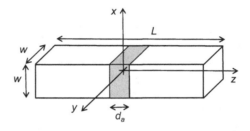

Figure 11.2 Schematic geometry of the laser cavity of a VCSEL.

where $U(x, y)$ is the normalized transverse profile of the electric field and $\beta = 2\pi\bar{n}/\lambda$ is the z-component of the propagation constant $\mathbf{k}$. The confinement factors can be written as:

$$\Gamma_{xy} = \frac{\int_{-w/2}^{w/2} \int_{-w/2}^{w/2} |U(x, y)|^2 dxdy}{\int \int_{-\infty}^{\infty} |U(x, y)|^2 dxdy}, \tag{11.3}$$

$$\Gamma_z = \frac{1}{L} \int_{-d_a/2}^{d_a/2} 2\cos^2(\beta z)dz = \frac{d_a}{L}\left(1 + \frac{\sin(\beta d_a)}{\beta d_a}\right). \tag{11.4}$$

It is also useful to write the axial confinement factor Γ_z as the product of a fill factor $\Gamma_f = d_a/L$ and of an enhancement factor Γ_{enh} given by:

$$\Gamma_{enh} = 1 + \frac{\sin(\beta d_a)}{\beta d_a}. \tag{11.5}$$

Note that for thin active layers composed by a quantum well $\Gamma_{enh} \simeq 2$ and that it is typically in the range between 1 and 2. Here, we have assumed that the active region is in the center ($z = 0$), in correspondence of the central maximum of the standing-wave pattern. If the active region is located at the field node, where $E(z) = 0$, the overlapping between the optical field and the active material is negligible and $\Gamma_{enh} \simeq 0$. Indeed, a more general expression of Γ_{enh} is the following:

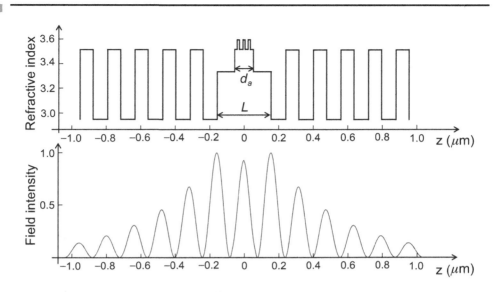

Figure 11.3 Upper panel: variation of the refractive index in the VCSEL described in the text. Lower panel: field intensity in the VCSEL.

$$\Gamma_{enh} = 1 + \cos\left(2\beta z_s\right)\frac{\sin\left(\beta d_a\right)}{\beta d_a}, \tag{11.6}$$

where z_s is the distance along the propagation direction between the center of the active region and the peak of the standing-wave pattern inside the cavity. To illustrate this important concept, Fig. 11.3 shows an example of a VCSEL where the active material in composed by three InGaAs quantum wells, with a thickness of 8 nm, separated by 2 GaAs barriers (8-nm thick), cladded on both sides by 8-nm-thick GaAs layers (emission wavelength: 980 nm). The spacers are composed by $Al_{0.2}Ga_{0.8}As$. The DBRs are formed by alternating quarter-wavelength layers of AlAs and GaAs. The figure also displays the standing-wave pattern, $|E|^2$, inside the cavity. We will use this particular structure as a case study in this section. In this example, the active region is placed in the center of the cavity in correspondence of the central anti-node of the standing-wave pattern ($z_s = 0$) so that the enhancement factor is of the order of 2. Moreover, the DBRs start with an AlAs quarter-wave layer close to the cavity. In this case at both ends of the inner cavity of length L, the electric field presents a maximum amplitude and the resonance condition for the emission wavelength is given by:

$$\langle \bar{n}\rangle L = m\frac{\lambda_0}{2}, \tag{11.7}$$

where $\langle \bar{n}\rangle$ is the spatially averaged refractive index and m is a positive integer. Since the active layer must be located in an anti-node of the standing-wave pattern, the minimum length of the cavity can be found from Eq. 11.7 with $m = 2$: $L = \lambda_0/\langle\bar{n}\rangle$. Note that the maximum total thickness of the active layer is of the order of $\lambda_0/4\langle\bar{n}\rangle$, since a larger thickness would correspond to active layers placed in regions with low optical intensity: This limits the number of quantum wells in the active region. In order to increase the gain, it is also possible to use a particular scheme, called *resonant periodic gain*, which makes

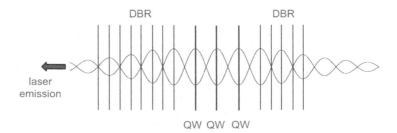

Figure 11.4   Schematic setup of a VCSEL with resonant periodic gain.

use of a relatively longer cavity, as shown in Fig. 11.4, where $m = 4$ so that the standing-wave pattern presents three anti-nodes and three quantum wells (or three double or triple quantum wells) separated by $\lambda_0/2\langle \bar{n} \rangle$ are placed at each anti-node.

Due to the very short length of the active medium, to achieve a reasonable threshold current the reflectivity of the bottom and upper mirrors must be very high, at least of the order of 99.5%. For this reason, metallic mirrors cannot be employed and DBRs must be used.

11.3   DBR for VCSELs

The DBRs used in VCSELs are obtained by alternating quarter-wavelength thick layers with low and high refractive index, as schematically shown in Fig. 11.5. The essential idea is that at the Bragg wavelength, the reflections at each index discontinuity add up in phase, thus leading to a large reflectivity. Generally more than 20 Bragg pairs are required for a mirror. In Fig. 11.5, r represents the reflectivity at each discontinuity and r_g represents the net mirror reflectivity, as in Section 10.5. If the wave is initially in the medium with refractive index n_2 and it is incident on the interface between medium 2 and 1 (at normal incidence), the electric field reflectivity is:

$$r = \frac{n_2 - n_1}{n_2 + n_1}. \tag{11.8}$$

Note that if $n_2 < n_1$ the reflected field has a π phase shift with respect to the incident field. As already mentioned, the condition for constructive interference is obtained by alternating layers with thickness $d_1 = \lambda_0/4n_1$ and $d_2 = \lambda_0/4n_2$, where λ_0 is the free-space wavelength. At the Bragg wavelength, the net electric field reflectivity of the mirror, assuming negligible absorption, can be written as:

$$R = \left(\frac{1-b}{1+b} \right)^2, \tag{11.9}$$

where b has two possible expressions. When the sequence of refractive indexes of the DBR shown in Fig. 11.5 is n_c (refractive index of the spacer layer), (n_1, n_2) repeated m times and finally n_s (refractive index of the substrate):

$$b = \frac{n_s}{n_c} \left(\frac{n_1}{n_2} \right)^{2m}, \tag{11.10}$$

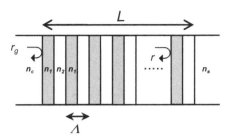

Figure 11.5   Schematic geometry of Bragg reflector.

while, when the refractive index sequence is n_c, (n_1, n_2) repeated m times, n_1 and finally n_s:

$$b = \frac{n_1^2}{n_c n_s} \left(\frac{n_1}{n_2}\right)^{2m}, \tag{11.11}$$

in this last case since the first and the last layers between the spacer and the substrate are made by the same material, the number of layer pairs of the DBR is $m + 0.5$. For example, if we consider a DBR formed by alternating layers of AlAs/GaAs and we assume that the waves are incident from $Al_{0.2}Ga_{0.8}As$ and are transmitted to air in the case of the top mirror and to GaAs in the case of the bottom mirror, the mirror reflectivities calculated according to the previous equations exceed 99.9% (neglecting absorption) for > 20 layer pairs.

The reflectivity of a Bragg mirror can be calculated as we have seen in Section 10.5. As shown in Fig. 10.11, when the number of layer pairs m is sufficiently high (corresponding to a high coupling parameter q_0), the mirror reflectivity is characterized by a broad spectral region of very high reflectivity around the Bragg wavelength λ_B. This region is called *stop-band* since it represents the wavelength interval where the periodic photonic structure does not allow transmission. The width of the stop-band is roughly given by [21]:

$$\Delta\lambda_{stop} \simeq \frac{2\lambda_B \Delta n}{\pi \langle n_g \rangle}, \tag{11.12}$$

where $\Delta n = |n_1 - n_2|$ and $\langle n_g \rangle$ are the spatial average of the group index given by Eq. 8.68. We can assume

$$\langle n_g \rangle = 2 \left(\frac{1}{n_1} + \frac{1}{n_2}\right)^{-1} . \tag{11.13}$$

For example, in the case of a GaAs/AlAs mirror with $\Delta n = 0.57$ and $\langle n_g \rangle \simeq 3.2$ at $\lambda_B = 980$ nm, we obtain $\Delta\lambda_{stop} \approx 110$ nm.

As we have already observed in Section 10.5, in the case of a wave incident from a high-index material, the mirror reflectivity can be written as:

$$r_g \simeq \sqrt{R} e^{2i\beta_0 L_{eff}}, \tag{11.14}$$

so that a Bragg mirror can be modelled as an effective mirror with a field reflectivity amplitude equal to $|r_g|$ placed at a distance L_{eff} from the first layer of the DBR, as shown in Fig. 10.10(b). L_{eff} is given by Eq. 10.80, where L is the length of the DBR (see Fig. 11.5).

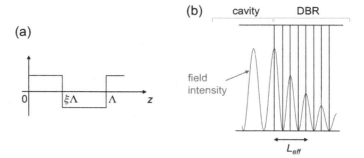

Figure 11.6 (a) Refractive index with a periodic square-wave modulation of period Λ and duty-cycle ξ. (b) Field intensity inside a Bragg mirror.

It is possible to shown that for a rectangular-shaped grating (i.e., for a grating where the modulation of the refractive index consists of a periodic square-wave function of period Λ), the coupling parameter can be written as:

$$q_0 = \frac{2\Delta n}{\lambda}\frac{\sin(\xi q\pi)}{q},$$ (11.15)

where q is the grating order and ξ the grating duty-cycle as shown in Fig. 11.6(a). If $\xi = 1/2$ and $q = 1$, the coupling parameter is given by:

$$q_0 = \frac{2\Delta n}{\lambda}.$$ (11.16)

In the case of low coupling (i.e., in the case of a DBR with low reflectivity) $\tanh(q_0L) \approx q_0L$ and the effective equivalent mirror is located at the center of the Bragg reflector $L_{eff} = L/2$, while in the case of high coupling $\tanh(q_0L) \approx 1$ so that $L_{eff} = 1/2q_0 = \lambda/4\Delta n$. It is interesting to note that the effective length gives also an approximated value of the penetration depth of the optical energy into the Bragg reflector, defined as the depth into the mirror at which the optical field intensity is equal to $1/e$ of its value at the input of the mirror, as shown in Fig. 11.6(b).

Taking into account the absorption losses inside a DBR, described by an absorption coefficient α_{DBR}, the mirror reflectivity can be written as:

$$R = R_m \exp(-2\alpha_{DBR}L_{eff}),$$ (11.17)

where R_m is the mirror reflectivity (with m pairs of layers) without absorption. Therefore, the maximum reflectivity (assuming $R_m = 1$) will be:

$$R_{max} = \exp\left(-\frac{\alpha_{DBR}\lambda_0}{2\Delta n}\right).$$ (11.18)

The total effective cavity length is given by:

$$L_{eff,T} = L + L_{eff,t} + L_{eff,b},$$ (11.19)

where $L_{eff,t}$ and $L_{eff,b}$ are the effective penetration depth of the top and bottom DBRs, respectively. This quantity is used to estimate the spacing between consecutive longitudinal modes:

$$\Delta\lambda = \frac{\lambda^2}{2L_{eff,T}\langle n_g\rangle}. \tag{11.20}$$

Exercise 11.1   Calculate the spacing between longitudinal modes in the VCSEL shown in Fig. 11.3, with an emission wavelength of 980 nm. Assume that the cavity between the two Bragg mirrors is one optical wavelength measured in the semiconductor. The mirrors start with an AlAs quarter-wave layer adjacent to the cavity. Refractive indexes: for AlAs $n = 2.95$, for $Al_{0.2}Ga_{0.8}As$ $n = 3.39$, for GaAs $n = 3.52$, and for InGaAs $n = 3.6$.

We can first calculate the thickness of the two $Al_{0.2}Ga_{0.8}As$ spacer layers. Between the two DBRs, the VCSEL is formed by two spacer layers, four GaAs layers, and three InGaAs layers, so that we must have (the thicknesses are expressed in nanometers):

$$2 \cdot 3.39 \cdot s + 4 \cdot 3.52 \cdot 8 + 3 \cdot 3.6 \cdot 8 = 980 \text{ nm,}$$

where $s = 115.2$ nm is the thickness of a spacer layer. The effective propagation length in both mirrors is (we assume a high coupling):

$$L_{eff} = \frac{\lambda_0}{4\Delta n} = 429.8 \text{ nm.}$$

The thickness L of the cavity between the two mirrors is $L = 286.4$ nm. The total cavity effective length is therefore given by:

$$L_{eff,T} = (286.4 + 2 \cdot 429.8) \text{ nm} = 1146 \text{ nm.}$$

The longitudinal mode spacing is given by:

$$\Delta\lambda = \frac{\lambda^2}{2L_{eff,T}\langle n_g\rangle} = 131 \text{ nm,}$$

where $\langle n_g\rangle$ is the average index of refraction of the mirrors given by Eq. 11.13: $\langle n_g\rangle = 3.21$. The calculated mode spacing is larger than the stop-band of the mirror (it is also larger than the gain bandwidth of the quantum well used as active laser material). Therefore, just a single longitudinal mode can oscillate in this VCSEL. Depending on the lateral size of the VCSEL, various transverse mode may oscillate and particular fabrication methods can be adopted to have a single transverse mode. ■

11.4   Threshold Conditions and Current Confinement

The threshold gain can be written as follows:

$$\Gamma g_{th} = \langle\alpha_i\rangle + \frac{1}{L_{eff,T}}\log\frac{1}{\sqrt{R_1R_2}}, \tag{11.21}$$

where R_1 and R_2 are the reflectivity of the top and bottom mirrors. Due to the small cavity length, absorption and scattering losses can be very small, thus leading to devices operating

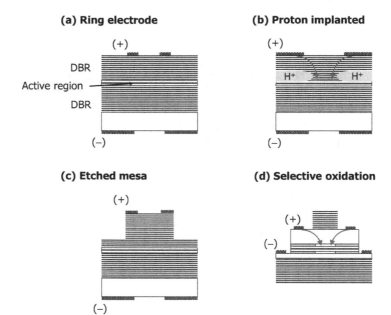

(a) Ring electrode

(b) Proton implanted

(c) Etched mesa

(d) Selective oxidation

Figure 11.7   Typical current confinement schemes in VCSELs.

with low threshold current density (down to ~ 350 A/cm^2). Since the transverse size is typically very small, in the case of single transverse mode operation (of the order of 5–10 μm), low threshold currents can be achieved. Typical threshold currents are of the order of 1 mA, with record values of about 0.06 mA.

In order to reduce the threshold current, it is crucial to use efficient current confinement structures. Figure 11.7 shows a few typical injection and current confinement schemes employed in VCSELs. The first one makes use of an annular electrode, which can limit the current flux near the ring electrode (see Fig. 11.7(a)). Laser radiation can be emitted from the inner part of the metallic ring. This structure is very simple but does not allow an efficient current confinement, mainly due to carrier diffusion. To limit carrier diffusion, proton implanted structures can be used (top or bottom emitting). In this case, the current is confined by the semi-insulating region generated after the implant, which funnels into the active area (see Fig. 11.7(b)). Several commercial devices are based on this structure, which guarantees a good current confinement. Proton implantation may introduce optical losses, and the implantation damage can affect the reliability of the laser. Optical losses introduced by this technique typically limit the laser efficiency to a maximum value of $\sim 20\%$. The ion implanted VCSEL allows a planar fabrication process. Both these schemes are examples of gain-guided VCSELs since the extension of the active region is determined by the current confinement, without any lateral photon confinement structures. In this case, the carriers injected in the active region can diffuse, thus decreasing the overall laser efficiency. With low current injection, single transverse mode emission can be easily achieved. Another largely used configuration is the so-called *etched mesa* (see Fig. 11.7(c)), where current is confined to the mesa width. In this case, the fabrication process in nonplanar and the etch depth is of the order of a few micrometers. Mesa structures can be obtained by using various dry etching techniques (e.g., reactive ion etching, based on the use of SiCl$_4$, which can produce etch depths of the order of 10 μm; chemically assisted ion-beam etching). The advantage of dry etching is that it allows the fabrication of structures with a very small diameter. Etching is stopped well above the active layer to maintain reliable device operation. The mesa structure provides an excellent current confinement and also an optical waveguide for the emitted photons. Another largely used structure is based on selective oxidation of a AlAs layer in the top Bragg reflector in order to generate a highly resistive layer of AlO$_x$ just above the active layer (see Fig. 11.7(d)). This selective oxidation gives rise to an aperture in front of the active layer for the current flow and to a waveguide for the emitted photons. These devices are characterized by small optical losses thus leading to a very high efficiency ($>$ 50%) and very low threshold currents ($<$100 μA).

11.5 Advantages and Applications

Since the first suggestion of a VCSEL device in 1977 and the demonstration of the first operation in 1979, VCSELs have become largely used laser devices for a wide range of applications, replacing the edge-emitting lasers in various applications where laser-optical fiber coupling is important, due to the small beam divergence and the symmetric mode profile. Coupling efficiency in single-mode optical fibers of the order of 90% can be achieved with top-emitting VCSELs. Emission wavelength and threshold conditions are relatively insensitive against temperature variations. Single longitudinal mode emission is easily achieved. A number of laser devices can be fabricated by fully monolithic processes thus leading to very low-cost chip production. The VCSELs allow simple optical on-wafer testing, similar to the testing of integrated electronic circuits, directly after growth (i.e., before the wafer is cleaved). Moreover, densely packed two-dimensional laser arrays can be formed. Due to the short cavity round-trip time, VCSELs can be modulated with frequencies in the gigahertz range, with important applications for optical fiber communications.

11.6 Exercises

Exercise 11.1   Consider a VCSEL with an active region composed by a sequence of quantum well of GaAs/Al$_{0.3}$Ga$_{0.7}$As, where wells and barriers have the same thickness $s = 10$ nm. The emission wavelength is 840 nm, and the active region has the following characteristics: $\sigma = 6 \times 10^{-16}$ cm^2, $N_{tr} = 1.2 \times 10^{18}$ cm^{-3}, $\tau = 4$ ns, $\gamma = 0.03$, $\eta_i = 0.95$, $\Gamma \simeq 1$. The refractive index of Al$_x$Ga$_{1-x}$As is $n = 3.3 - 0.53x + 0.09x^2$. Calculate:

a) the total thickness of the active region in order to optimize the gain;

b) d_1, d_2, and L (see Fig. 11.8);

c) the threshold current density;

d) the number of oscillating modes.

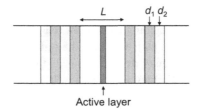

Figure 11.8   Structure of the VCSEL considered in Exercise 11.1.

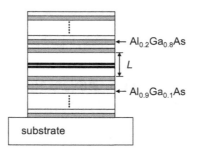

Figure 11.9   Structure of the VCSEL considered in Exercise 11.2.

Exercise 11.2   Consider the VCSEL shown in Fig. 11.9, emitting at 850 nm.

a) Determine the thickness d_1 and d_2 of the two layers in each period of the DBRs in order to have a peak reflectivity at 850 nm. The refractive index of $Al_{0.2}Ga_{0.8}As$ is $n_1 = 3.19$, and the refractive index of $Al_{0.9}Ga_{0.1}As$ is $n_2 = 2.90$.

b) Determine the total thickness, L, such that the lowest-order mode of the optical cavity has wavelength equal to the Bragg wavelength of the DBRs (i.e., 850 nm).

Exercise 11.3   We have to design a VCSEL emitting at 980 nm. The Bragg reflectors are composed by quarter-wave stacks of AlAs/GaAs. The top mirror has a power reflectance of 99.9%, including the air interface, and the bottom mirror must have a power reflectance of 99%. The cavity between the DBRs is one-optical wavelength long (measured in the material). It consists of 4 InGaAs quantum wells, with a thickness of 10 nm, separated by 3 GaAs barriers, with a thickness of 8 nm and cladded on each side by another 8-nm-thick GaAs layer. The spacers are made of $Al_{0.2}Ga_{0.8}As$. The two DBRs begin with an AlAs quarter-wave layer. The quantum wells are centered between the DBRs. Refractive indexes, AlAs: 2.95, GaAs: 3.52, $Al_{0.2}Ga_{0.8}As$: 3.39, InGaAs: 3.60. Determine:

a) the thickness of the two $Al_{0.2}Ga_{0.8}As$ spacers;

b) the number of AlAs/GaAs layer pairs in the two DBRs;

c) the longitudinal mode separation.

Exercise 11.4   Design the DBRs of a VCSEL with the following characteristics: emission wavelength 1.55 μm, power reflectance of the output coupler mirror 99%, power reflectance greater than 99.9% for the back reflector, alternating layers of InP (with refractive index 3.41) and InGaAsP (refractive index 3.52). The output coupler starts with the high-index material, while the other side terminates in air. The back reflector terminates with the higher index material on both sides. Determine the effective penetration depth into the mirrors.

Exercise 11.5   Repeat the same calculations of Exercise 11.4 in the case of a VCSEL designed to emit at 870 nm, with Bragg reflectors made by $Al_{0.2}Ga_{0.8}As/AlAs$ alternating layers (refractive index of $Al_{0.2}Ga_{0.8}As$: 3.39 and refractive index of AlAs: 2.95).

Exercise 11.6   Consider a VCSEL with a cavity length of 10 μm, with internal losses $\alpha = 15$ cm^{-1} and DBRs with power reflectance of 99%. Determine the threshold gain.

Exercise 11.7   Consider a VCSEL emitting at 980 nm with GaAs/AlAs Bragg reflectors. Waves are incident from $Al_{0.3}Ga_{0.7}As$ spacer layers and are transmitted to air or GaAs substrate in the case of the top or bottom reflector, respectively. In the case of the bottom mirror, a single low-index quarter-wave layer adjacent to the high-index GaAs substrate has to be added. Plot the power reflectance of the two mirrors as a function of the number of layer pairs.

Exercise 11.8   Show that the enhancement factor, Γ_{enh}, of the confinement factor of a VCSEL is unitary when $d_a = m\lambda/(2\langle \bar{n} \rangle)$, where m is a positive integer and $\langle \bar{n} \rangle$ is the spatially averaged refractive index.

Exercise 11.9   Consider a VCSEL with N active sections, with the same gain. If the i-th section is located between $z = z_{ia}$ and $z = z_{ib}$, demonstrate that the enhancement factor, Γ_{enh}, can be written as:

$$\Gamma_{enh} = 1 + \frac{1}{2\beta} \frac{\sum_{i=1}^{N} [\sin(2\beta z_{ib}) - \sin(2\beta z_{ia})]}{\sum_{i=1}^{N} (z_{ib} - z_{ia})}.$$

Exercise 11.10   A VCSEL has the following characteristics: average group velocity $\langle v_g \rangle = c/\langle \bar{n} \rangle$ where $\langle \bar{n} \rangle = 3.6$, total effective length 1.3 μm, average internal losses $\langle \alpha_i \rangle = 10$ cm^{-1}, thickness of the active region $d_a = 24$ nm, enhancement factor $\Gamma_{enh} = 1.8$, and power reflectance of the top and bottom Bragg reflectors $R_t = R_b = 99.5\%$. Calculate:
a) the photon lifetime;
b) the threshold gain.

Exercise 11.11   The amplitude reflection coefficient of a Bragg reflector can be written as $r = (1-b)/(1+b)$, where $b = (n_s/n_c)(n_1/n_2)^{2m}$. Show that the power reflectance $R = |r|^2$ can be approximated by the following expressions:

$$R \approx 1 - 4\frac{n_s}{n_c}\left(\frac{n_L}{n_H}\right)^{2m},$$

when $n_1 = n_L$, where n_L is the low refractive index, and by:

$$R \approx 1 - 4\frac{n_c}{n_s}\left(\frac{n_L}{n_H}\right)^{2m},$$

when $n_1 = n_H$, where n_H is the high refractive index.

Exercise 11.12   If an extra layer is added to the DBR of Exercise 11.11, the amplitude reflection coefficient at the central wavelength can be written as $r = (1 - b)/(1 + b)$, where $b = (n_1^2/n_c n_s)(n_1/n_2)^{2m}$. Show that good approximations for the power reflectance $R = |r|^2$ are:

$$R \approx 1 - 4\frac{n_L^2}{n_c n_s}\left(\frac{n_L}{n_H}\right)^{2m} \qquad \text{for } n_1 = n_L$$

$$R \approx 1 - 4\frac{n_c n_s}{n_H^2}\left(\frac{n_L}{n_H}\right)^{2m} \qquad \text{for } n_1 = n_H.$$

12 Quantum Cascade Lasers

12.1 Quantum Cascade Lasers

In this final chapter, we will analyze the quantum cascade lasers (QCLs), invented in 1994 by J. Faist, F. Capasso and coworkers [22], which are based on completely different physical processes compared to the semiconductor lasers we have considered so far. In the previous chapters, we have always considered lasers where light emission in the active material takes place by means of recombination of an electron in the conduction band with a hole in the valence band (*interband transitions*). QCLs are unipolar devices, since they use only one type of charged carriers (the electrons) and the radiative recombination takes place within a single band (the conduction band) between two bound states of a potential well (*intraband transition*), as first theoretically suggested in 1971 by Kazarinov and Suris in a seminal paper [23]. Once an electron has undergone a quantum jump between two levels in a particular quantum well with the emission of a photon, it can tunnel into an adjacent well so that the emission process can be repeated: This process is called *electron recycling*. Therefore, a single electron can produce the emission of multiple photons as it moves along the laser structure. This explains the name of this particular laser, since the emission is produced by a *cascade* process.

Figure 12.1 shows a simplified scheme of the conduction band structure for a basic QCL, where the laser transition is between subbands $|3\rangle$ and $|2\rangle$. The laser structure consists of consecutive stages composed by an electron injector and an active region. The number of stages typically ranges from 20 to 35 for lasers emitting in the wavelength region between 4 and 8 μm, but QCLs can have just a single stage or 100 stages. The tilt of the conduction band in Fig. 12.1 is produced by the applied electric field. The energy levels and wave functions can be calculated using the envelope function approximation. Electrons are injected into the upper laser level $|3\rangle$ by *resonant tunnelling*, which allows a highly selective injection, when the applied voltage exceeds a particular threshold value. In the particular structure shown in Fig. 12.1, which refers to an InGaAs/AlInAs laser, the thicknesses of the two quantum wells in the active region are 6.0 and 4.7 nm, separated by a 1.6-nm barrier, thus giving an energy separation of 207 meV (which corresponds to a wavelength of 6 μm) between the laser levels $|3\rangle$ and $|2\rangle$, and an energy separation between levels $|2\rangle$ and $|1\rangle$ of 37 meV, which is very close to the energy of an optical phonon mode of the material. In this way, electrons in the lower laser level $|2\rangle$ rapidly scatter by emission of an optical phonon to the energy level $|1\rangle$, with a relaxation time $\tau_{21} = \tau_2 = 0.3$ ps. The lifetime of level $|3\rangle$ is significantly longer, due to the larger energy separation between levels $|3\rangle$ and $|2\rangle$, such that the electron–phonon scattering processes between these two states are

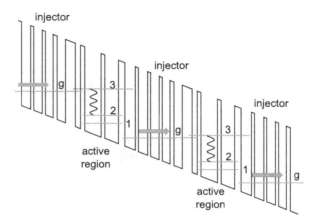

Figure 12.1 Schematic of the conduction band structure of a basic quantum cascade laser with laser transition between states $|3\rangle$ and $|2\rangle$ (wavy lines). $|g\rangle$ is the ground state of the injector. The arrows depict electron transport and injection into the active regions.

nonresonant. Electrons injected into level $|3\rangle$ can scatter into levels $|2\rangle$ and $|1\rangle$ by phonon emission, with scattering times $\tau_{32} = 2.2$ ps and $\tau_{31} = 2.1$ ps, thus giving rise to an overall lifetime of the upper laser level $\tau_3 = (\tau_{31}^{-1} + \tau_{32}^{-1})^{-1} = 1.1$ ps, which is much longer than the lifetime of the lower laser level, thus allowing the creation of a population inversion between these two states. For this reason, it is also important to prevent tunnelling from level $|3\rangle$ to states in the downstream adjacent injector region. This is obtained by proper engineering of the injector region so that at an energy corresponding to state $|3\rangle$ there is a region with very low density of states, called *minigap*, in the downstream injector. On the contrary, a dense manifold of states, called *miniband*, is present in the downstream injection region at energies corresponding to levels $|2\rangle$ and $|1\rangle$, in order to favor an efficient electron extraction from the active region after photon emission. The electrons injected in the downstream injection region gain in energy, relative to the bottom of the band and are injected into the following active region, so that the process described above proceeds in a stair-like cascade.

Due to the small energy difference between the laser levels $|3\rangle$ and $|2\rangle$, by proper engineering of the laser structure and materials, it is possible to obtain laser emission in the mid- to far-infrared wavelength regions from 3.5 to 24 μm and even more, up to 60 μm. For wavelengths $\lambda > 5$ μm, wells and barriers are typically formed with alloy compositions lattice matched to the InP substrate (e.g., $Al_{0.48}In_{0.52}As$ and $In_{0.47}Ga_{0.53}As$). In this case, the height of the quantum well barrier is the conduction band discontinuity between the two alloys, $\Delta\mathcal{E}_c = 0.52$ eV. For wavelengths shorter than 5 μm, the energy of the upper laser state increases, thus reducing the barrier for thermal activation of electrons. This may be a limiting factor in cw operation at room temperature, since it may introduce temperature-induced instabilities. This problem can be alleviated by using strained AlInAs/GaInAs heterostructures. By taking advantage of the strain, the height of the quantum well barrier can be of the order of $0.7-0.8$ eV, thus suppressing the electron leakage over the barriers. Another limiting factor is the so-called *thermal back-filling* of the lower laser level by electrons from the downstream injector. In order to prevent a thermal activated electron transfer from the injector into level $|2\rangle$, the energy separation between this level and the injector ground state should be larger than 0.1 eV.

12.2 Gain Coefficient

In Section 6.7, we have discussed the process of intersubband absorption; Eq. 6.87 gives the corresponding absorption/gain coefficient. In this equation, the electronic densities N_i are measured in cm^{-3}; in the following, we will use the carrier densities per unit area, $n_i = N_i L_p$, where L_p is the thickness of one period of active region and injector (such densities are therefore measured in cm^{-2}). In order to write the gain coefficient as a function of the photon energy, $\hbar\omega$, in Eq. 6.87, we have to write:

$$g(\omega - \omega_{mn}) = \hbar\, g(\hbar\omega - \mathcal{E}_m + \mathcal{E}_n) = \hbar\, g[\hbar(\omega - \omega_{nm})].$$

The squared modulus of the electric dipole moment, $|\mu_{nm}|^2$ can be written as $|\mu_{nm}|^2 = e^2|z_{nm}|^2$, where $z_{nm} = \langle n|z|m \rangle$ is the dipole matrix element between the states m and n. On the basis of Eq. 6.87, the gain coefficient can be written as:

$$g(\hbar\omega) = \frac{\omega\pi}{n_{eff}c\epsilon_0\hbar}e^2|z_{nm}|^2\hbar\, g_L[\hbar(\omega - \omega_{nm})]\frac{\Delta n}{L_p}, \qquad (12.1)$$

where n_{eff} is the mode refractive index, $\Delta n = n_m - n_n$ and $g_L(\hbar\omega)$ are the Lorentzian function:

$$g_L[\hbar(\omega - \omega_{nm})] = \frac{1}{\hbar}\frac{T_2/\pi}{1 + (\omega - \omega_{nm})^2T_2^2} = \frac{\gamma_{nm}/\pi}{\hbar^2(\omega - \omega_{nm})^2 + \gamma_{nm}^2}, \qquad (12.2)$$

with a peak value given by $1/\pi\gamma_{nm}$ (where $\gamma_{nm} = \hbar/T_2$) and a full-width at half maximum given by $2\gamma_{nm}$. With these definitions, the peak gain/absorption can be written as:

$$g_p = \frac{2e^2\omega_{nm}}{n_{eff}c\epsilon_0L_p}|z_{nm}|^2\frac{1}{2\gamma_{nm}}\Delta n = \sigma\,\Delta n. \qquad (12.3)$$

To take into account the shape of the mode inside the cavity, the confinement factor Γ has to be used, so that the modal gain can be written as:

$$g_m = g_p\Gamma = \sigma\Gamma\Delta n. \qquad (12.4)$$

We can also define a modal cross-section (measured in cm) as:

$$\sigma_m = \frac{g_m}{\Delta n} = \Gamma\sigma, \qquad (12.5)$$

which can be also written in terms of the emitted wavelength, λ:

$$\sigma_m = \frac{4\pi e^2}{n_{eff}\epsilon_0}\frac{|z_{nm}|^2}{2\gamma_{nm}L_p\lambda}\Gamma. \qquad (12.6)$$

12.3 Rate Equations and Threshold Conditions

In order to calculate the threshold gain and the threshold current density, we will use a rate-equation approach as in Section 8.2, in the case of semiconductor lasers. We will model a QCL as a three-level system, as shown in Fig. 12.2, where the electron states $|3\rangle$ and $|2\rangle$ are the upper and lower laser levels, respectively, and n_i are the electron densities (in units of cm^{-2}). We will assume that electrons are injected into the upper laser level from the ground state $|g\rangle$ of the injector with an injection efficiency η; the remaining electrons scatter directly into the lower laser level.

$$
\begin{cases}
\dfrac{dn_3}{dt} = \eta\dfrac{J}{e} - \dfrac{n_3}{\tau_3} - R_{st} \\[2ex]
\dfrac{dn_2}{dt} = (1-\eta)\dfrac{J}{e} + \dfrac{n_3}{\tau_{32}} + R_{st} - \dfrac{n_2 - n_2^{bf}}{\tau_2} \\[2ex]
\dfrac{d\phi}{dt} = \Gamma R_{st} - \dfrac{\phi}{\tau_c} + \Gamma\beta_{sp}R_{sp}
\end{cases}
\tag{12.7}
$$

where: J is the current density, $J = I/S$; ϕ is the photon density per unit area in the lasing mode (measured in cm^{-2}); R_{st} is the net stimulated emission rate. In the following, we will neglect the spontaneous emission term, $\Gamma\beta_{sp}R_{sp}$. The stimulated emission term, R_{st}, is given by:

$$
R_{st} = v_g g\phi = \frac{c}{n_{eff}}\sigma(n_3 - n_2)\phi.
\tag{12.8}
$$

The last term in the second rate equation takes into account the process of thermal back-filling from the injector back into the lower laser level $|2\rangle$ mentioned in Section 12.1. If n_s is the sheet carrier density in the injector ground state, n_2^{bf} is given by:

$$
n_2^{bf} = n_s \exp\left(-\frac{\Delta\mathcal{E}_{inj}}{k_B T}\right),
\tag{12.9}
$$

where $\Delta\mathcal{E}_{inj}$ is the energy difference between level $|2\rangle$ and the quasi-Fermi level in the injector, as shown in Fig. 12.2.

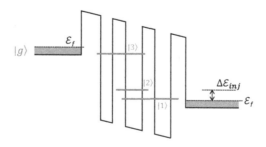

Figure 12.2   Three-level scheme of a QCL used in the rate equation approach.

We will assume steady-state conditions, so that the temporal derivatives in the rate equations are set to zero. At threshold $\phi = 0$ and, from the first rate equation, we have:

$$n_3 = \eta \tau_3 \frac{J}{e}, \tag{12.10}$$

which is then used in the second rate equation, where $n_2 = n_3 - \Delta n$, to calculate the population inversion $\Delta n = n_3 - n_2$:

$$\Delta n = \frac{J}{e}[\eta \tau_{eff} - (1 - \eta)\tau_2] - n_2^{bf}, \tag{12.11}$$

where:

$$\tau_{eff} = \tau_3 \left(1 - \frac{\tau_2}{\tau_{32}}\right) \tag{12.12}$$

is the effective lifetime of the population inversion. Indeed, assuming $\eta = 1$, from Eq. 12.11 we have:

$$\frac{d\Delta n}{\tau_{eff}} = \frac{dJ}{e}. \tag{12.13}$$

At threshold the spatially averaged gain must equal the spatially averaged total losses, which we will indicate as α_{tot}. The population inversion is clamped at the threshold value, Δn_{th}. From Eq. 8.32, we can write:

$$\Gamma \sigma \Delta n_{th} = \sigma_m \Delta n_{th} = \alpha_{tot}. \tag{12.14}$$

By using Eq. 12.11 in the previous equation, we can calculate the threshold current density, J_{th}:

$$J_{th} = \frac{e(\alpha_{tot}/\sigma_m + n_2^{bf})}{\eta \tau_{eff} - (1 - \eta)\tau_2}. \tag{12.15}$$

If we assume a unit injection efficiency ($\eta = 1$), we obtain:

$$J_{th} = \frac{e}{\tau_{eff}} \left(\frac{\alpha_{tot}}{\sigma_m} + n_2^{bf}\right) = \frac{e}{\tau_{eff}}(\Delta n_{th} + n_2^{bf}). \tag{12.16}$$

From this formula, it is clear that before reaching the threshold condition the current has to compensate both the losses inside the cavity and the electrons injected into level $|2\rangle$ by thermal back-filling. This last term strongly depends on temperature, as shown in Eq. 12.9. To minimize n_2^{bf} compared to α_{tot}/σ_m, a sufficiently high energy difference $\Delta \mathcal{E}_{inj}$ (typically > 0.1 eV) should be achieved by proper band gap engineering.

If we assume $\eta = 1$ and the back-filling process is neglected, from Eq. 12.11, the population inversion is given by:

$$\Delta n = \tau_{eff} \frac{J}{e}, \tag{12.17}$$

and the modal gain can be written as:

$$g_m = \sigma_m \Delta n = \left(\frac{4\pi e}{n_{eff}\epsilon_0} \frac{|z_{32}|^2}{2\gamma_{32}L_p\lambda}\right) \tau_3 \left(1 - \frac{\tau_2}{\tau_{32}}\right) \Gamma J_{th} \equiv g(\lambda)\Gamma J_{th}, \tag{12.18}$$

where we have introduced the gain coefficient $g(\lambda)$. With this definition, the threshold condition given by Eq. 12.14 can be written as:

$$g(\lambda)\Gamma J_{th} = \alpha_{tot}. \tag{12.19}$$

It is evident that a crucial parameter for the determination of the threshold conditions is the total cavity loss α_{tot}. A first contribution to α_{tot} is related to the mirror losses:

$$\alpha_m = \frac{1}{2L} \log\left(\frac{1}{R_1 R_2}\right), \tag{12.20}$$

where R_1 and R_2 are the reflectivity of the two laser end mirrors. A source of cladding losses is related to free-carrier absorption in the doped semiconductor regions: We will indicate this loss contribution with α_{fc}. Resonant absorption is taken into account by the thermal back-filling factor n_2^{bf}. Nonresonant losses (α_{nr}) of the active region are due to all the other transitions in the active region. Scattering losses (α_{sc}) are due to light-scattering processes in the cavity mainly related to waveguide sidewall roughness. Therefore, we can write:

$$\alpha_{tot} = \alpha_m + \Gamma\alpha_{nr} + (1-\Gamma)\alpha_{fc} + \alpha_{sc} = \alpha_m + \alpha_w. \tag{12.21}$$

The threshold current density given by Eq. 12.16 can be written as:

$$J_{th} = \frac{1}{\tau_{eff}}\left[\frac{n_{eff}\epsilon_0 2\gamma_{32}L_p\lambda}{4\pi e|z_{32}|^2}\frac{\alpha_m+\alpha_w}{\Gamma} + en_s\exp\left(-\frac{\Delta\mathcal{E}_{inj}}{k_B T}\right)\right]. \tag{12.22}$$

This formula can be improved to reproduce the temperature dependence of J_{th} by considering temperature dependent lifetimes, τ_i ($i = 1, 2, 3$):

$$\tau_i(T) = \tau_{i0}\frac{1}{1 + 2/\left[\exp\left(\mathcal{E}_{LO}/k_B T\right) - 1\right]}, \tag{12.23}$$

where τ_{i0} is the scattering time at low temperature and $\mathcal{E}_{LO}$ is the energy of the longitudinal optical (LO) phonons.

A simple formula giving the temperature dependence of the threshold current density can be written, as in conventional semiconductor lasers, by introducing a characteristic temperature T_0:

$$J_{th}(T) = J_0\exp\left(\frac{T}{T_0}\right). \tag{12.24}$$

Typical values of T_0 are in the temperature range between 100 and 200 K. A slightly different version of Eq. 12.24, which allows one to fit the dependence of J_{th} versus temperature over the entire temperature range of laser operation (from 4.2 K to above the room temperature) is the following:

$$J_{th}(T) = J_0\exp\left(\frac{T}{T_0}\right) + J_1. \tag{12.25}$$

12.4 Output Power, Slope-, and Wall-Plug Efficiency

Above threshold, the population inversion is clamped to its threshold value, $\Delta n = \Delta n_{th}$, and the photon density, ϕ, is no longer zero. From the rate equations 12.7, it is possible to calculate ϕ (we will neglect spontaneous emission). For simplicity, we will assume $\eta = 1$. From the first two rate equations, we have:

$$R_{st} = \frac{\tau_{eff}}{\tau_2 + \tau_{eff}} \frac{J - J_{th}}{e}, \qquad (12.26)$$

where J_{th} is given by Eq. 12.16. From the third rate equation, we have $\phi = \Gamma \tau_c R_{st}$, so that:

$$\phi = \frac{\tau_c \tau_{eff}}{\tau_2 + \tau_{eff}} \frac{\Gamma}{e} (J - J_{th}), \qquad (12.27)$$

where the photon lifetime is given by:

$$\tau_c = \frac{n_{eff}}{c \alpha_{tot}}, \qquad (12.28)$$

so that the photon density (per unit area) can be written as:

$$\phi = \frac{\tau_{eff} \Gamma (J - J_{th}) n_{eff}}{e \alpha_{tot} c (\tau_2 + \tau_{eff})}. \qquad (12.29)$$

To calculate the output power, we will proceed as in the case of conventional semiconductor lasers. The optical energy stored in the laser cavity can be written as:

$$\mathcal{E}_{st} = \hbar\omega (N_p S_p) \phi, \qquad (12.30)$$

where N_p is the number of periods (i.e., the number of stages of active regions and injectors), $\hbar\omega$ is the energy of the emitted photons, and the surface S_p is related to the volume V_p occupied by photons (typically larger than the volume of the active region occupied by electrons, $V = \Gamma V_p$). The photon loss rate due to mirror transmission is given by:

$$\frac{c}{n_{eff}} \frac{1}{L} \log\left(\frac{1}{R}\right) = \alpha_m \frac{c}{n_{eff}}, \qquad (12.31)$$

so that the output power is given by:

$$P = \mathcal{E}_{st} \alpha_m \frac{c}{n_{eff}} = \frac{\hbar\omega}{e} N_p \frac{\alpha_m}{\alpha_{tot}} \frac{\tau_{eff}}{\tau_2 + \tau_{eff}} (I - I_{th}), \qquad (12.32)$$

where we have used the current $I = J \Gamma S_p$. The slope efficiency is:

$$\eta_s = \frac{dP}{dI} = \frac{\hbar\omega}{e} N_p \frac{\alpha_m}{\alpha_{tot}} \frac{\tau_{eff}}{\tau_2 + \tau_{eff}} = \eta_{int} N_p \frac{\hbar\omega}{e} \frac{\alpha_m}{\alpha_{tot}}, \qquad (12.33)$$

where we have introduced the *internal quantum efficiency* η_{int}:

$$\eta_{int} = \frac{\tau_{eff}}{\tau_2 + \tau_{eff}}. \qquad (12.34)$$

It is evident that upon increasing the number of periods, N_p, large slope efficiencies and therefore large output powers can be achieved: Values in the range of W/A and W, respectively, are typical. The external quantum efficiency, η_{ext} is given by:

$$\eta_{ext} = \frac{d(P/\hbar\omega)}{d(I/e)} = \eta_s \frac{e}{\hbar\omega} = \eta_{int} N_p \frac{\alpha_m}{\alpha_{tot}}, \qquad (12.35)$$

so that the output power can be calculated as:

$$P = \eta_{ext} \frac{\hbar\omega}{e}(I - I_{th}),$$

(12.36)

as in the case of conventional semiconductor lasers.

12.4.1 Wall-Plug Efficiency

The wall-plug efficiency, which gives the conversion efficiency between electrical power, P_e, and optical power, P, is defined as:

$$\eta_{wp} = \frac{P}{P_e} = \frac{P}{UI},$$

(12.37)

where U is the voltage drop on the structure. In the case of a QCL, an equivalent and useful definition of η_{wp} is [24]:

$$\eta_{wp} = \frac{dP}{UdI} \frac{J_{max} - J_{th}}{J_{max}} = \frac{\eta_s}{U} \frac{J_{max} - J_{th}}{J_{max}},$$

(12.38)

where J_{max} is the maximum current density. So far we have assumed that electrons are injected into the upper laser level $|3\rangle$ from the preceding injection region quickly and almost exclusively and that they are efficiently removed from states $|2\rangle$ and $|1\rangle$. Electrons are injected into level $|3\rangle$ by *resonant tunneling*, and the electron injection into the active region is optimized when quasiequilibrium is reached between the population of the injector ground state and that of the upper laser level, which are therefore characterized by a common quasi-Fermi level. In this way, space-charge build-up in the active region induced by increased electron injection is avoided. A very good expression for the maximum current density is given by the following expression:

$$J_{max} = n_s \frac{e}{\tau_t},$$

(12.39)

where n_s is the free-carrier sheet density per period (i.e., the 2D electron concentration in the injector state) and τ_t is an effective transit time of the electron across a period of the active region at resonance. The concept of resonance is illustrated in Fig. 12.3, which shows the process of injection through a very thin barrier from the ground state of the injector, $|g\rangle$, to the upper laser level at resonance (the corresponding injection into the lower laser level is much less efficient). The resonance condition is obtained by applying an electric field, E_{res}, to the structure, which decreases the energy detuning between $|g\rangle$ and $|3\rangle$ states until it vanishes and the states $|g\rangle$ and $|3\rangle$ form a doublet with anticrossing energy $2\hbar|\Omega|$ (as shown in Fig. 12.3). The voltage drop on a single period is given, as depicted in Fig. 12.3, by $eU_p = \hbar\omega + \Delta\mathcal{E}_{inj}$, so that the voltage drop on the complete structure with N_p periods is given by:

$$U = \frac{N_p}{e}(\hbar\omega + \Delta\mathcal{E}_{inj}).$$

(12.40)

The threshold current density in the case of a N_p-period structure is obtained from Eq. 12.16 by using the total modal gain cross-section $N_p\sigma_m$:

$$J_{th} = \frac{e}{\tau_{eff}} \left(\frac{\alpha_{tot}}{N_p\sigma_m} + n_2^{bf} \right).$$

(12.41)

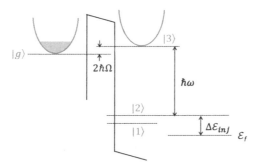

Figure 12.3 Schematic representation of the resonant tunneling process from the injector to the upper laser level.

By using Eqs. 12.35, 12.39–12.41 in Eq. 12.38, the wall-plug efficiency can be written as:

$$\eta_{wp} = \eta_{int} \frac{\alpha_m}{\alpha_{tot}} \frac{1}{1 + \Delta\mathcal{E}_{inj}/\hbar\omega} \left[1 - \frac{\tau_t}{\tau_{eff}} \left(\frac{\alpha_{tot}}{n_s N_p \sigma_m} + \frac{n_2^{bf}}{n_s} \right) \right]. \tag{12.42}$$

From Eq. 12.21, $\alpha_{tot} = \alpha_m + \alpha_w$. It has been shown that for QCLs operating both in the mid- and far-infrared spectral region, the waveguide loss, α_w, is proportional to the doping and can be written as:

$$\alpha_w = N_p \alpha_{fc} n_s + \alpha_{sc}. \tag{12.43}$$

In good quality lasers, α_{sc} can be neglected. Moreover, we will assume that, by proper laser design, $\Delta\mathcal{E}_{inj}$ is large enough so that back-filling process can be neglected. By using these assumptions, it is simple to calculate the value of the mirror losses, α_m, which maximizes the wall-plug efficiency:

$$\alpha_m = \sqrt{N_p \sigma_m n_s \alpha_w \tau^*} - \alpha_w, \tag{12.44}$$

where we have introduced the dimensionless upper state lifetime $\tau^* = \tau_{eff}/\tau_t$. By using the previous expression in Eq. 12.42 (with $n_2^{bf} = 0$), it is possible to calculate the maximum value of the wall-plug efficiency:

$$\eta_{wp,max} = \eta_{tr} \frac{1}{1 + \Delta\mathcal{E}_{inj}/\hbar\omega} \left[\frac{\sqrt{g^* \tau^*} - 1}{\sqrt{g^* \tau^*}} \right]^2, \tag{12.45}$$

where we have introduced the dimensionless gain cross-section $g^* = \sigma_m/\alpha_{fc}$.

It is interesting to note that the first QCL demonstrated by Faist, Capasso and coworkers in 1994 was characterized by a wall-plug efficiency $< 0.15\%$ in pulsed operation at 10 K and that η_{wp} remained of the order of 1% at room temperature for most QCLs for a decade. Then, the introduction of optimized laser structures allowed a significant increase of η_{wp} up to $\sim 50\%$. A first optimization consists in the reduction of the voltage drop in the active region, which does not contribute to the generation of light.

Another technique is based on the increase of the coupling strength, $\hbar\Omega$, since the tunneling rate between injector and active layer and other parameters that determine the laser efficiency strongly depends on this parameter. The increase of the tunneling rate gives rise to an increase of the maximum current density, J_{max}, thus leading to an increase of the wall-plug efficiency. A large tunneling rate presents other advantages: It reduces the electron density in the injector region, thus decreasing the leakage current from the injector to the lower laser level or the continuum energy levels; moreover, it increases the population inversion and therefore the gain.

The effects of the coupling strength on the electron transport characteristics can be understood by using the following expression (reported by Kazarinov and Suris) for the current between two subbands coupled by a tunneling barrier with a coupling energy $2\hbar|\Omega|$ (see Fig. 12.3):

$$J = en_s \frac{2|\Omega|^2 \tau_\perp}{1 + \Delta^2 \tau_\perp^2 + 4|\Omega|^2 \tau_3 \tau_\perp}, \tag{12.46}$$

where $\tau_\perp$ is the relaxation time of electron in-plane momentum, $\hbar\Delta$ is the energy level detuning from full resonance, which is related to the applied electric field E by the following relation:

$$\hbar\Delta = ed(E - E_r), \tag{12.47}$$

where d is the effective spatial distance between resonantly coupled injector state and upper laser level state, and E_r is the electric field required for full resonance. From Eq. 12.46, it is evident that upon increasing $|\Omega|$, the detuning term in the denominator, $\Delta^2 \tau_\perp^2$, does not significantly influence the resonant tunneling rate.

It has been recognized that interface roughness introduces a significant detuning to the energy levels in resonance (interfaces between adjacent semiconductor layers present roughness of the order of a few ångstroms, which is significant since typical semiconductor layer thicknesses in QCLs are of the order of $10 - 50$ Å). An increase of the coupling strength can be achieved by using very thin injector barriers, of the order of ~ 10 Å, between the injector and the active region. The obtained high coupling strength reduces the relative importance of the detuning of resonant tunneling induced by the interface roughness. Another advantage is that the upper laser level spreads even more into the injection region as a consequence of the very thin injector barrier: This effect increases the lifetime of the upper laser level and decreases the threshold current density. Detailed calculations based on a density matrix approach [25] show that the gain coefficient rapidly increases upon increasing the energy splitting $2\hbar|\Omega|$ up to ~ 20 meV and then it slowly decreases.

12.5 Numerical Example

In this numerical example, we will refer to a particular QCL described by Gmachl and coworkers [26]. The active region is composed by 3 InGaAs quantum wells closely coupled by thin AlInAs barriers. The energy separation between the upper and lower laser levels is $\mathcal{E}_{32} = 153.6$ meV (which corresponds to an emission wavelength $\lambda = 8.08$ μm) at the electric field value $E = 45$ kV cm^{-1}, corresponding to the measured threshold. The energy separation between the levels $|2\rangle$ and $|1\rangle$ is $\mathcal{E}_{21} = 38.3$ meV, designed to have an efficient depletion of the lower laser level by resonant LO-phonon scattering (indeed $\mathcal{E}_{LO} \approx 34$ meV). The energy separation between the upper laser level and the ground state of the injector is $\mathcal{E}_{3g} = 40$ meV. The structure is designed to have level $|1\rangle$ of the preceding active region in resonance with level $|3\rangle$ of the following active region in order to have a resonant carrier transport between consecutive active regions without a significant carrier relaxation in the injector. LO-phonon scattering time from level $|3\rangle$ to level $|2\rangle$ is $\tau_{32} = 3.1$ ps, from level $|3\rangle$ to level $|1\rangle$ is $\tau_{31} = 3.6$ ps, from level $|3\rangle$ to all states of the downstream injection region is $\tau_{3i} = 14.6$ ps (i.e., $\tau_{3i} \gg \tau_{31}, \tau_{32}$). Lifetime of state $|2\rangle$ is $\tau_2 \approx \tau_{21} = 0.3$ ps. The dipole matrix element of the optical transition is $|z_{32}| = 1.9$ nm. The sheet carrier density in the injector is $n_s = 1.6 \times 10^{11}$ cm^{-2}. The full-width at half maximum $2\gamma_{32}$ (measured value of the luminescence linewidth) is $2\gamma_{32} \approx 10$ meV. The length of one period is $L_p = 44.3$ nm, the effective refractive index of the waveguide is $n_{eff} = 3.27$, the number of periods is $N_p = 30$, the facet reflectivity is $R = 0.28$, and the length of the laser is $L = 2.25$ mm. The measured waveguide losses are $\alpha_w = 24$ cm^{-1}. The confinement factor, Γ, for the complete structure with N_p periods is $\Gamma = 0.5$. We assume an energy separation between the ground state of the injector and state $|2\rangle$ of the preceding active region $\Delta\mathcal{E}_{inj} = 150$ meV. Coupling energy $\hbar\Omega = 1.3$ meV, in-plane momentum relaxation time $\tau_\perp = 40$ fs.

Calculate the gain coefficient $g(\lambda)$, the threshold current density, J_{th}, the slope efficiency per facet, the external quantum efficiency per facet per stage, and an approximated value of the wall-plug efficiency.

The lifetime of the upper laser level is given by:

$$\tau_3 = \left(\frac{1}{\tau_{31}} + \frac{1}{\tau_{32}} + \frac{1}{\tau_{3i}} \right)^{-1} = 1.5 \text{ ps}.$$

The gain coefficient, $g(\lambda)$, can be calculated by using Eq. 12.18:

$$g(\lambda) = \left(\frac{4\pi e}{n_{eff}\epsilon_0} \frac{|z_{32}|^2}{2\gamma_{32}L_p\lambda} \right) \tau_3 \left(1 - \frac{\tau_2}{\tau_{32}} \right) = 59.1 \text{ cm/kA}.$$

The mirror loss α_m is given by:

$$\alpha_m = -\frac{1}{L} \log R = 5.6 \text{ cm}^{-1},$$

the threshold current density is (from Eq. 12.19):

$$J_{th} = \frac{\alpha_m + \alpha_w}{\Gamma g(\lambda)} = 1 \text{ kA/cm}^2,$$

in excellent agreement with the experimental value $J_{th,exp} \approx 1$ kA/cm^2. The effective lifetime τ_{eff} is given by:

$$\tau_{eff} = \tau_3 \left(1 - \frac{\tau_2}{\tau_{32}} \right) = 1.35 \text{ ps}.$$

The internal quantum efficiency can be calculated by using Eq. 12.34:

$$\eta_{int} = \frac{\tau_{eff}}{\tau_2 + \tau_{eff}} = 0.82.$$

The slope efficiency per facet can be calculated from Eq. 12.35:

$$\eta_s = \eta_{int} N_p \frac{\hbar\omega}{e} \frac{\alpha_m/2}{\alpha_w + \alpha_m/2} = 394 \text{ mW/A},$$

again in excellent agreement with the experimental value $\eta_{s,exp} \approx 400$ mW/A. The external quantum efficiency per facet and per stage is given by:

$$\frac{\eta_{ext}}{N_p} = \frac{e}{\hbar\omega} \frac{\eta_s}{N_p} = 8.5\%.$$

Therefore, external quantum efficiencies larger than 1 for the complete laser structure can be achieved by using a number of periods $N_p \geq 12$.

In order to calculate an approximated value of the wall-plug efficiency, we will first estimate the effective transit time of the electrons across a period of the active region at resonance, τ_t, by using Eq. 12.46 and Eq. 12.39:

$$\tau_t \approx \frac{1 + 4\Omega^2 \tau_3 \tau_\perp}{2\Omega^2 \tau_\perp},$$

where we have assumed zero detuning from the resonance condition. Using the numerical data, we obtain $\tau_t \approx 6.2$ ps. The population of level $|2\rangle$ due to thermal back-filling can be completely neglected, due to the high value of $\Delta\mathcal{E}_{inj}$, so that the wall-plug efficiency can be calculated by using Eq. 12.42:

$$\eta_{wp} = \eta_{int} \frac{\alpha_m}{\alpha_m + \alpha_w} \frac{1}{1 + \Delta\mathcal{E}_{inj}/\hbar\omega} \left(1 - \frac{\tau_t}{\tau_{eff}} \frac{\alpha_m + \alpha_w}{n_s N_p \sigma_m}\right) = 0.078,$$

which is very close to the measured value of 7%. In the previous expression, the modal gain cross-section σ_m can be calculated by using Eq. 12.6, $\sigma_m = 3.5 \times 10^{-9}$ cm.

12.6 Applications

A major application of QCLs is infrared spectroscopy, where the transmittance or absorbance spectrum of the infrared laser beam is measured after propagation through a sample. Mid-infrared spectral region, in the range between 3 μm and 15 μm, is particularly interesting since most chemicals have their fundamental vibrational modes in this spectral region. Chemical-sensing applications range from remote sensing of environmental gases and pollution in the atmosphere, to explosive detection and industrial process monitoring, to medical diagnostics such as breath analysis. For all these applications, it is important to use single-mode, narrow-linewidth, tunable mid-infrared lasers. This is required to achieve multiple-species detection with high sensitivity and selectivity. This can be accomplished by using QCLs, which naturally emit in the mid-infrared spectral region. Single-mode and tunable operation of QCLs has been demonstrated by using a number of different cavity configurations. For example by using distributed feedback (DFB) QCLs or by employing external cavity QCLs. Just as a recent example, a QCL source has been reported tunable between 6.2 and 9.1 μm obtained by integrating an eight-laser sampled grating DFB laser array with an on-chip beam combiner [27]. An important feature of the DFB QCL is that its emission wavelength can be continuously tuned by changing the heat sink temperature or the injection current, which changes the temperature of the device. In both cases, the effect is a change of the effective refractive index, which determines a shift of the resonance wavelength of the Bragg grating of the DFB structure, with a characteristic tuning rate around 0.1–0.2 cm^{-1}K^{-1}. This is particularly important since the DFB QCL can be continuously tuned across the absorption peaks of various molecules, thus offering very high detection sensitivity of trace gases. In order to cover a broad wavelength range, multiple DFB QCLs with different grating periods on a single chip can be employed.

Another emerging application of QCLs is free-space communication. The advantage is the possibility to use a wavelength in the highly transmissive windows of atmosphere between 3.3 and 4.2 μm or between 8 and 13 μm. Another important application is the use of QCLs as local oscillators in heterodyne detection systems for the far- and mid-infrared region. Advances in terahertz QCLs have led to a number of applications in this spectral region. Particularly, important is the use of QCLs as high-average power laser source for terahertz imaging.

Appendix: Fundamental Constants

	Symbol	Value
Electron charge	e	$-1.6021766208(98) \times 10^{-19}$ C
Electron rest mass	m_0	$9.10938356(11) \times 10^{-31}$ kg
Planck constant	h	$6.626070040(81) \times 10^{-34}$ J·s
		$4.135667662(25) \times 10^{-15}$ eV·s
Reduced Planck constant	$\hbar$	$1.054571800(13) \times 10^{-34}$ J·s
		$6.582119514(40) \times 10^{-16}$ eV·s
Boltzmann constant	k_B	$1.38064852(79) \times 10^{-23}$ J·K^{-1}
		$8.6173303(50) \times 10^{-5}$ eV·K^{-1}
Light velocity in vacuum	c	2.99792458×10^{8} m·s^{-1}
Vacuum permittivity	ϵ_0	$8.854187817 \times 10^{-12}$ F·m^{-1}
Vacuum permeability	μ_0	$4\pi \times 10^{-7}$ H·m^{-1}
		$1.2566370614 \times 10^{-6}$ H·m^{-1}
Avogadro number	N_A	$6.022140857(74) \times 10^{23}$ mol^{-1}
Bohr radius	a_B	$5.2917721067(12) \times 10^{-11}$ m
Rydberg constant	Ry	$1.097373156 \times 10^{7}$ m^{-1}
		$13.605693009(84)$ eV

References

[1] I. Vurgaftman, J. R. Meyer, and L. R. Ram-Mohan. "Band parameters for III–V compound semiconductors and their alloys." In *Journal of Applied Physics* 89.11 (2001), 5815–5875 (cited on 31, 159).

[2] P. Carrier and S.-H. Wei. "Theoretical study of the band-gap anomaly of InN." In: *Journal of Applied Physics* 97.3 (2005), 033707 (cited on 33).

[3] H. Q. Yang et al. "First-principle study of the electronic band structure and the effective mass of the ternary alloy $Ga_xIn_{1-x}P$." In: *Journal of Physics: Conference Series* 574 (Jan. 2015), 012048 (cited on 36).

[4] H. C. Casey and F. Stern. "Concentration-dependent absorption and spontaneous emission of heavily doped GaAs." In: *Journal of Applied Physics* 47.2 (1976), 631–643 (cited on 55).

[5] G. Bastard. *Wave Mechanics Applied to Semiconductor Heterostructures*. Wiley-Interscience, 1991 (cited on 123).

[6] C. Downs and T. E. Vandervelde. "Progress in infrared photodetectors since 2000." In: *Sensors* 13.4 (2013), 5054–5098 (cited on 154).

[7] J. Singh. *Electronic and Optoelectronic Properties of Semiconductor Structures*. Cambridge University Press, 2003. Chapter 9 (cited on 166).

[8] N. Grote. "Laser components." In: *Fibre Optic Communication, Key Devices*. Edited by Herbert Venghaus and Norbert Grote. Springer, 2017. Chapter 3 (cited on 178, 266).

[9] M. D. Sturge. "Optical absorption of gallium arsenide between 0.6 and 2.75 eV." In: *Physical Review* 127 (Aug. 1962), 768–773 (cited on 187).

[10] N. S. Köster et al. "Controlling the polarization dynamics by strong THz fields in photoexcited germanium quantum wells." In: *New Journal of Physics* 15.7 (Jul. 2013), 075004 (cited on 189).

[11] I. Akasaki. "Nobel Lecture: Fascinated journeys into blue light." In: *Reviews of Modern Physics*, 87 (Oct. 2015), 1119–1131 (cited on 210, 211).

[12] S. Nakamura et al. "High-power InGaN single-quantum-well-structure blue and violet light-emitting diodes." In: *Applied Physics Letters* 67.13 (1995), 1868–1870 (cited on 210).

[13] Y. Arakawa and H. Sakaki. "Multidimensional quantum well laser and temperature dependence of its threshold current." In: *Applied Physics Letters* 40.11 (1982), 939–941 (cited on 239, 261).

[14] M. Asada, Y. Miyamoto, and Y. Suematsu. "Gain and the threshold of three-dimensional quantum-box lasers." In: *IEEE Journal of Quantum Electronics* 22.9 (1986), 1915–1921 (cited on 239).

[15] N. Kirstaedter et al. "Low threshold, large T_0 injection laser emission from (InGa)As quantum dots." In: *Electronics Letters* 30.17 (1994), 1416–1417 (cited on 240).

[16] L. Goldstein et al. "Growth by molecular beam epitaxy and characterization of InAs/GaAs strained-layer superlattices." In: *Applied Physics Letters* 47.10 (1985), 1099–1101 (cited on 240).

[17] J. C. Norman et al. "Perspective: The future of quantum dot photonic integrated circuits." In: *APL Photonics* 3.3 (2018), 030901 (cited on 242).

[18] K. K. Nanda, F. E. Kruis, and H. Fissan. "Energy levels in embedded semiconductor nanoparticles and nanowires." In: *Nano Letters* 1.11 (2001), 605–611 (cited on 250).

[19] M. Grundmann, R. Heitz, and D. Bimberg. "Carrier statistics in quantum-dot lasers." In: *Physics of the Solid State* 40 (May 1998), 772–774 (cited on 260).

[20] L. V. Asryan and R. A. Suris. "Theory of threshold characteristics of semiconductor quantum dot lasers." In: *Semiconductors* 38 (Jan. 2004), 1–22 (cited on 260).

[21] P. Yeh. *Optical Waves in Layered Media*. J. Wiley & Sons, 1988 (cited on 293).

[22] J. Faist et al. "Quantum cascade laser." In: *Science* 264.5158 (1994), 553–556 (cited on 301).

[23] R. F. Kazarinov and R. A. Suris. "Possibility of the amplification of electromagnetic waves in a semiconductor with a superlattice." In: *Soviet Physics—Semiconductors* 5.4 (1971), 707 (cited on 301).

[24] J. Faist. "Wallplug efficiency of quantum cascade lasers: Critical parameters and fundamental limits." In: *Applied Physics Letters* 90.25 (2007), 253512 (cited on 308).

[25] J. B. Khurgin et al. "Role of interface roughness in the transport and lasing characteristics of quantum-cascade lasers." In: *Applied Physics Letters* 94.9 (2009), 091101 (cited on 310).

[26] C. Gmachl et al. "Recent progress in quantum cascade lasers and applications." In: *Reports on Progress in Physics* 64.11 (Oct. 2001), 1533–1601 (cited on 310).

[27] W. Zhou et al. "Monolithically, widely tunable quantum cascade lasers based on a heterogeneous active region design." In: *Scientific Reports* 6.1 (2016), 25213 (cited on 312).

Index